Energy Medicine

Jeremy P. Tarcher/Putnam

a member of
Penguin Putnam Inc.
New York

Energy Medicine

Balance Your Body's Energies for Optimum Health, Joy, and Vitality

Donna Eden
with David Feinstein

Illustrations by Brooks Garten

An Important Note: This book is not intended as a substitute for the medical recommendations of physicians or other health-care providers. Rather, it is intended to offer information to help the reader cooperate with physicians and health professionals in a mutual quest for optimum well-being.

Most Tarcher/Putnam books are available at special quantity discounts for bulk purchases for sales promotions, premiums, fund-raising, and educational needs. Special books or book excerpts also can be created to fit specific needs. For details, write Putnam Special Markets, 375 Hudson Street, New York, NY 10014.

Jeremy P. Tarcher/Putnam
a member of
Penguin Putnam Inc.
375 Hudson Street
New York, NY 10014
www.penguinputnam.com

Library of Congress Cataloging-in-Publication Data
Eden, Donna.
Energy medicine: balance your body's energies for optimum health, joy, and vitality / by Donna Eden with David Feinstein ; illustrations by Brooks Garten.
p. cm.
Includes bibliographical references and index.
ISBN 0-87477-945-6
1. Vital force—Popular works. 2. Mental healing—Popular works. 3. Self-care, Health—Popular works. I. Feinstein, David. II. Title.
RZ999.E327 1999 98-25809 CIP
615.8'9—dc21

Printed in the United States of America
1 3 5 7 9 10 8 6 4 2

This book is printed on acid-free paper. ♾

Book design by Deborah Kerner

Dedicated to the women of my clan:
Mama, Tanya, Dondi, Susie, Maude,
Sharon, Shawn, Chelli, Christy, Endiah, Tia,
Mozelle, Alice, Karen, Beverly, Mary, and Edith

And the men:
Daddy, Guy, Avon, Clarke, Danny,
Erinn, Ryan, Caz, Lehvi, Lawrence,
Seifert, Sol, Robin, Barry, Mark, Carl,
Ray, and David

Acknowledgments

Deepest in my appreciations is Mama, the first alternative healer I ever knew. I am also grateful to her doctors, Joseph Morgan, M.D., Howard Morningstar, M.D., and Janet Bates, M.D., who have restored my hope about Western medicine.

I could hardly overstate my personal gratitude and professional debt to Dr. John Thie for having had the courage during the 1970s, and the genius, to make available to the general public sophisticated self-healing techniques through his system of Touch for Health. I will forever be grateful to Dr. Thie, Gordon Stokes, Shanti Duree, Hazel Ullrich, Warren Jacobs, M.D., and the Touch for Health organization, which started me on my path.

Sandy Wand has been an amazing friend, healer, and source of boundless support. Paul Brenner, M.D., showed me what a physician can be. My editors, Irene Prokop and Wendy Hubbert, have been enormously patient, provocative, and very smart. My agent, Susan Schulman, has provided wise and empowering counsel. The institutional support of Innersource and its friends, particularly Laleah and Hugh Bacon, Jean Lamb, and Rodney Plimpton, are gratefully acknowledged. Stanley Krippner's brilliant critique of an earlier draft of the manuscript has strengthened the book substantially. My friends Carla Cunningham, Norman Friedman, Jane Maynerd, and Peg Mayo have also offered valuable suggestions on the developing manuscript. Richard Duree provided source material for several of the research studies reported in the book. Rik Jensen, our handyman with heart, has been a model of how to remain buoyant through floods and other disasters that could have derailed the entire project. And my daughters, Dondi and Tanya Dahlin, have always been a compelling inspiration toward wholeness and health.

My husband, David Feinstein, has been tireless in interviewing me and writing first drafts from those interviews, editing transcripts of tapes from my classes, conducting computer and library searches, and generally bringing left-brained organization to my right-brained way of being. His abilities to craft a phrase, suggest an analogy, find order amid complexity, and place an idea into its larger intellectual context—all the while retaining the spirit of *my* voice—permeate this volume. In short, this is the book *I* would have written if my mind worked like both of our minds combined.

Contents

Foreword • **xv**
Introduction: The Return of Energy Medicine • **1**

Part I: **Awakening the Two-Million-Year-Old Healer Within** • **13**

Chapter 1: Energy Is All There Is • **15**

Chapter 2: Energy Testing
Talking with Your Energy Body • **29**

Chapter 3: Keeping Your Energies Humming
A Daily Routine • **60**

Part II: **The Anatomy of Your Energy Body**
Eight Energy Systems • **93**

Chapter 4: The Meridians
Your Energy Transportation System • **95**

Chapter 5: The Chakras
Your Body's Energy Stations • **133**

Chapter 6: The Aura, Celtic Weave, and Basic Grid
Protecting, Weaving, Supporting • **172**

Chapter 7: The Five Rhythms
The Drummers to Which You March • **196**

Chapter 8: Triple Warmer and the Strange Flows
The Energetic Arms of Your Immune System • **224**

Part III: **Weaving It All Together** • **257**

Chapter 9: Illness
Your Body's Reset Button • **259**

Chapter 10: Pain
Nature's Idea of Tough Love • **277**

Chapter 11: Swimming in Electromagnetic Currents
Making the Best of an Ecosystem in Distress • **295**

Chapter 12: Setting Your Habit Field for Optimum Health and Performance • **317**

Epilogue: Journeys to the Outer Realms • **339**

Appendix: How to Find a Competent Energy Medicine Practitioner • **355**

Notes • **363**

Index • **373**

Figures

Figure 1. Heaven Rushing In
Figure 2. Spleen-Pancreas Energy Test
Figure 3. The Three Thumps
Figure 4. General Indicator Test
Figure 5. Cross Crawl
Figure 6. Wayne Cook Posture
Figure 7. Crown Pull
Figure 8. Spinal Flush
Figure 9. Neurolymphatic Reflex Points
Figure 10. Tracing the Central Meridian
Figure 11. Tracing the Governing Meridian
Figure 12. Tracing the Spleen Meridian
Figure 13. Tracing the Heart Meridian
Figure 14. Tracing the Small Intestine Meridian
Figure 15. Tracing the Bladder Meridian
Figure 16. Tracing the Kidney Meridian
Figure 17. Tracing the Circulation-Sex Meridian
Figure 18. Tracing the Triple Warmer Meridian
Figure 19. Tracing the Gall Bladder Meridian
Figure 20. Tracing the Liver Meridian
Figure 21. Tracing the Lung Meridian
Figure 22. Tracing the Large Intestine Meridian
Figure 23. Tracing the Stomach Meridian
Figure 24. Wellspring of Life Points
Figure 25. Alarm Points
Figure 26. Acupuncture Strengthening and Sedating Points
Figure 27. Spinal Suspension
Figure 28. Meridian Flow Wheel
Figure 29. Jet Lag Pressure Points
Figure 30. The Seven Major Chakras
Figure 31. Headache Isometric Press
Figure 32. Chakra Energy Test

Figure 33. Weaving the Aura
Figure 34. Rhythmic 8s
Figure 35. Rhythms of the Seasons Wheel
Figure 36. Neurovascular Holding Points
Figure 37. The Stomach Points
Figure 38. Acupuncture Tapping Points
Figure 39. Separating Heaven and Earth
Figure 40. Spleen and Large Intestine Drainage Points
Figure 41. Resetting the Ileocecal Valve
Figure 42. The Meridians and the Neurovascular Points
Figure 43. Muscle Meridian Chart
Figure 44. Dental Chart
Figure 45. Pain Zones
Figure 46. Meridian Beginning and Ending Points
Figure 47. Eye Pattern Release
Figure 48. Temporal Tap

Let the body think
Of the Spirit as streaming, pouring
Rushing and shining into it from
All sides.

—PLOTINUS

Foreword

Balance Your Body's Energies for Optimum Health, Joy, and Vitality

I have been asked for years about the nature of energy medicine—what is energy medicine? How is it different from allopathic medicine? What benefits does it offer a person that he or she cannot gain from conventional medicine?

Though these may appear to be questions that can be answered simply, they are not. They represent the fact that we have entered an energetic era in which we will redefine our understanding of ourselves. Soon we will come to know ourselves as multisensory, rendering obsolete the belief that we are five-sense creatures. No longer will we speculate about whether our thoughts and emotions have creative authority within our cell tissue; we will come to consider that form of perception as one of the central truths of life. We will view as ordinary such treatments as the laying on of hands and the use of healing oils and sounds. The belief that repressing an illness through drugs is of use to the healing process will come to be seen as harmful rather than as beneficial.

And when shall we expect such changes to occur? Dramatic shifts in medicine and healing are happening all around us now. Our society has shown respect for the holistic movement and energetic medicine by its growing interest in learning more about this field and choosing to engage in alternative healing methods for diseases. Research and studying the impact of thoughts and emotions in relation to the physical body increases continually, and again and again, the results indicate that the body/mind/spirit system is the accurate portrait of the human being.

The contribution Donna Eden has made with *Energy Medicine* will stand as one of the backbone studies as we lay a sound foundation for the field of holistic medicine. Her research is solid and the suggestions she gives for her readers to use in order to help themselves are simple and effective, the way help should be.

As someone who has been a part of the holistic field for fifteen years, I have met numerous people who either have wanted to know more about energy medicine as students or have needed to make use of its techniques. I feel extremely confident that this book will serve as a guide for people for years to come, because its information is based upon truth and truth does not change.

Looking at our lives these days, it would seem unwise not to acknowledge that we are residing in an age when the energetics not just of the body but of all of life are reshaping our world. The computer age represents our coming to rely upon energetic rather than vocal or written information. We are moving at a pace that is racing to keep up with the speed of thought—and thought drives the human body. Energetic medicine recognizes the power of our thoughts and of this world we are now residing in. We need texts of the quality achieved by this book to serve as guides along the way.

—Caroline Myss, Ph.D.

Energy Medicine

Introduction

The Return of Energy Medicine

OUR REMEDIES OFT IN OURSELVES DO LIE.

—WILLIAM SHAKESPEARE
As You Like It

You are a latticework of energies. The enormous implications of this single fact are the basis of energy medicine. I invite you to step into this introduction to that field and enter a domain beneath the world of appearances. Explore how invisible energies shape the way you feel, the way you think, and the way you live. Marshal these energies for your health and well being. Study your eternal dance with the unseen forces within and around you. And as you do, marvel with me as the dynamic energies that stream through your body, mind, and soul exquisitely reveal the genius of nature's "fashioning hand."[1]

The first practitioner of energy medicine is *you,* the one who inhabits the body being cared for. Using the principles of energy medicine, you can optimize your body's natural capacities to heal itself and to stay healthy. You can bring renewed stamina to a tired body, fresh vitality to a weary mind, and new bounce to a sagging spirit. You can manage your energies to more effectively meet stress, reduce anxiety, and free yourself of many ailments. And you can apply what you learn for yourself to benefit family members and other loved ones.

To cultivate these abilities, you will be learning a language your body already speaks and understands—the language of energy. Energy is the common medium of body, mind, and soul. Its wavelengths, rates of vibration, and patterns of pulsation form their shared vocabulary, much as fluctuations of tone and tempo form the vocabulary of music. The more fluent you become in sensing this shared vocabulary of body, mind, and soul, the more skillfully you can orchestrate their lifelong symphony.

You are, in fact, required today, when virtually everyone's lifestyle has become so alienated from the natural order, to live in conscious partnership with your body's energy systems if you are to live fully. And it is easier than you might think to learn how to mobilize inner forces that enhance your health, empower your mind, and literally brighten your spirit.

ENERGY MEDICINE IS TIMELESS

Energy medicine is safe, natural, and accessible. It is both contemporary and ancient. The term is being used in many ways today, ranging from the introduction of shamanic healing practices in modern cultures to the use of powerful electromagnetic and imaging technologies in modern hospitals. It embraces lawful principles and mysteries; routine procedures and artistry; matter and spirit. *Energy medicine* is the best term I know for describing the growing number of approaches where an understanding of the body as a system of energies is being applied for promoting health, healing, and happiness.

Energy medicine is the art and science of fostering physical, psychological, and spiritual health and well-being. It combines a rational knowledge and intuitive understanding of the energies in the body and in the environment. Cultivating your capacity to weave these energies can make you a more *conscious* and *sensitive* instrument of energy medicine than all the other technologies combined. By focusing on your body as a living system of energy, you begin to realize that powerful energy technologies are *already* inherent in your hands and in your being.

The term *energy medicine* is itself a double-entendre. Energy gives life to the body. Medicine is an agent that is used to heal or prevent disease. In energy medicine, energy *is* the medicine. And in energy medicine, this medicine *is applied* to the body's energies. Energy *heals,* and energy systems *are healed.*

The return of energy medicine is one of the most significant cultural developments of the day, for the return of energy medicine is a return to personal authority for health care, a return to the legacy of our ancestors in harmonizing with the forces of nature, and a return to practices that are natural, friendly, and familiar to body, mind, and soul. Regardless of the dazzling technologies that have already been introduced, such as electromagnetic imaging, or the undreamed-of advances that will come along, the essence of energy medicine will always be the *energy systems* that make up the subtle infrastructure of every body.

WORKING WITH THE BODY'S ENERGIES

I have always had an intimate relationship with the body's energies. I register them through my senses. Early in my career, licensed as a massage therapist, I would see and feel in my clients patterns of energy that told me about their physical problems. One of my first clients was a woman with ovarian cancer who came for a session with the hope that I could help her relax her body and prepare it for surgery that was scheduled in five days. She had been told to "get her affairs in order," as her immune system was so weak that her chances of surviving the surgery were limited. Metastasis was also suspected.

From looking at her energy, I was certain that the cancer had not metastasized. While her energy was dim and collapsed close to her body, the only place that looked like cancer to me was in her left ovary. In addition, the texture and vibration and appearance of the energy coming up through her ovary were responsive to my work with her. I could see and feel them shift, and by the end of the session, the pain that had been with her for weeks was gone.

I told her that her body was so responsive to what I had done that I wondered about her plan to have surgery. I was concerned that her immune system was indeed too weak, and I was confident that by working with her energy, not only would her immune system be strengthened, the tumor's growth could be reversed. While I carefully made my statements with the strong disclaimers required to avoid immediate arrest for practicing medicine without a license, she responded with horror to the implication that she cancel the surgery. I suggested that she at least delay the operation for two weeks. She scheduled a session with me for the next day and said she would discuss the surgery with her husband.

That evening I received a call from her husband. He was outraged and threatening. He called me a quack. He said I was putting his wife's life in jeopardy by giving her false hope, and he told me I would never have another chance to confuse her like this. He made it clear that she would not be coming back. When I began to respond, he hung up. I called back a short while later. She answered. Talking in hushed tones, she was clearly uncomfortable speaking with me. I said, "Okay, don't postpone the surgery, but please keep your appointment tomorrow. You don't have to pay. You have nothing to lose. I believe in what I am saying. In fact, I want you to bring your husband in with you. Find a way!" She did not believe he would come in, but the next day, they both arrived for the appointment.

I had her lie down on the massage table. My hope was to find a way to give this traditional and skeptical man, so poignantly fierce in his protection of his wife, an experience of healing energy that his senses could not deny. I could see a dark, dense energy at the site of his wife's left ovary, and it felt like my hand was moving through a muddy swamp. I asked the husband to place his hand a few inches above the area and begin to circle it, using a motion that tends to draw energy out of the body. To his great surprise, not only could he immediately feel that he was moving against something, within two minutes his hand was pulsing with pain. To his utter amazement, his wife reported that her pain diminished as his increased.

By the end of the session she was again pain free, felt better, and looked better. I showed them both, through the use of *energy testing* (to be explained later), that we had been able to direct healing energies from her immune system to the area of her cancer. I taught him a set of procedures to use with her every day. They decided to temporarily postpone the surgery and to ask for further medical tests before rescheduling it. After about ten days of these daily treatments from him and three more sessions with me, she went through the additional testing. The tumor was gone.

A PERSONAL JOURNEY INTO HEALING

I was, my mother tells me, born smiling, and the first energy I can recall feeling is the energy that wells into a smile. When I smile, it doesn't actually feel like *I* am smiling. It feels more like an energy is *smiling me.* I love to feel that energy come up through my face, my cheeks, my eyes, and permeate my entire being. Smiles and laughter are your birthright. If you can expand their energy within yourself, you will expand your joy. And you will bolster your health. While this book will focus on many kinds of energy, the joyous energy that wells into a deep smile is not insignificant. It is a natural and powerful healing force, and by clearing and balancing the various other energy systems in your body, you open a channel for it to shine through.

I write this book partially on the basis of my work with people's energies and partially on the authority of having overcome a series of personal health challenges. I was born with a heart murmur, contracted tuberculosis at age five, had terrible food allergies and hayfever, showed the first symptoms of what was later diagnosed as multiple sclerosis when I was sixteen, had a mild heart attack in my late twenties, had severe asthma in my early thirties, a breast tumor at thirty-four, and I have been hypoglycemic and struggled with severe PMS since I was twelve.

I was also given a fairly free spirit, and I took these difficulties in stride. But I learned quite early that conventional wisdom didn't always work for me, and I've had to use my body as a laboratory. Aspirins gave me headaches, sleeping pills kept me awake, fruits and vegetables made me gain weight. Nor were doctors particularly successful with this body that didn't seem to follow the rules.

By my early thirties, my health was extremely precarious. I retreated to Fiji to live a very basic life. Early in my stay, I was bitten by a poisonous insect. Because my immune system was already badly impaired, I had no resistance to the bite. I became very sick and went in and out of coma. It seemed I might die.

But the shamans of Vatukarasa, a nearby village, learning of my plight, came with a treatment for the bite. They buried me up to my neck in the sand and left me there for long stretches several times each day over a forty-eight-hour period. They believed that the toxins would be drained into the sand. I recovered. This was one of many incidents that turned the wheel, setting my course toward healing work.

My health returned while I was in Fiji. My family and I lived out in the wilds, far from the nearest town. We swam in the ocean every day. We ate breadfruit from the ground and fish from the sea. Nothing was processed. Nothing was canned. Everything was organic. There were no fumes from cars and no chemicals in our clothing. Life was lived at a slow pace. It was a world where you could just *be*. There was no competition and little stress. There was no radio, no newspaper, no television. After a time, I wasn't even sure the United States still existed.

By simply living naturally, I grew healthy. This experience was reminiscent of a formative period during my childhood. My mother contracted tuberculosis when I was four. She was put in a terminal ward, not expected to live. My father eventually brought her home, but our whole family was quarantined. I had tuberculosis by age five. My mother was told she would not survive if she wasn't on penicillin for the rest of her life. Instead, she switched to all natural foods and high amounts of vitamin C. We raised chickens so we could have fresh eggs. My father planted a garden so we could eat organic vegetables. Everything we ate was pure. We all got well. In the same way, everything I ate in Fiji was pure, and again I grew healthy.

However, when I returned to the States in 1977, I went into culture shock. My taste buds had become so sharp that I felt assaulted by the chemicals in the foods I ate. I could taste the packaging in which even the healthiest foods had been stored. I could feel the chemicals in clothing. I wanted to move to some tiny town, to get away from cities and everything that is polluted. I wanted to raise my daughters in a healthy world. But my marriage was ending, and I didn't know how I was going to make a living.

Even though healing work comes naturally to me, I had no idea that feeling and seeing energies could lead to a career. I had taken a premed curriculum in college, but my sensibilities were offended by an approach to health that was based more on what you could learn from cadavers than what you could learn from the body's living energies.

Shortly after returning from Fiji, through one of the more striking synchronicities of my life, I wound up in a Touch for Health instructor training course taught by Gordon Stokes and Shanti Duree. In the early 1970s, Dr. John Thie—after a collaboration with the founder of applied kineseology, Dr. George Goodheart—developed a health education system for laypeople called Touch for Health. Combining Chinese medicine with empirical techniques from the West, applied kineseology and its gregarious stepchild, Touch for Health, represent a potent synthesis of healing practices found in both cultures. Many of the techniques presented in this book are adapted from these two systems.

I learned about such approaches to healing after I encountered a woman wearing a T-shirt inscribed with a picture of a hand and the words "Touch for Health." When I asked her about it, her only reply was, "Oh, I'm so excited, I'm leaving next week to be trained as a Touch for Health instructor." I felt lightning strike. I still didn't know what Touch for Health was, but I heard myself say, "Me too." She gave me a phone number. I called the Touch for Health office to have them send me information. They made one of those serendipitous errors that changes a person's destiny. They sent a letter congratulating me for having successfully completed the basic Touch for Health classes, which were the prerequisites for attending the teacher's training. The next teacher's training began the following Tuesday. Off I went. Everybody else there already had a solid foundation in the Touch for Health system.

While I had never before been exposed to any training in alternative medical practices, I had the sense that I had come home to something that was deeply familiar. The training was ideal for me. Touch for Health gave me a structure that balanced my intuitive nature and a form with which to work with the energies I could see and intuit. Its use of muscle testing—which I also call energy testing—gave me a tool to demonstrate for a client or a student what I was seeing.

The training was residential and intensive. I completed it on a Tuesday and began teaching my first course that Friday. Another graduate of the training, Hazel Ullrich, and I charged ten dollars for an entire weekend class. I told the participants, "I've never taught a class before in my life, and that's why it is so cheap. I'm charging so little so I can have the space and feel the safety to make mistakes while I figure out how to teach this to you." I freely admitted my beginner's status and proceeded to learn how to teach energy healing.

I next studied therapeutic massage so I could be licensed to touch people. The Mueller College of Holistic Studies in San Diego had stringent requirements, and I learned a great deal about anatomy, physiology, and the many forms of therapeutic touch. The licensing exam for massage therapists in San

Diego in 1977 was, believe it or not, administered by the vice squad. Their interest was in preventing prostitutes from becoming licensed massage therapists. I'm not good at tests, so I felt nervous as the date approached. Alone in a room with a gruff, intimidating vice officer, I waited to see if he was going to ask me questions or ask me to massage him. Instead, he said, "Let me see your hands."

He looked at them briefly and said, "You pass." "How?" I asked. "You don't have nail polish and your fingernails are short," he replied. "You're not a prostitute."

Armed with this dubious certification of proficiency, I opened a private practice. My private practice and teaching have been my primary professional activities for over two decades. With my own physical illnesses and a lifetime of finding how to alleviate them as a backdrop, I have vigorously pursued ways to help people balance their own energies and heal their ailments. Many of my clients have been suffering with problems that have not responded to conventional medical treatment. Each of my twenty-two years working with them has brought me more deeply into the intricacies of energy medicine and has helped prepare me to write this book.

ENERGY KINESEOLOGY: MY BRAND OF ENERGY MEDICINE

I am ambivalent about the word "healer." It implies doing something to someone else, holding power over the other, separating the healer from the healed. Because the healing relationship feels sacred to me, I think of it more as being a minister—ministering to body, mind, and soul.

I do, however, appreciate the concept of the wounded healer. With my own personal history of health challenges, I could be the poster girl for "We teach what we need to learn." I also know that when you heal yourself, you discover what no one can teach you. It is an initiation into the very foundation of life, and it organically seems to follow that you have compassion for people who are frightened about their health and want to offer the harvest of your experience.

Since my introduction to Touch for Health, I have studied many natural healing approaches and taken something from each. Still I have never been attracted to set formulas. Each person is unique; each client takes me on a healing journey I have never before travelled. It is difficult to name my approach because I have borrowed from so many systems and drawn so deeply from my own instincts. I finally settled on the term *energy kineseology,* with a respectful nod toward applied kineseology's pioneering use of muscle testing.

Fresh understanding about energy and healing continually filters through me, so I evolve with each client I see. I also cherish and take confidence in knowing the body is designed to heal itself. Your body is engineered so that if you tap into its healing force, that force will lead you toward health. It is not just the personality or the soul wanting the body to get better. The *body* wants to heal, and every cell carries extraordinary intelligence and fortitude. While we all sometimes need outside help and direction, *healing is an inside job.*

HOW TO USE THIS BOOK

You can approach this book at any of several levels. You can go through it alone, practice its techniques with a friend, or use it as part of a study group. By reading the text and experimenting with the exercises, you will develop a solid overview of energy medicine to the extent it can be practiced by the nonprofessional. You can read the book without doing the exercises and return later to sections that are relevant for you. You can also use it as a reference guide. Individuals in the healing professions as well as individuals with an interest in self-healing can scan the chapters and refer to the table of contents, section headings, and index to locate information on specific topics. For instance, if you are buying the book because you suffer with chronic pain, you could begin by going directly to the pain chapter (Chapter 10) and consulting earlier chapters as needed. When a technique requires specific knowledge from previous chapters, you are referred to the relevant pages from those chapters.

Part I, "Awakening the Two-Million-Year-Old Healer Within," emphasizes my belief that we instinctively know a great deal more about our body's energies and how to optimize them than we usually realize or take advantage of. The first chapter addresses some of the elusive terms that arise in energy medicine—such as energy, subtle energy, soul, and spirit—and it recognizes the language of the body's energies as a language that can be learned. Chapter 2 introduces techniques for identifying the energies that operate within your own body as well as the energies that surround you. The chapter presents energy testing as a method for assessing your personal "energy signature." Energy testing also serves as a tool for determining the impact of the environment on your own energy field. In addition, it allows you to tailor the procedures in the remainder of the book to your unique needs. Chapter 3 introduces a "Daily Energy Routine" that can *unscramble* and bolster your energies for your own health and healing. It also shows you how to experiment with energy testing to gauge the benefits of each of its methods, and it closes with an enormously valuable approach for reprogramming your body's response to stress.

Part II, "The Anatomy of Your Energy Body," details eight major energy systems. Because my sensory apparatus translates subtle energies into pictures, I see in the human body a spectrum of energies that is often more colorful than the kaleidoscope of flowers in a spring meadow. Over the years I've found that I focus primarily on eight distinct patterns or energy systems. Descriptions of every one of these energy systems, it turns out, exist in the healing lore of one culture or another. Some cultures recognize many of them, others one or two. The eight energy systems include the meridians, the chakras, the aura, the Celtic weave, the basic grid, the five rhythms, the triple warmer, and the strange flows. I am not claiming there are only eight systems of energy in the human body, but I am claiming that within these eight systems, you can chart a systematic route to better health. If you keep these energy systems thriving, greater physical and emotional strength will follow.

Part III, "Weaving It All Together," shows how to apply what you've learned to what you encounter in your life. You will be focusing on ways of meeting your body's inevitable illnesses more effectively, relieving pain, and creating external and internal energy fields that optimize your health and happiness. The Epilogue invites you to reflect upon the mysterious realms that can be opened by an energy approach, and it considers the way such experiences can provide a window into your own soul's journey.

Integrated throughout the text are over a hundred procedures I have adopted or developed after more than ten thousand individual ninety-minute sessions and hundreds of classes. Self-care techniques are particularly important in energy medicine. While a powerful healer can on occasion effect a one-session cure, a good energy session almost invariably begins a process that requires ongoing support for a time if longstanding patterns are to be corrected. Unlike the primary tools of conventional medicine—surgery, drugs, and radiation—the techniques of energy medicine are gentler, more organic, less intrusive, and they lend themselves to becoming part of your lifestyle rather than being confined to the treatment setting. The procedures presented here have been designed for use with no special ability to perceive energy. Employing them, however, is a natural way to cultivate such sensitivities.

This is the kind of book where you are invited to use what draws you. As a manual covering eight major energy systems, it is far too detailed for you to expect yourself to master every exercise on your first reading. But my intention is straightforward. The book is designed to empower you. The first three chapters are themselves an introductory class in energy medicine. You could set the book aside with a sense of completion after reading them. You will have gained an overview of how energies can be optimized for your health in Chapter 1, picked up the invaluable tool of energy testing in Chapter 2, and mastered, in Chapter 3, a basic five-minute routine for keeping your energies balanced and vital. Or you could continue on and acquire more advanced information and skills as you proceed.

You can also randomly thumb through the book and find "how-to" boxes that address a potpourri of common complaints. These are not the challenges that usually bring people to me but the practical problems they mention incidentally, from how to prevent a cold to how to calm one's kids so they will go to bed at night. These boxes refer you to the specific pages where the techniques suggested are described. I offer them partly because they are so practical and partly because they give you a taste of how you can figure out ways to apply the techniques in the book to virtually any context.

References to laboratory or clinical research supporting many of the book's major assertions and suggestions are sprinkled throughout the text. When I entered the field two decades ago, anyone offering any form of energy medicine was going out on a scientific limb. Since that time the scientific literature has seen an explosion of studies related to the body's energies.[2] Although still a fringe area within medicine (even impressive research has generally not yet been reported in mainstream medical journals), energy medicine has reached a healthy adolescence.[3] Its rite of passage into American medicine occurred in 1992 when the National Institutes of Health established the Office of Alternative Medicine. While many younger health practitioners recognize that the future of energy medicine looks bright, the field's exuberance and unfamiliar paradigm make the old guard uneasy. The growing body of research sampled in the text hopefully provides a bridge that skeptics will at least find provocative.

Energy Medicine: Balance Your Body's Energies for Optimum Health, Joy, and Vitality lays out a systematic yet open approach for consciously working with your own energies, or the energies of those you care about, in order to achieve a healthier body, sharper mind, and more joyful spirit. Deciding which techniques are and are not appropriate for a self-help health book required a string of judgment calls on my part. I have generally not pulled my punches in presenting techniques that are potent and that can be helpful with serious health problems. On the basis of my experiences in teaching these techniques to laypeople in hundreds of classes, I have a good idea of how each method presented here impacts a wide range of people, and I have used my best discretion. Your part of the bargain is to exercise your own best discretion as you go through the program. If you need professional treatment, please consult a competent health practitioner! The techniques in this book will only complement the care you receive. But while sincerely stressing the value of professional intervention (see the appendix, "How to Find an Competent Energy Medicine Practitioner"), I do not want to undercut my primary message and deep conviction that there is much you can do to care for yourself. Ultimately it is you who is responsible for your health, and the more you know and the more you do for yourself, the better your health will fare. This book offers tools. May they serve you well.

Part I

Awakening the Two-Million-Year-Old Healer Within

BENEATH OUR CONSCIOUS INTELLIGENCE
A DEEPER INTELLIGENCE IS AT WORK—
THE EVOLVED INTELLIGENCE OF HUMANKIND.

—ANTHONY STEVENS
The Two-Million-Year-Old Self

Chapter 1

Energy Is All There Is

$E = mc^2$

—Albert Einstein

Your body is designed to heal itself. The ability of a body to maintain its health and overcome illness is, in fact, among nature's most remarkable feats. But you've been placed in a world that systematically interferes with this natural capacity, and your conscious involvement in your health is required if you are to truly prosper.

This book will show you how to work with the electromagnetic and more subtle energies that give your body life. These energies form the foundation of your health. They are your body's fuel and its atmosphere. You were spawned within Earth's electromagnetic, gravitational, and nuclear fields. You were raised under the sun's life-giving rays. Your own energy systems, such as your meridians (your body's energy pathways) and chakras (your body's energy centers), emit electromagnetic energy and light.[1] As you will see, energies that are more subtle than we have even known how to measure also dwell within and around you, and they are the key to energy medicine.

While our culture does little to help us look more closely, energy *really* is all there is. Even matter, as Einstein's elegant formula shows, is congealed energy. When you watch a log burning in a fireplace, you are seeing the congealed energy that is the log transform into the roaring energy that is the flame. The flame could then be transformed into mechanical energy, where it might propel a locomotive or run a generator. That generator might, in turn, produce electrical energy. Perhaps, as Einstein believed, there is only a single energy, "a unified field," but if so, it has countless faces.

Numerous cultures describe a matrix of subtle energies that support, shape, and animate the physical body, called *qi* or *chi* in China, *prana* in the yoga tradition of India and Tibet, *yesod* in the Jewish cabalistic tradition, *ki* in Japan, *baraka* by the Sufis, *wakan* by the Lakotas, *orenda* by the Iroquois, *megbe* by the Ituri Pygmies, and the *Holy Spirit* in Christian tradition. It is hardly a new idea to suggest that subtle energies operate in tandem with the denser, "congealed" energies of the material body.

THE INTELLIGENCE OF THE BODY AND ITS ENERGIES

The core premise of this book is that the body and its energies are intelligent, and that you can engage them in intelligent dialogue that fosters your health.[2] This becomes quite apparent when you are working in a healing context and connect with the energies that flow through the human body. Often, when I am with a client, I feel directed by the energies themselves, as if an amazing intelligence has taken over the session. I can do nothing better than pay attention and follow.

The biologist Lewis Thomas beautifully expresses his awe at the capacity of a single cell to make intelligent choices in his classic book *The Lives of a Cell.*[3] I feel that way about energy. The closer you look at the ballet of energy and protoplasm, the smarter each appears. In fact, the idea that a complex organ such as the heart carries a sort of consciousness has been gaining favor. In a remarkable book, *A Change of Heart,* heart recipient Claire Sylvia describes the way many people who have had a transplant may suddenly be obsessed with the thoughts, memories, dreams, tastes, and desires of the organ donor.[4] The more you work with the subtle energies in a person's body, the more clear it becomes that you are encountering and collaborating with an intelligent force.

SEEING SUBTLE ENERGIES

I tend to vibrate to the energies of other people. At times I feel like a tuning fork. I see and sense other people's energies as rhythms and vibrations, frequencies and flows, jolts and currents, colorful swirls and geometric patterns. Early in my life, I came to understand that the colors, shapes, movements, and textures I saw hold meaning.

As a fourth-grader, I once overheard a number of teachers gossiping about my teacher, Miss Proctor, belittling her intelligence, ridiculing her as strange and eccentric, wondering out loud how she ever got through college. I was shocked and totally puzzled. Were they blind? The strongest thing about Miss Proctor was a beautiful, pale, creamy yellow energy emanating from her body that said volumes to me. I intuitively knew it meant she was both wise and kind. She led with a sort of innocence, but it seemed obvious to me that she was an advanced and trustworthy being. Her peers, as I later came to understand, would have found it easier to respect her had she assumed a more sophisticated demeanor rather than being so spontaneous and carefree.

To this day, my appraisals of people are based on the subtle energies I feel and see emanating from them, not on their words, physical appearance, status, or personality. This quirk has proven invaluable in my work. I have learned over the years not only to understand that what I see and feel in a person's energies has meaning but also how to use my hands to weave these energies to improve a person's health, vitality, and clarity of mind. Matter follows energy. That is the fundamental law of energy medicine. When your energies are vibrant, so is your body.

SUBTLE ENERGIES IN HEALING

Leah was carried up the stairs to my office.[5] The man who brought her told me she had been to the Scripps Clinic, the Mayo Clinic, and many, many doctors. Leah was diagnosed with bronchitis, but no one could determine why she wasn't responding to established treatments, and it appeared she was dying. Her closest friend had died just before the onset of her illness, and she had been to a psychiatrist on a doctor's hunch that her bronchitis might be exacerbated by complicated grief, but to no avail. She was reduced to grabbing at straws, and I guess I was a straw.

When Leah tried to speak, she could only wheeze. Because she had already been tested by the best, I didn't start with the obvious systems, involving her lungs, but rather sank into her deepest energy system, which I call the basic grid. It appeared to me that two of the energy pathways in her basic grid were so seriously disrupted that her life force could barely flow through them. I touched the points that are at either end of one of these energy patterns (one near her hip and one on her forehead), using my own body like a jumper cable to connect the circuitry and rebuild the grid.

When this energy began to move through me, it at first felt jagged. Then I couldn't breathe. Leah was looking at me, not knowing what to think about this alternative healer who was now in some kind of agony. It went on and on. I was gasping for breath for about thirty minutes. At the moment I finally felt the energies hook up through my body, she gasped and then took her first normal breaths in months. My breathing returned as well.

I moved to the second blocked energy pathway. Very soon after I began to hold its points, a grief came through me that was as deep as any personal grief I've ever known. It was overwhelming. I began to cry and couldn't stop crying, regardless of how it might look to Leah. She told me later that had it not been for the fact that she could now breathe better, she would have been off the table and out of there in an instant because of my lack of professionalism. After what seemed a very long time—at the moment I felt the energies connect—*she* began to sob in uncontrollable, loud, heaving grief.

A puzzle was instantly solved for her. Leah and her best friend had travelled the world together. When her friend became ill with cancer, Leah had decided they would take this last journey together, and she had spent the final year of her friend's life caring for her. When her friend died, Leah's grief was not so much about the loss of her friend. They had consciously grieved together during that final year, and she had been irritated with the psychiatrist who insisted her physical problems were based on unresolved grief. No! She had fallen into a pit of existential despair over the thought that no one would ever be there for her as she had been for her friend. At a level below her conscious awareness, she had given up on life. Her sense of emptiness and isolation had buried itself in her lungs and manifested as bronchitis. With her basic grid reset and her immune system given a boost, I could only hope for the best, as she was contending with a major illness. Still, she had been carried in, and now she walked out.

Several days later, I came to my office to find a long string of Guatemalan tribal bells hung on my door with a note from Leah telling me that she had been getting "better and better." She also felt renewed in spirit, and she had a vision for her future. She wanted to study energy kineseology and then go to Guatemala, where she and her friend had dreamed of settling. She came to see her healing crisis as a gift that opened her to a new sense of purpose. After studying with me, she moved to Guatemala and began to do healing work. For several years I received cards from her each Christmas with stories of people she had successfully treated.

How to Nip an Illness in the Bud

If you are just starting to feel sick, you can often reverse an illness at the onset with a few well-chosen energy techniques. Any or all of the following will help. The first three can be done without a partner. (time—5 to 20 minutes):

1. *Do the Separating Heaven from Earth exercise (page 249).*
2. *Do the Hook Up (page 119) and the Three Thumps (page 63).*
3. *Massage your neurolymphatic points to detoxify your body (page 118).*
4. *Have a friend do a Spinal Flush on you (page 79).*
5. *Have a friend clear your chakras (page 167)*

ENERGY WORK IS SOUL WORK

The medical institutions that treated Leah's bronchitis without correctly assessing how her symptoms reflected the predicament of her soul failed to help her. Illness and healing are way stations on the soul's journey. The soul, however, is not an easy concept to fathom, and conventional medicine has hardly even attempted to incorporate soul-level concerns into its understanding.

To approach the soul, Renaissance theologians would have an initiate study the vastness of the night sky as a metaphor of the vastness within. It is an illuminating exercise. The soul is the source of the most subtle energies of your being. Yet this subtle energy gives form to everything else about you, from your cells to your sense of self. If spirit, as it is often defined, is the all-pervasive, intelligent energy of creation, soul is its manifestation at the personal level. Soul and Spirit are the unfathomable, vital mysteries of our existence. We can, however, experience them directly, whether through the energies of love, contemplation, healing, or mystical experience. William Collinge, a researcher of subtle energies, describes energy as "the bridge between Spirit and matter":

> *Einstein showed through physics what the sages have taught for thousands of years: everything in our material world—animate and inanimate—is made of energy, and everything* radiates *energy. . . . He concluded that the continuously unfolding and dynamic nature of the universe could only be understood as the work of a higher guiding intelligence of another dimension.*[6]

I think of soul as the spark of Spirit that infuses the body with life and the brain with consciousness. When the soul leaves, the brain dims, the body dies. To work with a person's energy is to touch a soul as well as a body. Ironically, the deeper you enter into the life of your personal soul, the more fully you identify with your roots in the life of a universal, unifying intelligent Spirit. And the better your body will fare.

When all your energies are brought into harmony, your body flourishes. And when your body flourishes, your soul has a soil in which it can blossom in the world. These are the ultimate reasons for energy medicine—to prepare the soil and nurture the blossom.

How to Overcome Despair Through Spiritual Connection

Whenever you are in despair, feeling alone, sick at a crossroads, suffering with physical pain, in anguish because of an illness that will not heal, or otherwise unable to find the answers you ache to find, the following can be an uplifting practice. It can put you in touch with the sacred dimension of your life, infuse you with the knowledge that you are not alone, and give you the odd comfort of realizing that all you see is only a glimpse of a larger picture.

If at all possible, do this out in nature, perhaps under the stars, or in another place that feels sacred. For me, this is like a prayer. It connects me with the spiritual realm. I call it Heaven Rushing In *(time—2 minutes or longer):*

1. *Stand tall. Take a moment to ground yourself by spreading your fingers on your thighs, breathing deeply, feeling your feet on the ground, and being conscious of your connection to the Earth as the energy pours out of your fingers, down your thighs, and into the ground. You are preparing yourself to make a sacred connection.*

2. *Take a deep breath in, open your arms wide, and bring them into a prayer position in front of your chest.*

3. *With another deep breath, open your arms wide, lifting them up. Look to the heavens (see Figure 1). Reach toward heaven as heaven reaches back to you. Release your breath. Bask in the knowledge that you are not alone in this universe and that you are worthy of this blessing from the heavens. You may feel a tingle, a buzz, or heat in your hands. You have been inviting healing energies from the cosmos.*

4. *Scoop this energy into your arms and bring your hands into the middle of your chest. There is a vortex here called Heaven Rushing In, and "heaven" rushes into your heart with healing, with a glimpse* *of your true nature, and with a peek into who you are in the larger plan. Even when you do not receive guidance or inspiration, know that they will unfold in their perfect time.*

5. *If there is a specific area in your body that needs healing, place your charged hands over that area and let the energies stream in.*

IN AND OUT OF BALANCE

Energies from the Sun and from the Earth penetrate your every cell, and they give shape to your energy body. Your energy body, in turn, becomes a distinct, self-regulating universe, a force within your body and a force in the environment. It is continually interacting with the energies around it and moving its own energies to warm you, cool you, activate you, calm you, and establish a cycle of repair and rejuvenation. In this exquisite alchemy, energies are built up, stored, spent, transformed, harmonized, and brought into balance.

Balance is a pivotal concept within energy medicine, just as homeostasis is a pivotal concept within biology. All systems move toward an energetic balance, a state of internal stability and harmony with other energies. At the same time, every expenditure of effort and every interaction with the environment upsets this balance. You are always moving toward balance and always disturbing that balance in living and growing.

When one of your body's energy systems is chronically out of balance, or when several systems are not in harmony with one another, your body does not work as well. Your energy body is always adjusting the energies available to it to restore its balance. Several facts of modern life make it more difficult than perhaps ever before in history for your body to maintain the energetic balances that would best preserve and nurture it, ranging from our multiple and highly refined psychological stresses to our immersion in polluted air, processed foods, and artificial electromagnetic energy. This book will describe how to zero in on precisely the energies in your body that need balance, and it will show you how to balance them by dissolving energies that

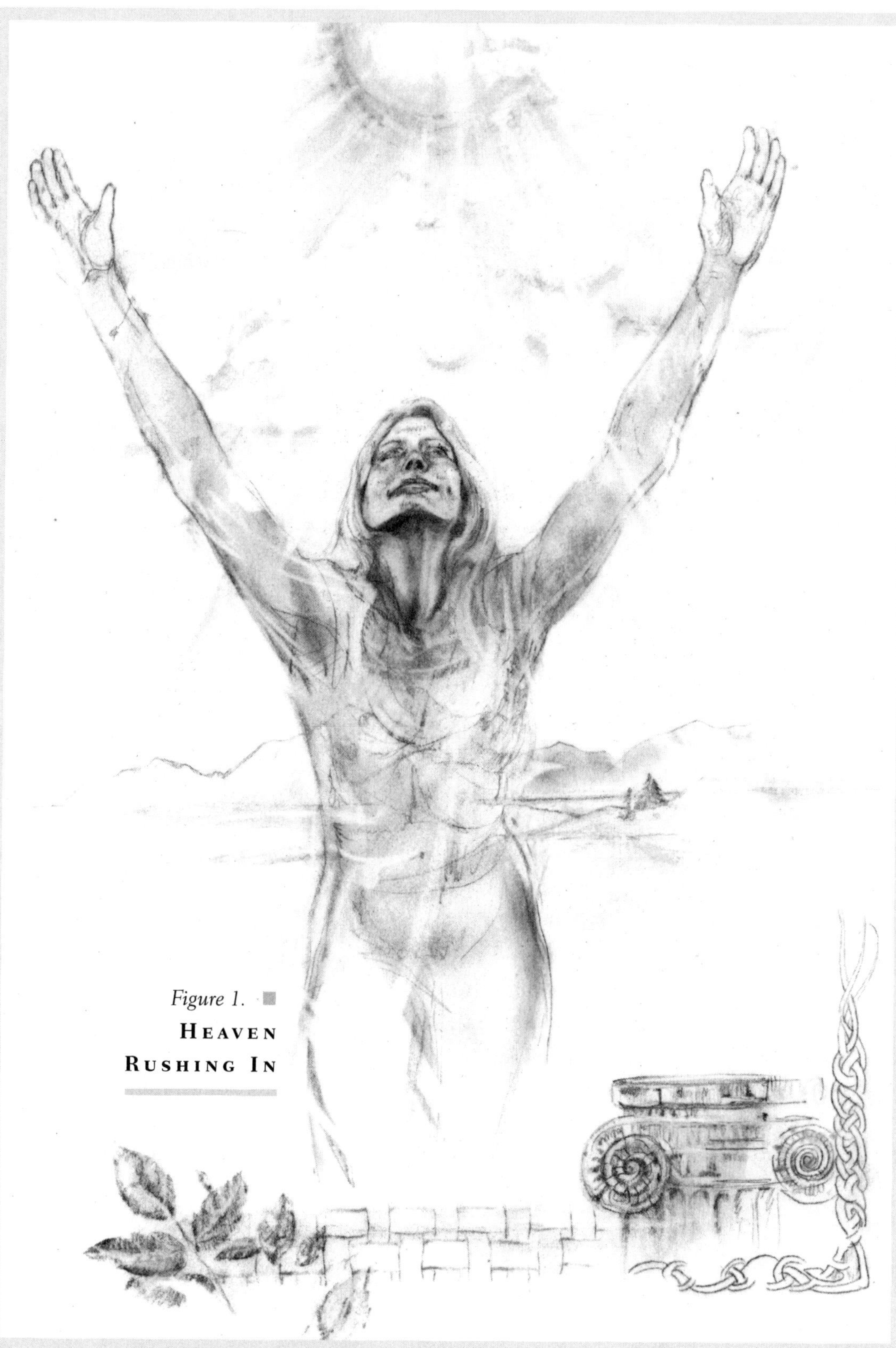

Figure 1. **Heaven Rushing In**

are stagnant, boosting those that are faint, and disbursing or sedating those that are excessive.

AWAKENING THE TWO-MILLION-YEAR-OLD HEALER WITHIN

Having taught healing classes all over the world, I firmly believe that virtually anyone can learn to influence their energies to improve their health. You can do this! Instinctive knowledge of how to heal themselves was a possession of your ancestors, and it still resides deep within you.

Evolutionary psychologists have shown that survival strategies are passed along in our DNA.[7] Healing energies are embedded in your tissues, and healing abilities are coded in your psyche. Carl Jung spoke of the archaic component of the human psyche as the "two-million-year-old man" who lives within each of us.[8] Two million years ago, our ancestors, with brains only slightly larger than those of the gorilla, were carrying out many cultural activities. Not only were they "hunting, building shelters, making tools, treating skins, [and] living in base camps,"[9] they were also healing one another. For most of humanity's history, *natural* healing was the only game in town.

Our ancestors had to depend on themselves entirely for their own and their clan's survival. They couldn't go to the neighborhood clinic for an antibiotic or a kidney transplant. They were their own authorities, which forced them to be far more finely attuned than we are to subtle energies. They sensed the energy in a potential food, a wild animal, or a diseased compatriot, and they knew how to regulate their own energies to maximize their strength and proficiency. They could not have survived the nasty and brutish world of millions of years ago if they had not intuitively understood what was required to maintain, in a hostile environment, the most complex physical structure that had yet evolved. That your ancestors knew how to heal themselves by working with their energies and that their hard-earned survival strategies are deeply imprinted in your own urges, passions, and reasoning should bolster your confidence in your own latent abilities. You need only knock on the door to arouse the two-million-year-old healer dwelling within you.

How to Rid Yourself of a Tension Headache

The traditional headache points in Oriental medicine have been used for thousands of years and are effective (time—about a minute):

1. *Massage the traditional headache points at the indents where your neck meets your head, move up to the ridge of the skull, and massage all across the back of that ridge, rubbing hard with a circular motion.*

2. *Place your fingers at the bottom of your neck, next to your spine. Push in on the back of your neck with a circular massage.*

3. *Maintaining pressure, drag your fingers out to the sides of your neck.*

4. *Repeat, moving up a notch higher. Work your way up to the top of your neck.*

IS THE BODY'S ENERGY LANGUAGE A LANGUAGE THAT CAN BE LEARNED?

Our healing abilities still reside in our ancestral bones, our healing hands, our compassionate hearts, and our intuitive brains. I believe that as infants we see and feel the energies that surround people and that we innocently absorb information about their health, mood, character, and soul. But our culture does not talk about or reinforce such perceptions. By the time we are two or three, these capacities atrophy from lack of use, just as empathy, which is an inborn response, will fade if not mirrored by the infant's caregivers.

Nonetheless, I have time and again watched this capacity to see or feel subtle energy rebuild itself. Occasionally someone will even begin to hear or taste the subtle energies that pulse through the body. Animals are naturally attuned to these energies,[10] and humans can cultivate this ability. I once saw a demonstration by blind people in Java who had learned to attune themselves so accurately to the energies of inanimate objects that they were able to flawlessly walk through mazes constructed of bamboo poles, moving with the precision of bats navigating by sonar.

While some people have more healing aptitude than others, just as some have more natural ability in the arts or physics or athletics, I have never worked with a person who had no healing ability at all. And I have many times witnessed people in my classes see energy fields for the first time or perform a successful healing, instinctually innovating a valid method that had never been taught to them.

I saw an amazing healing while I was teaching a large class in San Diego. Within the first half hour, a man in the back of the room had a heart attack and fell to the floor. I couldn't get back to him as a sea of people crowded around. A physician who was sitting nearby did get to him. He began to administer CPR as the paramedics were called. The doctor had clearly become the person in charge, but he couldn't get the man's heart going, and before the paramedics had even arrived I heard him say, "He's gone." My heart jumped into my throat. At that moment, as I anxiously watched the whole scene unfold, a boy of about sixteen suddenly seemed electrified and made his way over to the man. The boy picked up the little fingers of each of the man's hands, put them between his teeth, and chomped down on the sides, at the tips. The man bounced up off the floor, and his heart started beating! Unbeknownst to the boy, those points are precisely at the end of the heart meridian. He said he had no idea what inspired him to do what he did.

One of the first classes I ever taught was at a retirement home. A man well into his eighties had been paralyzed on his right side by a stroke a few months earlier, and he was extremely depressed. He wondered why his life had to be dragged out, bemoaned the fact that he had survived the stroke, and wished he were dead. With no feeling or control on his right side, he couldn't do the physical exercises designed to improve the flow of his energies. Reaching for a way to help him, I asked him to sit in front of a mirror and *imagine* seeing himself doing exercises that cross energy over from each brain hemisphere to the opposite side of the body. I encouraged him to "see" the energy radiating light, color, and power to the frozen side. He did this every day for several days. Feelings began to return to the paralyzed side of his body. He had been told he would never have feeling or movement there again. By the time the class ended twelve weeks later, he had enough movement in his right arm and leg that he could begin to perform rather than only imagine the exercises. He was so excited to see that his mind had reactivated his body that his whole self-concept shifted. His was not just a dwindling body and mind; his mind could make his body stronger. Simply imagining he was doing the exercises had led to improvements in his physical condition that exceeded anyone's hopes. This makes little sense until you understand that subtle energies can be influenced by the mind.

MEASURING SUBTLE ENERGIES

In recent years scientists have introduced the term *subtle energy* to describe previously undetected though apparently intelligent forces in the environment and in the body.[11] If a "green thumb" is a force that helps plants grow, if prayer is a force that speeds healing in the person being prayed for, if a healer's hands transmit a force that relieves a patient's symptoms, each of these forces is comprised of "subtle energy."

Believed to exist outside the electromagnetic spectrum, subtle energies operate within a domain that has until recently eluded even the most sensitive scientific instruments. Now, however, there are devices that are sensitive at least to the denser electromagnetic counterpart of subtle energies. Electrodes attached to the hands and the feet, for instance, can provide a complete assessment of the energy flowing through each meridian and the corresponding internal organs it feeds.[12] Changes measured in the ion flow or light emissions emanating from the chakras and meridians correspond with the energetic shifts that follow meditation, acupuncture, qi gong, and other energy healing treatments.[13] Such devices can even anticipate illness. Shifts in meridian activity, for instance, have predicted changes in physical conditions that have occurred hours, days, and sometimes even weeks later.[14]

Thought also emits a subtle energy. William Tiller and his colleagues at Stanford University built a gas discharge generator that registers electron activity, and they have found that with mental focus alone, people can increase the electron activity within the generator.[15] Independent studies originating in the School of Engineering at Princeton University also suggest that thoughts affect subtle energies. Not only could certain individuals use their minds to influence instruments that generate numbers on a random basis but also the presence of organized groups imposes a degree of order on the output of *random* number generators. At ten separate gatherings, ranging from business meetings to scientific conferences to religious events, the effect was strongest during periods when the group's attention was focused, when the group's cohesion was high, or when the group's members were sharing a common emotional experience.[16] Other studies have produced similar findings.[17]

Among the most dramatic applications of subtle energies in healing are demonstrated by individuals who consistently provide accurate medical diagnoses from a distance. When I look at a person's energies, I can often recognize whether the energy holds a particular disease. Some practitioners need not even be present. In research conducted by the neurosurgeon C. Norman Shealy, a patient's name and date of birth were given to the medical intuitive Caroline Myss, who had no contact with the patient. Myss would psychically "enter" the patient's body and systematically examine the relative health of each organ system in the patient's body. On one sample of fifty patients, Myss's clairvoyant diagnosis matched Dr. Shealy's medical diagnosis in 93 percent of the diseases identified. Myss's statements were specific, such as "left testicle malignant, spread to left kidney," "venereal herpes," and "schizophrenia."[18] A growing body of evidence suggests that certain subtle energies are, like quantum fields, "nonlocal" in their effects—their influence is unaffected by distance. Thus there is a plausible explanation within physics for the explosion of scientific findings supporting the existence of telepathic communication, the influence of prayer on healing, and feats such as Myss's long-distance diagnosis.[19]

People who are suffering and experiencing themselves as victims of their own body become empowered when they are able to direct subtle energies to effect their healing. A pathway opens that they are, not surprisingly, inclined to follow. In discovering that you have the ability to help yourself feel better and think and function more effectively by becoming attuned to the subtle energies within and around you, I believe you will want to cultivate this capacity. While technologies for reading subtle energies are clearly on their way, all the tools you really need are built into your own physiology. By introducing you to *energy testing,* the following chapter shows you how to read subtle energies with no special equipment or extrasensory perception required.

Chapter 2

Energy Testing

Talking with Your Energy Body

THE CHINESE PHYSICIAN CAN DETECT IMBALANCES IN MERIDIANS BY FEELING THE PULSES, BUT THIS IS A SENSITIVE TOUCH, AND IT MAY TAKE TEN TO TWENTY YEARS TO DEVELOP PROFICIENCY WITH IT. TOUCH FOR HEALTH USES MUSCLE TESTING TO DETECT THESE SAME IMBALANCES.

—JOHN THIE
Touch for Health

Your body is a cascading fountain of energy systems, remarkably complex, exquisitely coordinated, and entirely unique. That is why no expert can tell you exactly what you must do to thrive. Not only does every person carry a distinguishing energy, so too does every cell, every organ, and every system of the body. Each speaks its own local language, and they also share a universal language.

Energy testing is a precise tool for translating these languages, phrase by phrase. Through energy testing you can assess your own unique, ever-oscillating energy body or energy field. This chapter is a primer on energy testing. It also coaches you to intuitively sense your own energy field and the fields of energy around you.

Illness shows up in your energies before it manifests in physical symptoms. In provinces of China where the doctor's primary job was seen as maintaining health by keeping people's energies balanced, the physician was paid only as long as the person was well. If the doctor had to treat an illness, he had failed, so the patient didn't pay. Western medicine's practice of waiting for physical symptoms to appear before intervening in a person's health is rather crude by comparison. It is much easier and much wiser to treat the imbalance while it is still only a disturbance in the energy field than to wait until it has progressed into a physical symptom that is far more traumatic, entrenched, and difficult to reverse.

Anyone who has successfully adapted to a modern society has pushed a great deal of visceral wisdom beneath the threshold of awareness. When you were growing up, your culture affirmed your intellect while negating the primal knowing of your body. We have all been instinct-wounded. Becoming civilized is, to a large extent, learning *not to do* what your body wants you to do.

Still, your body carries an intelligence that exceeds the understanding of your intellect. Over the years, I have been figuring out ways for people who do not consciously register subtle energies to begin to detect, connect with, interact with, and work in partnership with them. In learning ways to determine how subtle energies are affecting you, you will be cultivating your own latent capacities to see, feel, and influence your body's health. This is not a passive read. But each ounce of effort wisely applied to your energy systems can yield a ton of reward.

THE ENERGY BODY

Your energy body is the subtle counterpart of your physical body, and it is more responsive than your physical body to many kinds of influences. Because the energy body holds the blueprint of the physical body's health, it is the focus of energy medicine. Treatments that affect your energy body reverberate throughout your entire system.

The energy body reveals itself rudely and unmistakably in phantom limb pain, one of medicine's most disconcerting puzzles. A person may feel terrible and inexplicable pain in a part of the body that is no longer there. Some 300,000 people in the United States have had an arm or leg amputated, and a majority of them report itching or pain in the area where the limb was.[1] Because of "their vivid sensory qualities and precise location in space," notes Ronald Melzack, a research psychologist at McGill University, phantom limbs "seem so lifelike that a patient may try to step off a bed onto a phantom foot or lift a cup with a phantom hand. . . . A person may feel a painful ulcer or bunion that had been on a foot or even a tight ring that had been on a finger."[2] Phantom limb pain is reported by people born without a limb as well as by people who have lost a limb, and medical science has generally been unable to relieve either group of its discomfort.

Some neurologists believe that the brain is coded from birth with a map of the body and that the limb remains on the map even if it never develops or is severed. The sensations in the phantom limb, according to this theory, are activated by nerves at the stump of the missing limb.[3] This theory would better account for my observations in working with phantom limb patients if it were complemented by an understanding of the energy body. I believe a phantom limb is the energetic "double" of the missing limb.

A man who had a finger amputated and preserved in a jar in the heated basement of his mother's house had no pain for several years. He then went back to the doctor who had performed the amputation, complaining of a feeling of extreme cold in the area of the missing finger. Following the doctor's advice to check on the condition of the preserved finger, the man learned that the basement where the jar was stored had become very cold due to an undetected broken window. After the finger was taken out of the cold, the pain left.[4] An assortment of similar reports suggests that the energy of the phantom limb somehow remains connected to the energy of the physical limb. This becomes more conceivable as an increasing number of nonlocal effects, such as distance healing and the healing power of prayer, are being documented.[5]

The following three cases involve phantom limb pain and provide insight into the nature of the energy body. Phantom limb pain wouldn't be seen as such a puzzle if it were understood that an "energy limb" remains after the physical limb has been lost. Just as phantom pain is undeniable to those who have it, its responsiveness to energy work is undeniable as well.

Unbearable Phantom Pain. A good-looking man who had lost both of his legs in Vietnam was brought to me in a wheelchair. No one had been able to help with the pain at the end of where his right foot had been. He vividly recalled the scene of stepping on a land mine and watching the bones and flesh of his right foot explode into pieces. The pain he now had was massive. The sensations were so similar to the original shock that he could never get relief from the traumatic memories. The relentless pain also led to terrible nightmares. His left foot wasn't as painful. It sometimes itched, but it was a mild discomfort compared with the area of his right foot.

As he sat there with his friend, he cried and said, "The pain is so excruciating, and the way it keeps me tied to my past is so bad, that I sometimes think of taking a gun to my head." I could see the energy still there in both of his absent legs, and I could feel it with my hands. I followed the energy along his absent legs to the end of where is feet had been. It was palpable. My hand began to hurt terribly. I asked him if my hand was in the area of the worst pain, and it was. The most painful areas were at the sides of his feet, which happen to be at the end of the bladder meridian. I said to him, "This may sound crazy to you, but I believe I can hold some points in midair where your feet were and help you."

I moved my hands to the ends of his legs, where his feet had been, and held the points on the bladder meridian. As the two men watched these strange conjurings, it must have seemed to them that I was just holding air. But I was not! I felt and saw the meridian lines as strongly as if his legs were still there. At first it was painful for him to have me touch the area of his absent right foot. After a couple of minutes, he reported that not only was the foot being relieved of the pain, but another chronic pain in his back, just above his waist, was also lifting. Interestingly, that area is also governed by the bladder meridian.

His kidney and liver meridians were also involved, and I held those points as well. The kidney meridian governs fear, and holding these points opened a powerful catharsis. Fear and terrors buried deep inside, which had kept him locked up emotionally, just plummeted out of him. He cried and cried. I held on to the points until the emotions had worked their way through him. As the fear subsided, he became euphoric. By the time I had finished holding his liver points, he was pain free. I showed the friend, who lived with him, how to hold points and which points to hold. The man and his friend never returned, but on my invitation they called me about once a month. The friend told me that after the session, the man began to lift out of his deep depression. While the phantom pain would return every now and then, they knew how to deal with it.

Itchy Fingers and Indigestion. A man whose hand had been mangled in a machine at the factory where he worked was missing his index finger and the tip of his little finger. He had some feeling where the fingers had been, though this was not a problem for him. He actually thought these sensations were "kind of funny," except when his phantom fingers itched. He was referred to me because he was having terrible digestive difficulties. After a trail of unsuccessful medical treatments, he went to a psychotherapist to see if the problem might be related to underlying stress or unconscious fears, perhaps because of his accident. As he told me this, I was realizing that his missing index finger is at the beginning of the large intestine meridian and that the small intestine meridian begins at the little finger. I began to wonder if the meridians that flow along his missing fingers might be related to his difficulties with digestion. I told him this. He didn't know anything about meridians, and it seemed far-fetched to him that there might be some kind of connection between his digestion and his missing fingers.

Nonetheless, I traced meridians up and down both his arms to try to loosen the energies in his small and large intestines, concentrating especially on the missing portions of his index and little fingers. I used a technique designed to open energy that is blocked on the acupuncture points themselves. It's a little like squeezing a toothpaste tube: you just keep pushing the energy up and out. As I held the points that activate the large intestine meridian, the area of his missing index finger started to throb so painfully that he wanted me to stop. To take the throbbing away, I held the "sedating points" on that meridian, which meant that I appeared to be holding the air at the spot where his index finger had been. The pain went away, but the energy still felt stagnant to me. I tried again, this time starting at the other end of the meridian and moving the energy in the opposite direction. This time it worked. The bloated feeling in his intestines did not return for several days, but then it did. In a second session, I showed him where to hold, where to massage, and where to trace his meridians. He did these techniques daily, and his digestive problems cleared and this time did not return.

Congenital Phantom Pain. A woman brought her infant son to a session with me, hoping he would sleep through it, but the boy was extremely restless. He had been born without an arm. As his mother held him, the boy screamed and was inconsolable. All I did was to stroke the length of where the arm would have been, sometimes with soft strokes, sometimes with bold strokes. He began to calm down, to coo, and finally fell asleep. His mother said that he often cried and couldn't be comforted. I spent a little time teaching her how to move and harmonize the energies in his "phantom limb," and from then on she was able to provide comfort whenever his phantom limb pain would return.

These and similar cases add another twist to the mystery of phantom limb pain. While it may seem quite odd to see me hold meridian points in thin air, people quickly set that concern aside when chronic pain begins to disappear or other medical conditions associated with those meridians resolve themselves. Imagining that an energy limb exists, whether or not the actual limb is there, gives you a good opening image for visualizing your own energy body.

THE UNIQUE ENERGIES WITHIN

Everyone's energy field shifts from one moment to another. You are distinctly aware of the energetic changes that occur within you when someone you are attracted to or someone you are afraid of walks into the room, and you can learn to become aware of ever more subtle energies flowing within and around you.

Your energies share common features with everyone else's energies: everyone is surrounded by a colored hue called an aura, everyone has seven major energy centers called chakras, everyone's meridians flow along similar pathways. But these universal statements are only dim abstractions compared with the rich textures, uncommon shapes, intricate weavings, and assorted colors that distinguish each of our energy fields.

Changes in Your Energy Precede Changes in Your Body. Your energies are, in fact, so unique, and they so potently shape your physical body, that you begin to look like your energy body just as surely as two people who share a life for fifty years sometimes begin to resemble each other. As your energy field pulses through your body, it gradually adds its complexion to the physical structure that was determined by your genes. Matter follows energy. If you want to age well, energy well. Your body's health reflects the health, clarity, and flow in your energy body, and the program you are embarking on will show you how to maintain both.

Individuals who are sensitive to energy fields are able to identify shifts in the electrical activity of another person's body. Their descriptions correspond with readings from standard instruments for measuring galvanic skin response, brain waves, blood pressure, heartbeat, and muscle contraction. Moreover, their reports of changes in the energy field, like the readings of the instruments described in the previous chapter that gauge subtle energies, slightly *preceded* the changes in the physical measures.[6] Changes observed in the energy field predicted measurable physical changes. Keeping your energy body vibrant keeps your physical body vibrant.

How to Restore Your Energies While Taking a Bath

I love to take baths. A number of exercises can turn a bath into a pool of energy restoration. Here are a few of my favorites:

1. *Fill the bathtub with an eight-ounce box of baking soda. The baking soda cleanses your aura.*

2. *Slide back into the tub and deeply relax as you hold the neurovascular points on the front of your forehead for at least a minute (page 90).*

3. *You can leisurely massage your own lymphatic points (Figure 9).*

4. *On top of your feet, in the spaces between the bones that correspond with the spaces between your toes, are your "gait reflexes." Energy tends to clog there. With your fingers below each foot and your thumbs on the gait reflex, massage and smooth the energy down each of the five gait reflexes with your thumbs. End with a foot massage.*

5. *If you feel a cold coming on, let your knees fall out so they are supported by the sides of the tub. Cross your arms and push your elbows into the inside middle of your thighs. Push your thumbs into your biceps. Sitting this way for two or three minutes helps the congestion drain out of your lungs, and by the time you get out of the tub, you just may have stopped that cold from coming on.*

Detecting Your Body's Energies. One of the simplest experiments for detecting your body's energies uses your hands. Your hands can perceive energy as surely as your eyes perceive light:

1. Rub your hands together, briskly, for about 10 seconds.
2. Pull them away from each other, about six inches, and hold them there. Many people have the sense when they do this that their hands are up against something that is almost palpable.
3. Experiment with holding your hands still or moving them ever so slightly toward and away from each other.

The subtle energy you may feel between your hands is similar to the energy that continually moves, swirls, and spins throughout your body. If you can't quite feel the energy, rub your hands together again. This time cup them. Sense the "edges" of the energy you are holding, and pat them as you encircle this ball of energy between your hands.

Try a second experiment.

1. Again, rub your hands together briskly.
2. Place your left hand in front of your chest, about four inches out from your body.
3. Begin to make slow counterclockwise motions over the area (as if the clock were facing out from your chest).
4. Attune yourself to any subtle sensations in your palm or fingers or in your chest. You are moving through the energy field of your heart chakra, and its energies are being received by your hand. Just as we all dream each night but don't necessarily recall our dreams, we all register such sensations, but they don't necessarily filter into our awareness.
5. Place your right hand in the same area and circle clockwise with it. Notice if you detect similar or different sensations.

THE ENERGIES AROUND YOU AFFECT THE ENERGIES WITHIN YOU

The amount of solar radiation hitting the Earth fluctuates with the phase of the moon and seems to influence human behavior. Crisis calls to 911 increase on the day of the full moon; psychiatric records indicate a sudden rise of psychotic behavior on the day of the full moon; suicide rates increase with the waning moon; and death rates peak two days after the full moon. Mutual funds also fluctuate in a pattern that is related to the lunar cycle.[7] Periods of calm in the Earth's magnetic field correspond with a heightening of sensitivities such as telepathy, clairvoyance, and precognition.[8]

The energies of other people also impact your own energies and mental state, as expressed in comments such as: "I love being in her energy," "You could have cut the energy with a knife," "The energy at the meeting was amazing." You can feel another person's energies with your hands:

1. After rubbing your hands together and shaking them off, slowly circle your left hand counterclockwise about four inches above a friend's chest. Remain alert for any subtle sensations in your hands.
2. Follow this with a clockwise motion using your right hand. Register the sensations in each direction.

The impact of people's energies on one another has been scientifically studied. The energy produced by the heart of one person, as measured by an electrocardiogram, has an effect on the heart activity and brain waves of another person, as measured through electrodes attached to the skin. The effect is strongest when the two people are in physical contact, but it can still be detected when the people are sitting as far as three feet away from each other.[9]

The ability to sense subtle energies often becomes acute in people who have been blinded. Shortly after losing his sight in an accident at the age of eight, Jacques Lusseyran had a realization that there was another way to see: "I was aware of a radiance. . . . I saw light and went on seeing it though I was blind. . . . Light threw its color on things and on people. My father and mother, the people I met or ran into in the street, all had their characteristic color which I had never seen before I was blind. Yet now this special attribute impressed itself on me as part of them as definitely as any impression created by a face."

This "sight" proved indispensable when in 1941, at age seventeen, Lusseyran helped form the French Resistance. He was put in charge of perhaps its most sensitive activity—recruiting—because of his special "sense of human beings." He was a prime mover in the Resistance's extraordinary accomplishments, and his downfall came only after he underestimated the meaning of a color he saw in one individual. Upon meeting Elio, the infiltrator who caused Lusseyran's arrest, along with the arrests of many in the Resistance's leadership, he reflected that despite Elio's impeccable credentials and presentation, "Something like a black bar . . . slipped between Elio and me. I could see it distinctly, but I didn't know how to account for it."[10]

Whether or not your mind registers your own energy and the fields of energy that surround you, you *experience* them all the time. For example, consider what it is like if you walk into your child's room or a coworker's office when he or she is angry at you; or eager to see you; or depressed. Or consider the experience of delivering a speech. Before a word is uttered, a welcoming audience greets you with an entirely different energy from that of an unfriendly group. You probably know people who cause you to feel happy, talkative, and at ease the moment you step into their field of energy. You have also probably stepped into other energy fields that cause you to feel self-conscious, tired, or inarticulate. When someone's field is in an extreme state, the experience for me is like being hit by an intense weather front.

The energies in your environment can help you or harm you. There is evidence, for instance, that pets make an atmosphere more conducive for healing.[11] On the other hand, extended exposure to high-intensity electromagnetic fields has been associated with increased incidence of Alzheimer's disease, depression, suicide, leukemia, and cancers of the blood, brain, colon, prostate, nervous system, lymph system, lung, and breast.[12]

Rooms have an energy all their own. Think of your grandmother's kitchen. How does your body resonate to its energy? Consider a classroom; a garage; a library; a theater; a cathedral; a jail cell. You may also find yourself pulled toward specific kinds of energy and not toward others. Some people, for instance, are drawn particularly to the energy of art. Others to the energy of horses. Some to the energy of those in need. Others to the energy of cyberspace. Some to the energy of mountain climbing. Of course, other factors are also involved, but each experience has a distinctive energy. Think about the kinds of energies to which you are particularly attuned.

How to Escape the Grip of Fear

When you are caught in fear, anxiety, or a phobia, tapping a point on the triple warmer meridian, which governs the fight or flight response, can alleviate the fear. The next time you feel afraid (time—about a minute):

1. *Start tapping on the back side of your hand halfway between your wrist and fingers, aligned with the point where your fourth and fifth fingers converge. Tap for about a minute.*

2. *If you still feel fear, tap on the other hand.*

Even for a longstanding phobia, repeatedly using this technique begins to alter the underlying energetic pattern.

"WEARING" AN ENERGY FIELD

Occasionally I will see in a stranger's energy something so unique and special that I have an impulse to run up and gush about a glorious color in the person's aura or a chakra that stands out like a light show. I generally restrain myself from such effusive remarks, but I often cannot refrain from giving the person a smile, reflecting back the good feelings the person's energies have just brought to me. When I am with a client, however, it is my job to share what I see. If the person appears, for instance, to be in great despair and hopelessness, I might mention a part of the energy field that reveals an undeveloped capacity or other strength that might show a way out of the despair.

Any of a multitude of features may stand out. A person might have a shape or combination of colors in a particular chakra that I've never seen before but whose meaning is intuitively clear to me. I recall being stunned when I first saw one young man's energy field. His mother had brought him to a session. He was a junior in high school, and he was considered to be very slow, perhaps retarded. When he walked into the room, however, I could hardly meet his eyes because my own eyes were so strongly drawn to the beautiful geometric shapes I saw throughout his energy field. He was enfolded in an explosion of shapes that far exceeded anything I had ever seen in its complexity, design, and aesthetic splendor. My first thought was, "This can't be a retarded person. This is the energy of a genius!"

Never before had a client's energy field been so totally different from what I was expecting. I was so distracted that I could not hide my astonishment. I took a piece of paper and said, "I just want to show you what I am seeing." As he and his mother watched, I began to draw the different shapes I was witnessing in his energy field. Then it was *they* who were astonished. His mother explained that he spent a great deal of his time making drawings of geometrical figures shaped like the ones I was drawing. The figures, with their complex interweavings, were remarkable. I wondered for a moment if he might be a mathematical savant, but I quickly dismissed this. His energies didn't focus into a laser beam as I have seen in the energies of savants. Rather, his was a huge field. I suspected that he had enormous intelligence that just hadn't been measured yet.

I first saw him in April, toward the end of the academic year. The school didn't want him to return the following year. His mother knew he wasn't stupid, but she had no notion that her son might in some ways be brilliant. His problem was that he was totally out of synch with his own energies. No one had understood his unique way of experiencing the world, and he was working with all his might to do things the way other people do them. Most of us become slightly impaired in our thinking when our energies are somewhat disrupted. He was mentally disabled by the jangle between his energies and the way he was trying to live. I am actually unnerved by the way we humans judge one another. Until we've walked a mile in another's energy moccasins, we hardly have a clue as to what it would be to live from within that energy field.

The only thing that seemed to put the boy in synch with his energies and make him feel a little better was to draw the geometrical shapes he saw all around him, which is why he would doodle so much of the time. He saw these shapes distinctly and constantly. His mother had taken him to an eye specialist because the shapes sometimes intruded into his vision. They had been told that something was organically wrong with his eyes.

His body and his energies were in such discord that he had to move through life very slowly. His speech was slow, his movements were slow, his responses were slow, all adding to the school's impression that he was retarded. So the entire session with me was devoted to getting his body into better synch with the world and with his own energies. Think of a man whose movements are restricted because he is wearing his sweatshirt as his pants and his pants over his head. That's how this young man was living in his energy field. Most of us are shown by our upbringing and our culture how to wear our clothes, and we are subliminally oriented toward our energy field as well. My job was to teach him how to wear what he had, however unusual it might be, and to invent energy exercises that repatterned his movements and actions to correspond with his energies. Many of these exercises, I might add, emulated the way he spontaneously aligned himself with his energy field by drawing the geometrical patterns he saw within it.

Once he was provided with a bit of validation for the world as he experienced it, he claimed it full measure. It was as if he was suddenly in tune with the field that surrounded him. His speech and movements sped up markedly. His improvement in school was immediate and conspicuous. Now, with his "energy trousers" no longer over his head, he could, both figuratively and literally, run and jump and skip and play instead of having to blindly inch his way long. In fact, while we never focused on it directly, chronic fatigue that had plagued him for years also serendipitously vanished within our first few sessions.

Because of the geometric shapes I kept seeing in his energies, I asked him if he was smart in math. He said he didn't think so, but on further questioning I learned that he'd not had much of a chance at it. I encouraged his mother to find someone who had an adventuresome approach to geometry and to see where it would go. They soon found both a geometry class for him and a geometry tutor. Within a few weeks, he was teaching the teacher and tutoring the tutor. His special mental abilities had simply not been identified, but there they were, blooming. After his amazed teachers started to talk with their colleagues, a major university recruited him and gave him a scholarship for his prowess in geometry. He was, in his special way, brilliant beyond any human I'd ever met, and it was obvious in his energy field.

SENSING SUBTLE ENERGIES

I am certain that when babies are born they are far more open to the realms of energy than we are as adults. Have you ever noticed the way infants will often look intently just above your head or to the side of your face? There is no question in my mind that they are seeing the energies that surround you. Babies see energy, feel it, sense it, know it. But because the brain has so much to learn, and the realm of subtle energies is rarely spoken of and rarely validated, these sensitivities become dormant. They fall out of the loop in the learning process.

I have on occasion, however, had an opportunity to encourage a pregnant woman or the parents of a baby to talk with their child about energy from the beginning. Regardless of whether or not the parents see energy, I ask them to imagine the energies that animate all of life and talk about them. The parents may be speaking only about their imaginings, but they are attuning themselves to what is a reality for the child. From the offspring of these parents, I now know a handful of older children who can still see energy in vivid colors and are able to talk about it freely and easily. I have also witnessed many adults as they began to see and accurately sense energies after experimenting with techniques presented in this book. They are opening to a deeply ingrained but forgotten ability.

A woman dragged her physician husband to one of my weekly evening classes. He found the idea of energy healing implausible and the concept that a person could see color in people's energies ludicrous. But as he delivered his sometimes sarcastic comments about "seeing" colors, he spoke in one of the deepest and most beautiful voices I had ever heard. When people paired up to practice techniques with one another, he would always work with his wife, back in a corner of the room. On the last day of the class, as people were practicing with partners, a high-pitched, squeaky voice from the back of the class exclaimed, "Purple, I see purple!" It was the physician. He was so surprised by seeing the purple in his wife's energies that his voice jumped three octaves.

When you learn the native language of your own unique energy field, you become able to read it, hear it, and converse with it. One of the trickiest things in cultivating new sensibilities about your own and others' energies is that they often don't show up in the ways you expect them. Like synesthesia, where certain people can *smell* colors or *see* sounds, the perception of energy may just slide into one of your normal sensory channels. I've known people who can hear, smell, or taste specific energies, rather than see or feel them as I do. My own sense of taste has actually also grown stronger. I can usually taste which of the five elements is most dominant in a client's energies—I may have a metallic taste in my mouth while working with a metal energy person—and this alerts me to imbalances. I've now known three people who can *hear* energy move. One of my colleagues began to smell energies, and the scents became so overwhelming that she had to get out of the work for a time. You cannot know in advance how you will register subtle energies. We all have different strengths and different ways of knowing. As you expand into the language of energy, the only thing you can predict for sure is that it will speak to you in the way it chooses, not necessarily in the way you hope for or expect.

My friend and associate Sandy Wand often sees symbols when she is working on someone. She never knows where they will take her. But she has learned to describe them to her clients, often having no idea of what they mean but trusting that they will eventually begin to make sense. I had not told her of a frightening experience I had in which I thought I was dying. Lying in my hotel room bed in London, I felt that all my energy instantly dropped, like a runaway elevator, down into my root chakra, stopping with a jerk. Suddenly, I couldn't see the room anymore. All I could see was a deep blue-black that felt like hot ink boiling within my root chakra. It began to rise, filling my entire body. It felt like a poisonous fluid was inside me.

When I returned home, still feeling poisoned, Sandy gave me a session. After working on me for a few minutes, she said, "Well, this isn't going to make a lot of sense, but you know how squid squirt out ink to protect themselves? I'm getting an image that, like a squid, your root chakra has been squirting out energy in order to protect you." The piece of the puzzle she gave me that I didn't have was that while I thought this deep blue-black was an energy of death, she saw it as quite the opposite. This was an energy of life protecting itself. Squid squirt out ink so nothing can seem them or get to them. This provided me with an extremely useful insight. If I didn't start protecting myself better, *then* I might be dead. Since I wasn't succeeding all that well at setting boundaries, my energy system was trying to set a boundary for me. The blue-black ink was putting out a force to both contain my own energies and keep out the energies of others who might harm or drain me. Sandy has a gift, but not so many years ago, she didn't know she had it. Open to the unique ways energies reveal themselves to you, and your own natural capacities to work with energy can flourish as well.

ENERGY TESTING: BIOFEEDBACK WITHOUT THE GADGETS

As you develop your sensitivities to the subtle energies swirling within and around you, it will be extremely useful to have a tool that is tangible and palpable rather than to be left to rely totally on intuition. Enter energy testing. Energy testing, developed as "muscle testing" by George Goodheart, the founder of applied kineseology, and meticulously refined by his protégé Alan Beardall, is a concrete and very tangible procedure.[13] It allows you to determine whether an energy pathway is flowing or blocked, whether an organ is getting the energy it needs to function properly, or whether an outside energy (such as the energy of a particular food or a suspected toxin) is harmful to your system.

When I was first learning energy testing, a problem whose solution had eluded me all my adult life was solved. After years of experimentation, I had not been able to figure out why I couldn't manage my hypoglycemia or my weight, even though I was diligently taking the steps that should have controlled both. During my Touch for Health training, my spleen meridian consistently tested weak. The pancreas, which controls hypoglycemia and often weight, is on the spleen meridian. By working with the spleen meridian, I was able to counter my lifelong hypoglycemia. I was impressed. I also lost seventeen pounds without changing my diet. I was *really* impressed. Over the years now, energy testing has been priceless within my personal life.

I prefer the term "energy test" to the more common term "muscle test" to emphasize that the objective of a test is not to determine the *strength* of a muscle but rather how the body's energies are flowing through it. While MRIs, EEGs, and CAT scans provide vital and often life-saving information, I have yet to find a medical device that can reliably make the *subtle* determinations afforded by energy testing. Built into each of us, however, is all the equipment we need to ascertain which energies are good for a person and which are not. Energy testing is always available, night or day, requires no instrumentation, and if you practice it regularly, it can begin to feel almost instinctive. It is a tool that lets you ask your body for information about what it needs. And it lets your body answer in a language with a small enough vocabulary that you can master it.

In fact, energy testing is quite easy to learn. Deceptively easy. As a result, many people apply and misapply it casually and often inaccurately. Misapplied, it may reveal more about the beliefs of the tester, the fears or hopes of the person being tested, or other factors that have nothing to do with the information being sought. Many people have encountered energy testing by an inexperienced practitioner, have been given flamboyant explanations of its capabilities, or have seen it used more like a parlor trick than a tool for self-knowledge. I want to lift energy testing in your mind to its rightful place, which is between science and art. It is an art to learn how to energy test reliably. Once you have, it becomes a dependable barometer of your body, your energies, and your environment.

If I had the power to impact the medical profession in just one way, it would be that physicians add energy testing to their diagnostic tool kit for determining the choice and dosage of medications. Iatrogenic illnesses—disorders induced by medical treatment—are one of the most serious problems in medical care today. Energy testing could significantly reduce their incidence. And if I could choose just a few ways for this book to impact your lifestyle, one of them would be that you use energy testing to determine what foods, vitamins, and supplements you should and should not take into your body. In maintaining health, knowledge is power, and because each of us is so unique—we are simply not cookie-cutter people—energy testing is vital in my work. My use of the term "energy kineseology" (using the muscles to gauge the energies) to refer to my own brand of energy medicine emphasizes the important role of energy testing.

Energy testing will allow you to assess the state of your own or another's energies, identify imbalances, and tailor the procedures presented in the remainder of the book to suit your own unique needs or those of someone you care about. A woman with multiple chemical sensitivities and a history of life-threatening allergic reactions needed an antibiotic for a strep infection. When she went to the drugstore to get the prescription filled, she asked if she could bring it back if she proved to be allergic to it. The pharmacist of course said that would not be possible, but the prescription was expensive, and she was adamant that she not wind up buying medicine she couldn't use. She knew how to energy test, and she convinced another of the pharmacists to energy test the medication on her. She taught him the spleen meridian test, and when she tested strong, she immediately said, "Okay, I'll take it, but I want to buy only half the prescribed amount. I'll buy the other half if I do not build up an intolerance."

The first pharmacist was so amused that the story got back to me. It had an unexpected but instructive ending. After several days with the medication working effectively, the woman's neck began to itch, her stomach and ankles swelled, and her heart began to fibrillate. She recognized this as a severe allergic reaction and had her granddaughter energy test everything she had recently ingested. The antibiotic now tested weak. After she stopped taking it, the symptoms soon subsided. As often happens for people with hyperallergenic tendencies, she had built up an intolerance to the medication. Does this mean that the original energy test was inaccurate? No. The body is a dynamic system that is always in flux. Intolerances can and do develop. Because energy testing is quick, free, and always available, it is an extraordinarily useful tool for obtaining current information about your body's changing needs.

How Reliable Is Energy Testing? A 1984 article in the journal *Perceptual and Motor Skills* is the first published laboratory study that supports the potential value of muscle testing, or energy testing.[14] The investigator, Dean Radin, later commented: "To my surprise I found that in double-blind tests, people were slightly weaker when holding an unmarked bottle of sugar than they were when holding an unmarked bottle of sand of the same weight."[15] Other experiments have failed to show a correlation between muscle testing and nutritional needs.[16] A smattering of subsequent studies has also yielded contradictory results. One showed significant agreement among three examiners when the test used the *piriformis* or the *pectoralis* muscles but not when the test used the *tensor fascia lata* muscle.[17] Because there are so many ways that energy testing is conducted and so many nuances within a single energy test, all that can really be said about the research validation of energy testing at this point is that the jury is still out.

While research confirmation always lags behind clinical experience, some nuances that must be incorporated into both clinical work and research studies have been established. For instance, different testers consistently reported the same results on the same clients when the instructions were to "push against me as hard as you can," but their results did not match nearly as often when the instructions were merely to "resist."[18] Other studies have investigated the physiology of the test. For instance, muscles that show weakness in an energy test register different voltage levels than muscles that are simply tired, so the test is measuring an internal shift that is different from fatigue.[19] Moreover, energy testing correlates with electrical activity in the central nervous system, so the information gathered during an energy test reflects brain activity, not just the state of the indicator muscle.[20]

While the research is inconclusive, I have taught energy testing to thousands of individuals during the past two decades, and I have feedback from the hundreds of them—who return to classes, write letters, and schedule private sessions—that energy testing has proven a useful and accurate way of obtaining information about the body's needs. My experience has consistently been that the energy test corresponds with what I see in a person's energies. Information from the test also suggests where to work on the body, and the subsequent results have again and again confirmed for me the validity of the tests. The client frequently confirms the test as well, as when the test shows a weakness on the bladder meridian and the client then remarks, "I'm just recovering from a bladder infection!"

A BIOLOGICAL BASIS FOR ENERGY TESTING

Your nervous system is a phenomenally sensitive thirty-seven-mile-long antenna that reverberates to the subtle and not so subtle energies of the world in which you live. Everything from the food you eat to the people you encounter carries its own frequency and impacts your nervous system. While most of these vibrations exist below the threshold of your awareness, your body resonates to some and tenses against others. As a result, you will embrace the energies of some foods or some people while rejecting the energies of others. This sensitivity of your nervous system to the energies that come into its range is the basis of energy testing. The frequency of the substance being tested affects your nervous system, and this is reflected by the resistance in the muscle used in the energy test.

Energy testing can discriminate among subtle energy frequencies that cannot be detected by existing scientific instruments. Every energy system has a frequency, and many can be measured. In the human brain, for instance, the four basic frequencies are known as delta, theta, alpha, and beta waves. These frequencies are the rates at which electrical charges move through the brain's neurons. They may have a stabilizing effect, as in the delta waves of sleep, where signals move through the brain cells at about four cycles per second, or the charge may be arousing, as in the beta waves of normal waking consciousness, where signals move through at twelve cycles per second and above. Many disorders, such as depression, anxiety, obsessiveness, epilepsy, and attention deficit disorders are associated with abnormal brain frequencies and are responsive to a form of biofeedback called "neurofeedback" where the intervention is to activate a corrective frequency rather than to use a drug or other physical intervention.[21] The various tapping techniques you will be using in subsequent chapters also work in part by introducing a corrective frequency.

Because energy testing detects the vibrational impact of a substance on your nervous system, subtle distinctions may be identified that could not be revealed by a blood test. For instance, the vibration of a food on your nervous system may be very different depending on whether the food has been processed or is in its natural form. I personally test strong for raw milk, weak for pasteurized milk, and very weak for skim milk. Natural foods are balanced within themselves, and when we remove a portion of that food, this balance is corrupted and the food's vibration is altered. Your body then has to assimilate a vibration that has been skewed, which may upset its own balance or at least prevent it from gaining the full nutritional value of the food. It is very hard to be smarter than Mother Nature, and energy testing can tell you if your body's vibration is in harmony with the vibration of a food or vitamin.

Learning to Energy Test. You can learn energy testing in a few minutes, but to master it, you need to incorporate it into your kinesthetic skill bank, like learning how to shift the gears of a car. As you begin to learn the simple steps of energy testing, please understand that it is only with practice that you will be able to control for extraneous influences and attune yourself to the subtle distinctions that make an energy test accurate. The pressure you use in energy testing a child, for instance, is different from the pressure you use to energy test a typical adult, which is yet again different from working with a high school football player. But when your energy aligns properly with the energy of the person you are testing, the test will work regardless of the other person's strength.

Being energy tested reinforces the link between your brain and subtle energies in your body, establishing new levels of internal communication. New areas of self-awareness begin to unfold. Many people find they intuitively know what the result of an energy test will be before they apply pressure to the other person's arm. It's not like guessing before the test but rather launching a communication in which your awareness is working in tandem with subtle energies.

While energy testing is usually done with a partner, you can perform an energy test on yourself, and I will describe that procedure as an alternative. For learning the process, however, it is better to have a partner. I cannot state too strongly how valuable it will be for you to push through any shyness or reluctance to involve another person in your learning here. Whether it is someone already close to you or just a casual acquaintance, it will be a gift for you both. We tend to touch people for affection, for sexual gratification, or in anger. Another very important reason to become comfortable with touching one another is for healing. It is an altogether different kind of touching. Beyond opening a door to new perceptions, feelings, and understanding, healing touch can save your life!

Energy Testing with a Friend. Every muscle, every meridian, and every organ in your body can be energy tested. A meridian is a fixed energy pathway that distributes energy to and from at least one organ. We will begin with a single test that you will be able to use in many contexts. This test determines the way the energy is flowing along the spleen meridian—the energy pathway that passes through the spleen and the pancreas. The spleen is involved with the immune system. It also determines if the body will be able to metabolize a particular food, emotion, thought, energy, or other external influence. The spleen and the pancreas are both involved with food metabolism, blood sugar levels, and the mood swings associated with them. Both organs and their shared meridian influence your general energy level and are extremely responsive to stress. For all these reasons, the spleen-pancreas energy test can answer many questions you may have about the way your body will respond to something you are considering ingesting or otherwise bringing into your life. It is also an excellent gauge for assessing your body's overall health.

Because preconceptions can affect an energy test, it is better not to try to guess what the results will be. Subtle energies are responsive to your thoughts, so clear your mind as well as you can prior to conducting an energy test. If either participant is thirsty, begin with a drink of water. Water conducts electricity, and dehydration interferes with the flow of energy in your body. Also ask if the person being tested has any injuries that might be aggravated by having pressure applied to the arm being tested. To energy test the spleen meridian:

1. Both of you take a deep breath. With your exhalation, release your expectations.
2. The person being tested places either hand straight down the side of the body with the thumb touching the side of the leg, fingers pointing down.
3. The tester slips an open hand between the body and the arm, just above the wrist (see Figure 2).
4. The tester then asks the other person to hold his or her arm firm, elbow straight. I often use words such as, "Don't let me pull your arm away from your body."
5. With an open hand and with pressure from one to two seconds, the tester slowly pulls on the arm.

With neither person straining, the arm will either pull away relatively easily or it will feel locked in place. When pressure is applied, a muscle with energy flowing through it may also move a fraction of an inch, but it will bounce right back. Do not struggle so much to hold your arm firm that you involve other muscles; if you are the tester, do not struggle to pull the other's arm away. This is not a competition, nor is it about muscle strength. If the energy is flowing freely, the arm will stay locked in place or give a bit but bounce right back when you pull on it.

If the energy test shows the muscle to be weak, strengthen it by vigorously tapping or deeply massaging the spleen points that are illustrated in Figure 3 (page 51) and then retest. (In fact, if you find yourself overly susceptible to exhaustion, infection, or illness, you can keep the spleen meridian strong by routinely tapping these points. After you tap them, the energy test will probably indicate that the meridian is strong. If the test still shows it to be weak, your energies are probably quite scrambled. Don't be alarmed. Go directly to the following chapter, which offers a sequence of techniques for unscrambling your energy field, and then return here to complete this chapter.)

Energy Testing Without a Partner. Because I sometimes wanted someone to energy test me when on one was available, I hit upon a solution

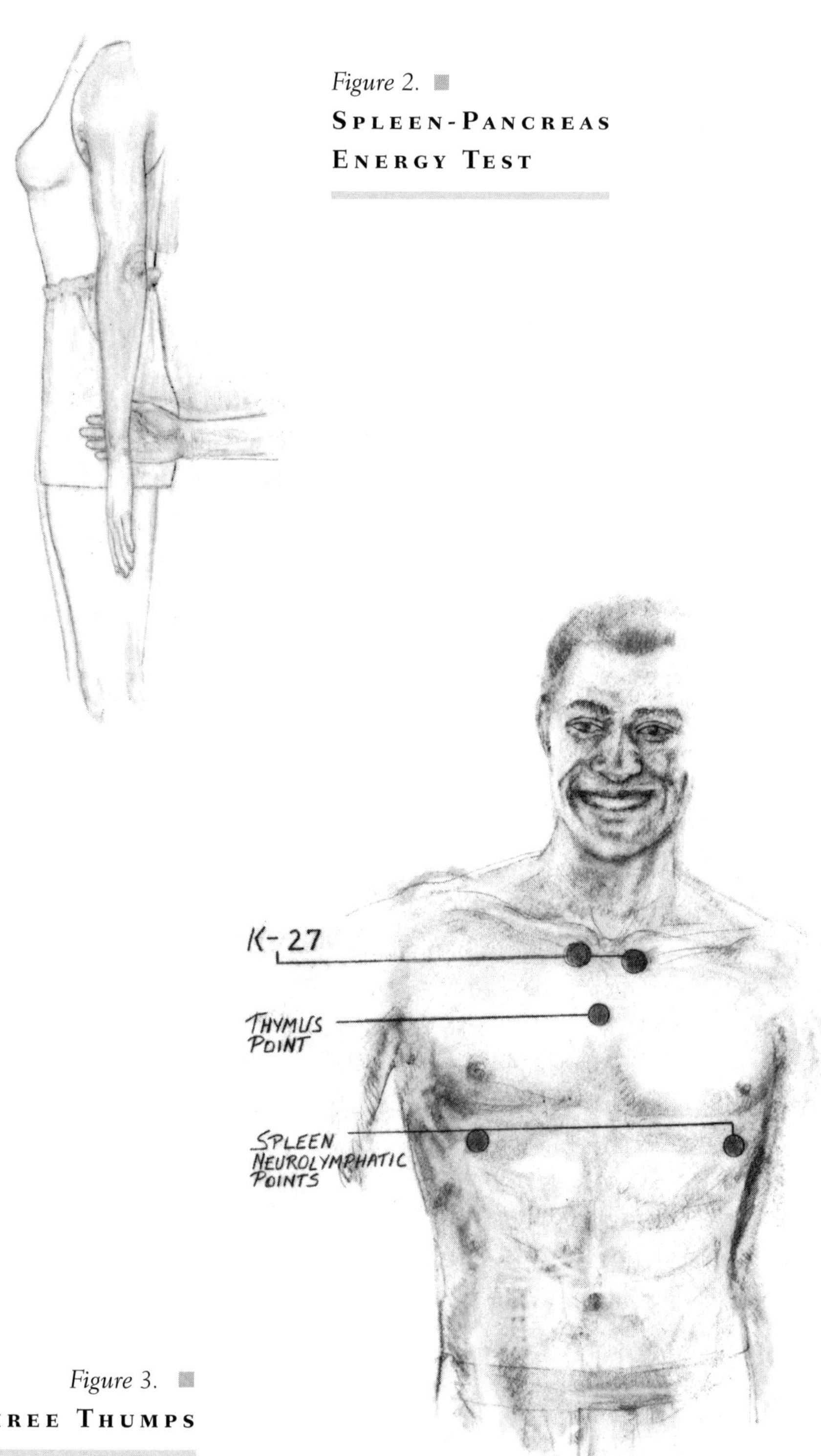

Figure 2.
SPLEEN-PANCREAS ENERGY TEST

Figure 3.
THE THREE THUMPS

while I was first experimenting with energy testing. I went to a sporting goods store and found a barbell that I could lift if I held it straight out in front of me when I was holding a comforting thought but could not lift when I was holding a depressing thought (as suggested by the experiments described on page 27, thoughts generate subtle energies that can affect matter, and comforting as well as depressing thoughts affect the muscles.)

I placed the barbell on a dresser that was the height of my shoulders. I would put my arm out in front of me, grab the barbell, and try to lift it. If I wanted to energy test for a specific food or vitamin, I would hold the substance in one hand and try to lift the barbell with the other. The energy of the food would affect my energies as decisively as an uplifting or a depressing thought. I could find out whether the substance was having a positive or negative impact on my energies by my ability or inability to lift the weight. Because barbells exert a steady pressure downward, they can provide a reasonably objective measure of what is being tested.

Numerous other ways for energy testing without a partner have been devised, and it would be convenient to have a system that doesn't require something as cumbersome as a barbell. But because most of the alternatives require the person to simultaneously exert and resist pressure, to be both the tester and the tested, it is very difficult to assure an unbiased result unless the person is highly experienced. I know of no other self-test that is more reliable than the barbell method. The critical factor, however, is finding a barbell with the correct weight—a weight you can lift while holding it straight out in front of you and thinking a positive thought but not while thinking a negative thought. Sometimes people start using a five-pound or other weight they happen to have handy; it usually doesn't work for them. You will probably need to shop, as I did, for the precise weight you need. I recommend going to a sporting goods store and energy testing yourself with different weights until you find one that works for you.

Practice Energy Testing at Your Next Meal. Although there is a different energy test for each of the fourteen meridians and seven chakras, learning the spleen-pancreas test can serve you well in innumerable contexts. The test is based on the *latissimus dorsi* muscle, and it is particularly attuned to food metabolism. You can, however, use the test as an *indicator* to find out about almost anything that is going on in the body.

Your next meal can become a personal workshop for practicing energy testing. Energy test each of the foods you are planning to eat. Touch the food and allow your other arm to be energy tested. If you lose your strength upon touching the food, your body's vibration is not in harmony with the food's vibration. This may mean the food is never good for your body, the food is not good for your body right now, or you have an allergy to the food.

Testing the food at various points in time will give you information about whether it is always, sometimes, or never good for you. Because everyone's chemistry is unique, nutrition is an entirely individual matter. One person's vitamin is another's poison. If the energy of a food, vitamin, or supplement doesn't match the energy of your body, you will not absorb and metabolize it though all the experts in the world say you need it. Even good food is toxic if its vibration triggers your immune system to tie itself up in a defensive reaction. Such food allergies often go undetected, but not without accumulating damage. Energy testing can help you to know what your body needs at a given moment, and it can help you develop a superb nutrition program for your unique body.

How To Eat Smarter

For many people, learning to energy test for food changes their relationship to what they eat. Test to see if your body is going to like a particular food. You can get children involved with picking out foods you buy at the grocery store or energy test them before you cook a meal. It can be a game. They will enjoy it, at least until the first time they energy test weak on something they really want. But often, even if they are begging for something else, seeing their arm lose its strength on an energy test of a junk food will bring laughter and a reduction in the tension.

Of course they may want to eat it anyway, but I have seen kids become less interested in foods that consistently cause them to lose their energy. The best results happen if they make the connection that losing strength on an energy test may mean that the food being tested may reduce their energy level and leave them feeling bad. But you have to be honest. A healthy body may be able to metabolize a certain amount of "junk" food, and it would undermine the process to overpower the child with your strength or your mind. Use the guidelines presented in this chapter to be sure the energy test is accurate, and particularly that it isn't being influenced by your beliefs or your child's desires.

Finally, let children draw their own conclusions about how eating food that energy tested weak makes them feel. When they can authentically relate energy testing to their life—that it isn't just a trick or a frivolous game—energy testing becomes a source of empowerment, a way to get useful information and biofeedback. They might also like turning the tables on Mom and Dad and energy testing you on that cheesecake or cigarette. And why not? You may be interested to find out for yourself whether that daily glass of red wine really is good for your heart. You can also use energy testing to assess the likely reaction of your own body or your children's bodies to vitamins and other food supplements.

HOW THE PHYSICAL ENVIRONMENT AFFECTS YOUR ENERGY FIELD

The moment you come into contact with an external energy, before you are even dimly aware of its impact on you, your own energy field is already responding and adjusting to it. By simply bringing something near you and doing an energy test, you can discern its immediate impact on your subtle energies. I have many times arranged to meet a client at the supermarket for the purpose of energy testing to see which foods elicit a positive response and which do not. Even different brands of the same food or vitamin may affect you in different ways. Your dietary needs can also change from one stage of your life to the next, but a tremendous amount of useful information can be garnered from a trip to the grocery store.

These first energy tests are experiments for you and your partner, and you will both become more proficient in your ability to perform accurate energy tests as you proceed. For now, play with energy testing. Have your friend test you when you walk into your favorite part of the house and then when you walk into your least favorite spot. Have a third person think "negative" thoughts while in your proximity. Find out how that affects your energy. Then have someone give you a genuine smile. Test again. Explore how the television affects you when you are two feet from it. Eight feet. Experiment with different setups of your computer or your office. Discover how your energies are affected when you focus on a particular piece of art or music.

As you explore your environment using energy testing, do not be concerned about temporarily bringing something that weakens you into your own energy field. Because we are continually entering fields of energy that affect us, the body—at least when it is relatively healthy—adjusts to the initial impact of a disturbing energy field by quickly rebalancing itself. In fact, one definition of health is how readily your body can adapt to a spectrum of environmental conditions. You can use energy testing to explore any element of your environment that stirs your curiosity. By having fun with it, you will be developing a strong aptitude with this invaluable tool.

How to Energy Test an Infant, a Pet, or Someone in a Coma

"Surrogate testing" allows you to test someone who is not able to offer resistance in an energy test. If someone is too sick to use his or her own strength or is mentally impaired or too young to follow your instructions, surrogate testing can provide valid information. This will even work with a pet.

It can also be used when you cannot get an accurate test on someone. For instance, because the energetic links with family members are so complex, you may find that you can get accurate results on everyone but your spouse or your child. Or sometimes a person is muscular, macho, and uses auxiliary muscles not to look weak on the test. Surrogate testing bypasses these difficulties.

1. *Stand to the side of the one who cannot be tested. Hold hands.*

2. *Having someone energy test your arm will give information about the one whose hand (or paw) you are holding.*

This seems bizarre at first, but it is simply based on the way energy flows. If you want to know, for instance, if a particular food is causing an adverse reaction in an infant, you can be a surrogate for the infant. With the food on the infant's skin, place your hand over the infant's hand or abdomen and have someone energy test your other arm. If the test shows weak, the energy of the food is not compatible with the infant. If this seems too strange, please don't take my word for it. You can easily demonstrate its validity for yourself by first having yourself energy tested on a number of substances and then having someone surrogate test you on the same items.

FINE-TUNE YOUR ENERGY-TESTING ABILITIES

I have long sensed that energy testing does more than simply provide information. It feels rather like the early part of the treatment, as if performing the test somehow engages my energies and the client's energies within a healing context. Energy testing immediately directs healing forces to the area being tested. It is as if the energy test focuses your subtle energies in a way that prepares them for the work to follow. I have found that the identical procedure yields better results if I have energy tested it first. How can this be?

Because subtle energies are affected by thoughts and intention, and because they are nonlocal, the energies of healer and client start mingling as soon as an energy test is even contemplated. We cannot be truly objective in any test. Not only does the act of observing influence the observed in physics, subtle energies are ultrasensitive to many influences. Yet we can learn to control for our hopes, fears, and expectations and create a container within which an energy test provides information that is very useful and quite precise. In fact, energy testing with a partner can reveal a deeper level of information than can any inanimate instrument. The complex interactions of tester and tested evoke two-way feedback that can make the energy test an incomparable dance of exploration for the purpose of obtaining trustworthy information. Having now trained scores of students whom I know to be capable of providing reliable energy tests, I can tell you with confidence that this skill can be yours, and it is a skill worth developing. The following tips will increase your accuracy. Practice them now, or proceed with the book and return here when the time is right.

1. Maintain a Beginner's Mind. Your mind affects your energy field, instantly and potently. To factor out preconceived ideas about what the test will reveal, approach energy testing as a contemplative practice, coming into a centered, meditative "beginner's mind." If you are concerned that your hopes or preconceived ideas, or your partner's, may be getting in the way of the test, the following procedure can energetically disengage them.

2. Disengage Your Expectations. This approach for neurologically disconnecting preconceived ideas from an energy test, taught to me by Gordon Stokes, my first Touch for Health instructor, may seem implausible, but over the years I have found that it works. It can be used by the tester, the tested, or both.

a. Take the thumb and middle finger of one hand and place them in the two indents where the back of your neck meets your head.
b. With your other hand, proceed with the energy test.

These indents are called the headache points in traditional Chinese medicine, but holding them does more than relieve headaches. The energy that travels up your nervous system along your spine passes through these points. When you are the tester, holding the points on yourself breaks the energetic circuitry with the other person and brings you back into your own energy loop, disengaging you and your beliefs from the other person. When you are being tested and place your hands on these points, it disengages you from your own thoughts about the matter and seems to allow the test to reveal a deeper level of information. Keeping your own beliefs and expectations from interfering is the most important single step you can take to ensure an accurate energy test.

3. Focus Your Intention. Your intention can influence a test's outcome. Rather than having to work around the effects of intention, however, you can use the fact that expectations influence energy testing and resolutely set your intention for valid results. To demonstrate, find a food or other substance that tests strong on a friend. Repeat the test on the same substance without telling your friend what you are up to. On the second test, decide in advance that the test will show weak. Do not pull harder. Just see if this shift in your intention influences the results. It often does. So set your intention for an accurate test.

4. Establish a Resonance Between Tester and Tested. You can create an energetic resonance with your partner:

a. Both take a deep breath in tandem.
b. Together release the breath along with all thoughts.
c. Test once the breath is released.

5. Stay Alert. Be sure the person is ready before beginning the test. Pull the arm smoothly. Apply pressure only long enough to determine if the arm's resistance holds (generally between one and two seconds). If a muscle is locked in place or gives a fraction of an inch and then bounces back, the test has shown that the energy is flowing through the muscle. Do not overwhelm the muscle "just to be sure." Along with learning the mechanics of energy testing, I am also interested in helping you cultivate a strong and reliable intuition. Energy testing is a superb means to that end because it provides tangible information about subtle energies. With each energy test you perform, you receive feedback that attunes your intuition to the flow of energies within you and around you as well as to the subtleties of energy testing itself.

6. Practice Under Double-Blind Conditions. You can develop confidence in your ability to do an accurate energy test by practicing under conditions where you will get immediate feedback. You will need two other people. One will supervise the test, and one will be tested. Find some substances that the person being tested *knows* are healthy and agreeable to his or her system, such as organic apples or mint tea. Take other substances that you *know* would be bad to ingest, such as a bottle of ammonia. The person supervising the experiment takes a substance and puts it into the energy field of the person being tested in a way that neither you as the energy tester nor the person being tested can see, perhaps against the back of the person's shirt. Energy test. Do not be surprised if you do not get a perfect score at first. The double-blind test can be a great forum for learning the intricacies of energy testing, and it provides a benchmark as you continue to build your proficiency.

Energy testing provides you with a method for assessing your own energies and for evaluating how the environment, including the food you eat, is affecting them. The following chapter will take you a step further by teaching you how to begin to balance your personal energy field to optimize your health.

Chapter 3

Keeping Your Energies Humming

A Daily Routine

I SING THE BODY ELECTRIC.

—WALT WHITMAN
Leaves of Grass

When a client's energies are humming, an exquisite, pulsating, captivating scene is laid out in front of me, as if ribbons of energy are moving up and down the body, weaving intricate patterns. The energy pathways are open and spacious rather than dense and congested. Like an endless waterfall, other energies spill up and over the top of the head, and a field of energy surrounds and caresses the body.

In contrast, an energy field that is out of balance appears disordered or scrambled. It may resemble the static on a television screen. The energies may be running backward from their natural, organic direction. The patterns may appear frenzied and chaotic, or they may seem stagnant, without flow.

The good news about understanding your body as an energy system is that you can learn skills for keeping your energies flowing in a vibrant harmony that fosters your good health. The bad news is that modern life assaults you every day with conditions that tend to drain your energies and push them out of balance.

Much in our world today lends to these disruptions. We are continually exposed to radiation and electromagnetic fields that scramble our energies in ways never dreamed of a hundred years ago. Medicines that give short-term relief may disorganize our energies, creating long-term problems. The pesticides killing off the insects that threaten our foods wind up *in* our foods and bewilder our energy systems. The disorienting effects of many such influences, combined with the complex stresses we routinely encounter and the fast pace at which we live our lives, can have the cumulative impact of leaving a person's energy field disrupted. Even energies that naturally work in harmony can become antagonistic.

Back to the good news. There is much you can do to reverse these disturbances. Throughout history, people have devised exercises and techniques that bring the body's energies back into balance. Cultures separated by time and distance, with no communication between them, often used very similar methods. In my own work, I have several times spontaneously "invented" a technique or exercise that I later learned was part of the tradition of an indigenous culture. While we are each as energetically unique as a thumbprint, we all share a similar physiology. You will be learning techniques that are beneficial for *every* body as well as how to adjust the procedures to your *particular* body. You can literally weave your energies so they become stabilized, strengthened, and harmonized as you move through the turbulence of modern life.

How do you know your energies are moving in a healthy, organic direction? If you feel fabulous, *they are!* If not, *they're not.* It's that simple.

When your energies are rhythmic and clear, they support your health. The more coherent the wave patterns in an individual's electrocardiogram, for instance, the more efficient the functioning of the nervous, hormonal, and immune systems appear to be.[1] In one pilot study, a coherent electrocardiogram signal correlated with a slowed growth rate of cultured cancer cells and a faster growth rate of healthy cells.[2] Healers often have the effect of synchronizing a client's energies. During healing sessions, the brain waves of the healer and of the recipient also enter a state of coherence and synchrony, becoming unified in a single energy field. This unity has been shown to raise the level of hemoglobin in the recipient's blood, reduce the severity of pain, lower anxiety, and heal wounds more rapidly.[3]

If your energies are flowing naturally, activities such as walking or running vitalize and strengthen you. One of the first things that happens when you are extremely tired or feeling ill is that activities that would normally replenish you begin to drain you. This is because your meridians have begun to run backward. The natural direction of each of the fourteen meridian pathways is illustrated in Figures 10 through 23 (pages 101–109). Your body is designed so that walking or running pumps the meridian energies in their organic direction; however, if the energies have reversed their direction, your activity is now going against the flow of the energy, so it depletes rather than replenishes you. Your body forces you to slow down so it can rebalance, release toxins, and regenerate itself in the energy-restorative magic of sleep. By accelerating your fatigue whenever you push onward, your body is trying to force you to rest when you need rest.

How to Release Tension

The face can hold an immense amount of tension. Here are two quick tension-release techniques (time—15 seconds each):

1. *Place the pads of your fingers against your cheekbones. Push upward and hard. Gradually push your fingers along the curve of the cheekbone toward your ears. This brings blood into your face and opens the spaces that had been tense.*

2. *Lay one hand over your opposite shoulder. Push in and pull your fingers forward over the shoulder. Do three times on each side.*

If you keep ignoring your body's demand that you slow down, your body will eventually make the decision for you. Perhaps you will "catch" a cold or the flu, injure yourself, or become depressed. Your body grows ever more insistent that you *must* slow down. Yet we live in a world that is governed by the artificial rhythms of clock and calendar rather than the natural rhythms of body, planet, and heavens. What to do?

Understanding the language of your body helps you better hear its needs and communicate with it so it can stay in better synch with your lifestyle. Rather than being at odds with one another, your choices and your body can stay in alignment, even when you need to trudge on when tired. While it is almost impossible in today's world to live primarily by your body's natural rhythms, you can literally tell your body not to reverse its energies at times when you are unable to rest. This doesn't mean you can forever go without sleep. You may still be drawing on your reserves when you make this request, but you won't be wasting energy by fighting your body's mechanisms to get you to stop when you cannot stop.

The following simple techniques can benefit nearly anyone living in the stress-producing, polluted, nature-alien, energy-scrambling environments that mark our technological progress. I also suggest that you combine these methods into a five-minute "energy routine," as detailed later in the chapter, and that you use it every day. The daily routine builds positive habits into your energy field. The techniques are simple yet potent, and they are cumulative. Each technique is followed by instructions for energy testing its effects. First learn the technique and *sense* its effects on your body. Then experiment with the energy test.

1. THE THREE THUMPS

Certain points on your body, when tapped with your fingers, will affect your energy field in predictable ways, sending electrochemical impulses to your brain and releasing neurotransmitters. By tapping three specific sets of points, a set of techniques I call the Three Thumps, you can activate a sequence of responses that will restore you when you are tired, increase your vitality, and keep your immune system strong amid stress. You will come to know instinctively when you need these points tapped. Also do not be too concerned about finding the precise location of each point. If you use several fingers to tap in the vicinity described, you will hit the right spot.

Thump Your K-27 Points

Acupuncture points are tiny centers of electromagnetic and more subtle energies; they are arranged along fourteen meridians. The K-27 points are the twenty-seventh pair of acupuncture points on the kidney meridian. Both acupuncture, which employs needles, and acupressure use the same points. Tapping or massaging your K-27 points is a simple exercise that will:

- Energize you if you are feeling drowsy
- Focus you if you are having difficulty concentrating

They are juncture points that affect all of your energy pathways. Working them sends a signal to your brain to adjust your energies so you can feel more alert and can perform more effectively. To tap your K-27 points, do the following (time—about 30 seconds):

1. Place your fingers on your collarbone (clavicle). Now slide them inward toward the center and find the bumps where they stop. Drop about an inch beneath these corners and slightly outward to the K-27s (see Figure 3). Most people have a slight indent here that their fingers will drop into.
2. With the fingers of both your hands turned toward your body, cross your hands over one another, with the middle finger of each hand now resting on the opposite K-27 point. Crossing your hands is not essential, but it does assist the energy to cross over from the left brain hemisphere to the right side of the body and from the left hemisphere to the right side.
3. Tap and/or massage the points firmly while breathing deeply, in through your nose and out through your mouth. Continue for about twenty seconds. If your situation doesn't allow you to use both hands at the same time, use your thumb and fingers to tap or massage both points at once.
4. After you have tapped or massaged K-27, you can boost the effect by hooking the middle finger of one hand in your navel and resting the fingers of your other hand on the K-27 points. Keeping your fingers hooked into your navel, pull upward for two or three deep breaths. You will feel a stretch below your belly.

Tapping the K-27s "flips" your energies around if they have started flowing backward. "Flowing backward" literally means that you are going *one way* and your energies are going *the other way.* It is so simple and unobtrusive to tap or massage your K-27 points that you can use this technique in a classroom or a business meeting. Or suppose you are driving late at night and you are drowsy. If you take one hand off the steering wheel and tap K-27, you will probably feel the energy come up into your eyes and wake you up a bit. You can lock in this alertness by pulling off the road and doing a cross crawl (see next section), exaggerating your movements. If you do not feel an immediate response from tapping the K-27 points, your body probably isn't going to let you override your exhaustion much longer. You should *immediately* pull over and rest. Techniques for dealing with extreme situations where circumstances insist that you push onward are presented later (see homolateral crossover, page 233, and Separating Heaven and Earth, page 249).

Tapping the K-27 points can not only keep you awake but can also help you think more clearly. I have countless times seen children in school, sitting at a desk, looking first at what the teacher is writing on the blackboard and then looking back to the desk. By the time some children's eyes reach the desk, their mind is blank or the information has become totally muddled. Their energy circuitry has been disrupted by this up-and-down movement of the eyes. This is the mechanism that allows eye movement desensitization to disengage a traumatic stress response from a traumatic memory, but it can also keep you from absorbing new information. By tapping K-27, however, the energy circuitry will hold strong even as the child's eyes move up and down. This technique is also helpful with dyslexia and other learning disabilities, as well as any situation, such as tennis or reading, where your eyes are rapidly shifting from one spot to another.[4]

Energy Testing the K-27 Thump. In Chapter 2 you learned to energy test the spleen meridian for determining your response to foods and environmental conditions. If the spleen meridian is chronically weak, however, it will not serve well in a general purpose test. The "general indicator" test is another energy test that may be used in virtually any situation because it does not isolate a specific meridian but rather indicates a more general disturbance in the energy field.

The basis principles for the tests are the same; the only difference is in how you hold your arm. To do the general indicator test:

1. Place either arm straight out in front of you, parallel with the ground, and then move it 45 degrees to the side. Your elbow should be straight and your hand open, palm facing the floor (see Figure 4).
2. Have your partner place the fingers of an open hand on your arm just above the wrist. As you hold firm, your partner pushes down slowly for up to two seconds and only hard enough to determine if there is a "bounce."

Figure 4. GENERAL INDICATOR TEST

You can use the general indicator test to demonstrate how the body's energies respond to fatigue. When you are particularly tired:

1. Walk forward seven or eight steps, stop, and have your partner test you, using the general indicator test. Does it show weak or strong?
2. Now walk backward seven or eight steps, stop, and retest.
3. If walking forward weakens you, more than half of your meridians are running backward. If this is the case, tap or massage your K-27 points and repeat number 1.

People are often bewildered when, during a demonstration, they lose their energy by walking forward but *retain* their strength when walking backward. If your meridians are reversed, you are actually in better harmony with your energies when you walk backward than when you walk forward, so you will test strong. Tapping or massaging the K-27 points radiates energy to all the meridians, flipping them into their forward direction, allowing you to walk forward or backward, as well as to do other activities, without losing your energy.

Thump Your Thymus Gland

Tapping the area over your thymus gland is a simple technique that will:

- Stimulate all of your energies
- Boost your immune system
- Increase your strength and vitality

This technique can help you if you are feeling bombarded by negative energies, catching a cold, fighting an infection, or if your immune system is otherwise challenged. Your thymus gland is located about two inches below your K-27 points in the center of your chest. It is your immune system's surveillance gland. If you override your thymus's intelligence by ignoring your body's needs, your thymus's surveillance mechanism becomes confused. If you do this consistently, it becomes sluggish and is thrown into disarray. Thumping the thymus wakes it and aligns it (time—about 15 seconds):

1. Move your fingers down a couple of inches and into the center of your sternum after tapping K-27.
2. As you breathe deeply, firmly tap your thymus point with the four fingers of each hand for about 20 seconds (see Figure 3).

Many people automatically tap their thymus point when under stress. Have you ever noticed how people tend to hold their midchest when an emotional shock comes their way; or how, if they are feeling faint, they pat this spot? So picture Tarzan and thump your own thymus before moving into a challenging situation.

Energy Testing the Thymus Thump. The test for the thymus is simple. Lay the fingers of one hand over your thymus and have your partner energy test you with your other arm.

Thump Your Spleen Points

Once, after running on empty for far too long, I found myself teaching a class to 120 people in another country and feeling extremely ill. I felt I was about to pass out. Without thinking, I instinctively marshaled whatever energy I had left and tapped my spleen "neurolymphatic reflex points." On the stage, in front of everyone, and with no explanation, I began tapping hard and furiously. Suddenly I could feel new life stirring within me, climbing up the trunk of my body, and pulsing outward. As I watched the class watching me, I realized how ridiculous I looked, and I began to laugh. I continued to laugh and tap, tap and laugh. Suddenly I knew I was going to be all right.

The class saw my recovery with their own eyes, as I changed in an instant from looking white, pasty, and shaky to centered, vibrant, and full of color. This proved to have been an effective demonstration. Two people from that class have since written to tell me that using this technique has pulled them out of a physical crisis. One is a young woman who worked in a medical care facility. She had pushed herself so beyond her physical limits that she began to wobble one day and was afraid she was about to fall in front of a patient. She started to tap her spleen points and quickly recovered. After she explained what had occurred, her patient took the technique into his repertoire and later described it as one of the most significant benefits he had received from his treatment.

Tapping the neurolymphatic points on your spleen meridian is a quick way to:

- Lift your energy level
- Balance your blood chemistry
- Strengthen your immune system

As contrasted with acupuncture points, neurolymphatic points are part of the lymph system. Because the spleen is central to the functioning of the immune system, tapping your spleen's neurolymphatic reflex points serves to synchronize your body's rhythms, harmonize its energies and its hormones, remove toxins, fight infection, combat general malaise during or after stress, counter dizziness, modulate blood chemistry, and better metabolize food. To tap your spleen points (time—about 15 seconds):

1. Find the points by moving your fingers down from your thymus, out to your nipples, and straight down to beneath your breasts. Then move them down over the next rib (see Figure 3).
2. Tap firmly with several fingers for about 15 seconds, breathing deeply in through your nose and out through your mouth.

Energy Testing the Spleen Thump. Since the spleen thump affects the spleen meridian, you can use either the general indicator test or the spleen-pancreas test you learned in the previous chapter. To experiment with the effects of the spleen tap, have a friend energy test you when your blood chemistry may be out of balance, as when you are feeling emotional, confused, irritable, dizzy, nauseous from hunger, or going through PMS. If the test shows weak, do the spleen tap and retest.

2. THE CROSS CRAWL

The Cross Crawl facilitates the crossover of energy between the brain's right and left hemispheres.[5] It will help you:

- Feel more balanced
- Think more clearly
- Improve your coordination
- Harmonize your energies

You may find the cross crawl to be particularly valuable if, for instance, you are feeling physically or mentally exhausted for no apparent reason, are feeling worse rather than better after you exercise, are becoming lethargic and unmotivated, or are off-balance from carrying a suitcase, a shoulder bag, or a child.

Prior to beginning the cross crawl, tap your K-27 points to ensure that your energies are traveling in their natural direction. The cross crawl is as simply as marching in place (time—about a minute):

1. While standing, lift your right arm and left leg simultaneously (see Figure 5).
2. As you let them down, raise your left arm and right leg. If you are unable to do this because, for instance, of being confined to a wheelchair, simply lift your knees to the opposite elbows, or twist your upper torso so your arm passes over the midline of your body.

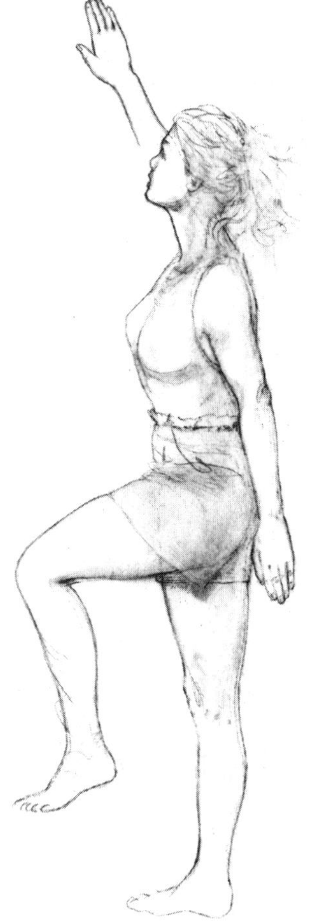

Figure 5. CROSS CRAWL

3. Repeat, this time exaggerating the lift of your leg and the swing of your arm across the midline to the opposite side of your body.
4. Continue in this exaggerated march for at least a minute, again, breathing deeply in through your nose and out through your mouth.

The Cross Crawl is based on the fact that the left hemisphere of your brain needs to send information to the right side of your body, and the right hemisphere needs to send information to the left side. If energy from the left or right hemisphere is not adequately crossing over to the opposite side of your body, you cannot access and utilize your brain's full capacity or your body's full intelligence.

When energy is unable to cross over, it slows down dramatically. It begins to move in what is referred to as a homolateral pattern—straight up and down the body—and the body's ability to heal is severely diminished. A homolateral pattern is natural in newborn babies. Their energy does not yet cross over: the right hemisphere governs the right side of the body; the left hemisphere the left side. Crawling establishes the patterning that allows the energies to cross over from each hemisphere to the opposite side of the body. This is one of the reasons an infant's learning curve increases exponentially with the ability to crawl, and children who do not crawl often develop learning disabilities. Modeled after observations of the importance of crawling for infants, the Cross Crawl helps your whole system function more effectively, and it also promotes the healing process.

Human evolution arranged for us to spend a considerable amount of time cross-crawling every day. Walking, running, and swimming are all natural ways of performing a Cross Crawl. However, in our sedentary lifestyles, we often do not put in the time with these activities that nature intended. In addition, when we do walk we often are not using a cross-crawl pattern. For instance, a shoulder bag prevents your arms from swinging and cuts through the major meridian lines in your shoulder. The Cross Crawl allows you to quickly—within a minute if you are reasonably well balanced—reestablish the crossover pattern. So if you have to carry something over your shoulder, you might do a cross crawl when you get to your destination. If the Cross Crawl tires instead of energizes you, it probably means that your energies are locked in a homolateral pattern, unable to cross over from one side of your brain to the other. If that is the case, skip over to the homolateral crossover (page 233) and use that procedure prior to doing the Cross Crawl.

Even when you are sitting, you can gain some of the benefits of the Cross Crawl. Postures that help your energies to cross over will feel natural to you when you need them. When you cross your arms in front of your chest for instance, the reason may not be to close yourself off to other people, but rather to get your own energies balanced. Sometimes people become self-conscious when they notice their arms are crossed because they fear someone will read their body language as meaning they are closed off or defended. The body language experts who have given us pat interpretations of the meaning of certain postures have actually done a disservice to our natural instincts. On the contrary, if you come into the presence of another person whose energy is throwing you off, crossing your arms can reconnect your own energies and actually allow you to be more authentic and present with the other person. A theme in my work, and throughout this book, is to stay attuned to the truth of your body's messages to you. Over the next few days, notice the effects spontaneous postures have on your energies.

I heard about an extreme example of the potency of the Cross Crawl years ago when a friend's eight-year-old son was playing with several other boys on a tractor. My friend's son fell in front of the tractor, which then rolled over his head. The other boys panicked and shifted the gears, accidentally putting the tractor into reverse, and running over his head a second time. Some of the contents of his skull literally spilled out. He wasn't expected to live. My friend was well versed in the use of the Cross Crawl. She rode with her son in the ambulance, and no doubt in the desperation of her helplessness as she looked at his horrendous injuries, she lifted his arms and legs in a Cross-Crawl movement on the way to the hospital. At the hospital, the doctors were surprised he hadn't died immediately. The family was Catholic, and he was given last rites.

Throughout the commotion, my friend just kept Cross-Crawling her son. When the boy didn't die, the staff sewed up his head. He was in a deep coma, and the doctors doubted that he would ever regain consciousness. My friend spent much of her time at the hospital, Cross-Crawling her son several times every day. She was certain that the procedure was keeping his brain alive. He stayed in the coma for six months. One day as she went to Cross-Crawl him, he woke up and Cross-Crawled for himself. His optic nerve had been severed, so he was blind, but his mental functioning returned. Whatever the fortunate combination of factors that contributed to his remarkable recovery, I can attest that I have seen the Cross Crawl perform lesser neurological marvels dozens of times.

Testing the Cross Crawl. Do a few jumping jacks. Have your partner energy test you. The jumping jacks probably weakened you. Many exercises that are designed to tone your muscles are not attuned to your energies. Jumping jacks, for instance, prevent your energies from crossing from each brain hemisphere to the opposite side of your body. If an activity weakens you, do a Cross Crawl and retest. You are probably strengthened. It is possible to alter most exercises so they are in alignment with your energies. For example, a form of jumping jack that will be in harmony with your energies as well as your muscles begins with your feet spread and your hands at your sides. Bring your feet together and your hands over your head on the jump. After several of these "reverse" jumping jacks, test again. It may be very useful to have a friend test you after any exercise you do routinely. If it weakens your energies, use the principles of the cross-crawl pattern to find a way to modify the exercise so it leaves you strong. Or simply, as a precaution, Cross Crawl after you complete any exercise.

How to Pick Yourself Up at Your Droopy Time of the Day

Take a break from your work and do an energy exercise that activates your respiratory system. Separating Heaven and Earth (page 249), for instance, pulls in extra oxygen, releases carbon dioxide, stretches the body so energy can more readily flow through it, and opens the joints, releasing trapped energy. Also valuable for the blahs are the Three Thumps (page 63) and, particularly if you're having trouble taking in information, the Wayne Cook posture (below).

3. THE WAYNE COOK POSTURE

I use the Wayne Cook posture when I am overwhelmed, am hysterical, cannot get clarity on a situation, cannot bring order into my life, must confront someone, or am upset after someone has confronted me. This procedure is named to honor Wayne Cook, a pioneering researcher of bioenergetic force fields, who invented the approach I have modified into the form presented here. It can help you to bring a sense of order into your affairs, to better comprehend the world you are facing, and to literally put one foot in front of the other. Even when the upset is so intense that you are unable to quit crying, are finding yourself snapping and yelling at others, sinking into despair, or feeling you are beyond exhaustion, it will move stress hormones out of your body. Almost immediately, you will begin to feel less crazy and overwhelmed. The Wayne Cook posture can help you:

- Untangle inner chaos
- See with better perspective
- Focus your mind more effectively
- Think more clearly
- Learn more proficiently

To do the Wayne Cook posture, sit in a chair with your spine straight (time—about 2 minutes):

1. Place your right foot over your left knee. Wrap your left hand around your right ankle and your right hand around the ball of your right foot (see Figure 6a).
2. Breathe in slowly through your nose, letting the breath lift your body as you breathe in. At the same time, pull your leg toward you, creating a stretch. As you exhale, breathe out of your mouth slowly, letting your body relax. Repeat this slow breathing and stretching four or five times.
3. Switch to the other foot. Place your left foot over your right knee. Wrap your right hand around your left ankle and your left hand around the ball of your left foot. Use the same breathing.
4. Uncross your legs and place your fingertips together forming a pyramid (see Figure 6b). Bring your thumbs to rest on your "third eye," just above the bridge of your nose. Breathe slowly in through your nose. Then breathe out through your mouth, allowing your thumbs to separate slowly across your forehead, pulling the skin.
5. Bring your thumbs back to the third eye position. Slowly bring your hands down in front of you, pulling them together into a prayerful position while breathing deeply. Surrender into your own breathing.

Wayne Cook demonstrated the effectiveness of this technique for treating dyslexia and stuttering. The procedure connects the energy circuitry in a manner that allows a smooth flow throughout the body. Stress causes the reptilian section of the brain to dominate the forebrain. Your thinking brain turns off. Long before the forebrain evolved, the autonomic nervous system kept our predecessors alive during times of threat. Reactions were automatic, unpremeditated. They had to be. The teamwork between the forebrain and the autonomic nervous system, however, is not one of nature's showcase achievements. The Wayne Cook posture helps the energies between the two work in better harmony. It can also help with a wide range of psychological problems, including confusion, obsessiveness, compulsivity, disorganization, depression, and excessive anger. The Wayne Cook posture also strengthens the body's energy integrity, making it less vulnerable to outside influences such as pollution and toxic energies in the environment.

When your energy is scrambled, you speak with a scrambled tongue. The listener often does not receive the information you are trying to convey because it gets caught in this energetic chaos. A speaker's energies can put audience members to sleep or rivet their attention. The energetic impact is instant, and it is often outside the listener's awareness. If you were to energy test the members of an audience when a speaker's energies are scrambled, you would find that most of the audience would also test weak. Scrambled energy is contagious. When a person's energies are clear and centered, it is also contagious.

For instance, have you ever had the experience of going to the theater to see a play you loved so much that you couldn't wait to go home and persuade someone to come see it with you the next night? The next night, however, the play simply didn't have the power of the previous evening. The first night the actors spoke in a smooth, congruent manner. On the second night, for whatever reasons, usually a combination of stress and exhaustion, their energies were scrambled. Practicing the Wayne Cook technique could have helped the actors on the second evening.

Because I live near a theater town, I have often demonstrated how scrambled energy can cause performers to lose their audience, and I have taught many how to use the Wayne Cook posture to unscramble their energies before a performance. It makes a profound difference. Even if you are not an actor, there is almost certainly a place in your life for this technique. Use

Figure 6.
WAYNE COOK POSTURE

it and you will, among other things, be a better listener or audience member. As any musician, actor, or comedian will tell you, great audiences create great performers. Or if you are in a relationship with a partner who doesn't seem to hear you, or if you have difficulty speaking your own truth, this is an excellent technique to use prior to any confrontation or important discussion.

If you are in a board meeting or some other setting where you would not be comfortable holding the Wayne Cook posture but definitely need to get clear, you can cross your arms, knees, and ankles, and/or wrists. Breathe slowly. This will give you at least some of the benefits of the Wayne Cook technique.

Testing the Wayne Cook Posture. If you feel tired while reading or are having difficulty comprehending the words, your energies will test weak. This is because when your brain recognizes fatigue, it in its wisdom tries to slow you down. Preventing you from comprehending new material is a way of slowing you down. Your energies become scrambled, making it difficult for you to read, make sense of new material, or perform linear tasks.

Use the Wayne Cook posture when you have been having trouble reading or concentrating and then retest. In fact, read a line from right to left (reading the words backward: "backward words the reading"). Energy test. If your energies are scrambled or reversed, reading backward makes you strong while reading forward makes you weak. It is similar to the way walking backward can strengthen you when you are exhausted (page 66).

4. THE CROWN PULL

Energy naturally accumulates at the top of your head, but it can become stagnant if it doesn't move out through your crown chakra, the energy station at the top of your head. The "Crown Pull" releases this energy. It clears the cobwebs from your mind and brings a calm to your nervous system. It can also often take away a headache or a stress-induced stomachache. The Crown Pull:

- Releases mental congestion
- Refreshes the mind
- Opens the crown chakra to higher inspiration

The crown chakra is your gateway to the higher energies of the cosmos, the spiritual amniotic fluid that surrounds and supports each of us. Over the years, I've had many people tell me that the crown pull has helped their overly active intellect surrender to transcendent sources of information. The crown pull clears your head and mind, making space for the energy in your skull to move more freely. It reminds you to attend to the realm of the crown chakra, which governs your spirituality.

A growing number of people are hungry to receive this inspiration and guidance and to feel more fully connected with nature or god or however they conceive of the larger picture.

While doing the Crown Pull, breathe deeply, in through your nose and out through your mouth (time—about 15 seconds):

1. Place your thumbs at your temples on the side of your head. Curl your fingers and rest your fingertips just above the center of your eyebrows (see Figure 7a).
2. Slowly, and with some pressure, pull your fingers apart so that you stretch the skin just above your eyebrows.
3. Rest your fingertips at the center of your forehead and repeat the stretch.
4. Rest your fingertips at your hairline and repeat the stretch.
5. Continue this pattern with your fingers curled and pushing in at each of the following locations:
   a. Fingers at the top of your head, with your little fingers at the hairline. Push down with some pressure and pull your hands away from one another, as if pulling your head apart (see Figure 7b).
   b. Fingers at the center of your head, again pushing down and pulling your hands away from one another.
   c. Fingers over the curve at the back of your head, again using the same stretch. Repeat each of these stretches one or more times.

Testing the Crown Pull. Test the crown chakra by touching anywhere on top of your head with one hand and having your partner energy test your other hand. You will probably test weak at times when your head feels jammed. Use the Crown Pull at such a time and then retest.

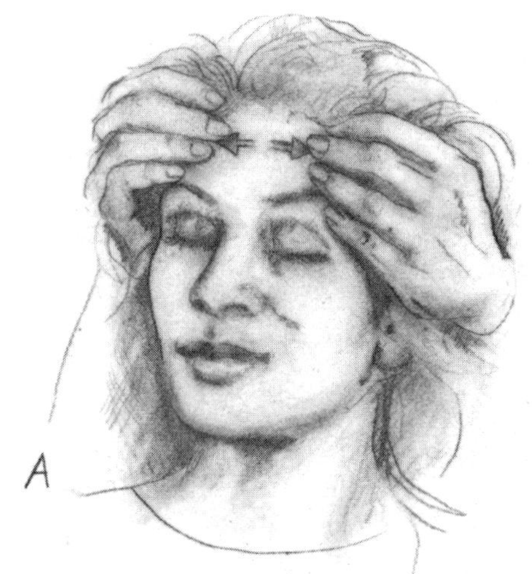

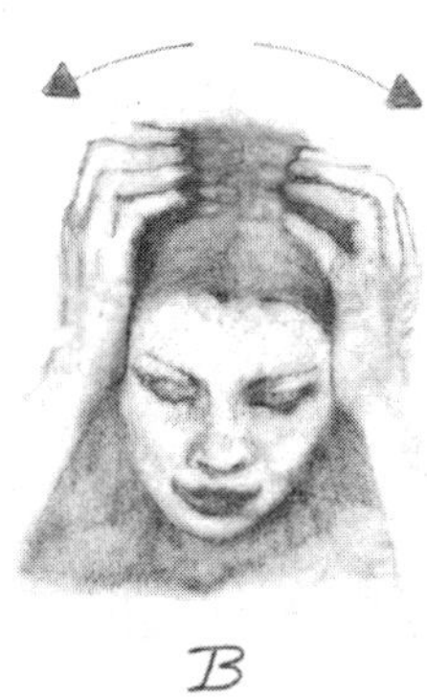

Figure 7.

Crown Pull

5. THE SPINAL FLUSH

If you wake up tired, doing the spinal flush will bring fresh energy to you; if you are sore in the evening, it will relax you. The Spinal Flush works with your lymphatic system. The lymph has no pump of its own, but you pump it whenever you move your body. You can also pump it by massaging your neurolymphatic reflex points. Located mainly on your chest and back, these points regulate the flow of energy to the lymphatic system. When the neurolymphatic reflex points become clogged, every system in your body is compromised. The Spinal Flush will:

- Energize you
- Send toxins to your body's waste removal systems
- Clear stagnant energies from your body

Congested neurolymphatic reflex points feel sore when massaged. For that reason, they are not hard to locate. And there are so many of these points so close to one another that you won't miss them. Massaging them is a way to clear them and allow the energy that has been blocked to flow again. It is a particularly nice procedure to do with a friend (time—about a minute, but you may be pleading "Don't stop!"):

1. Lie face down, or stand three or four feet from a wall and lean into it with your hands supporting you. This positions your body to remain stable while your partner applies pressure to your back.
2. Have your partner massage the points down either side of your spine, using the thumbs or middle fingers and applying body weight to get strong pressure. Massage from the bottom of your neck all the way down to the bottom of your sacrum (see Figure 8).
3. Have your partner go down the notches between your vertebrae and deeply massage each point. Staying on the point for at least 5 seconds, your partner moves the skin up and down or in a circular motion with strong pressure.
4. Upon reaching your sacrum, your partner can repeat the massage or can complete it by sweeping the energies down your body. From your shoulders, and with an open hand, your partner sweeps all the way down your legs and off your feet, two or three times.
5. Do not worry about a point being missed. Each of your meridians will be covered by simply going between all the notches. Rather than knowing which meridians are associated with which points, simply ask for special attention on any points that are sore.

If you are moving through intense emotional or physical stress, or if you have been exposed to environmental toxins, doing the Spinal Flush will clear your lymphatic system. You know lymph as the clear fluid that comes out of a cut. It is there to remove foreign matter. Lymph plays a key role in your immune system by helping fight conditions ranging from colds to cancer. It creates antibodies and produces one-fourth of your white blood cells.

The lymphatic system has been called the body's other circulatory system. While blood is pumped by the heart, movement and massage pump the lymph. Lymphocytes are specialized white blood cells produced in the lymph nodes, found in your neck, armpits, abdomen, and groin. Your lymph system also carries proteins, hormones, and fats to the cells and eliminates dead tis-

Figure 8.
SPINAL FLUSH

sue and other waste products. There are twice as many lymph vessels as blood vessels in your body.

The Spinal Flush not only cleanses the lymphatic system, it stimulates the cerebrospinal fluid, clearing your head as well. It is a quick rebalance, and of all the energy techniques I've seen, it probably delivers the most benefit for the least effort in the greatest number of situations. As a cold is coming on, the Spinal Flush can stop it in its tracks. I regularly recommend it to couples, both as a way of lovingly caring for one another and as a way to head off problems. If you sense that an interchange is headed toward an argument, tell your partner, as lovingly as you can, "Up against the wall!" and firmly work the neurolymphatic reflex points. This simple technique immediately reduces built-up stress and takes an edge off of emotional overreactions.

Flushing Without a Partner. If you do not have a partner available to work the neurolymphatic reflex points on your back, massage the many points you can reach that are shown in Figure 9. You can clear many of the points on your back by reaching over your shoulders or around your waist and pushing into the notches yourself. Wherever you have tenderness, work with these points for several seconds. While the tenderness will not necessarily go away immediately, you are clearing the congestion. Neurolymphatic massage is great to do on a daily basis. You will feel a difference. Again you can energy test before and after any of these procedures to demonstrate to yourself its effects.

6. ZIPPING UP

When you are feeling sad or vulnerable, the central meridian, the energy pathway that governs your central nervous system, can be like a radio receiver that channels other people's negative thoughts and energies into you. It's as if you are open and exposed. The central meridian runs like a zipper from your pubic bone up to your bottom lip, and you can use the electromagnetic and more subtle energies of your hands to "zip it up."[6] Pulling your hands up the central meridian draws energy along the meridian line. The Zip Up will help you:

- Feel more confident and positive about yourself and your world
- Think more clearly
- Tap your inner strengths
- Protect yourself from negative energies that may be around you

To Zip Up (time—20 seconds):

1. Briskly tap K-27 to assure that your meridians are moving in a forward direction.
2. Place your hand at the bottom end of the central meridian, which is at your pubic bone (see Figure 10, page 101).
3. Take a deep in-breath as you simultaneously move your hand with deliberation straight up the center of your body, to your lower lip. Repeat three times.

This is the natural direction that the central meridian flows. By tracing it in this manner, you strengthen the meridian, and the meridian strengthens you. You can zip up the central meridian as often as you wish. Again, remember to breathe deeply as you do so, and you will begin to feel centered, in control, and in your own power. The energies of a healer's hands emanate an electromagnetic force, and tracing your meridian with your hand moves the energy in the meridian.

The central meridian is highly sensitive to other people's thoughts and feelings as well as to your own. It also directly affects each of your chakras. When you are feeling good, it is as if this "zipper" is zipped all the way up and you are protected.

I like to demonstrate for an audience the tangible way that people's energies affect one another by first establishing, using an energy test, that a volunteer's central meridian is strong. I then ask the audience to bring negative thoughts to mind. Almost always, an energy test will show that this has weakened the volunteer's central meridian. Then I will have the person zip up and "mentally lock it in" while the audience not only thinks negative thoughts but ups the ante by sending negative energies toward the volunteer. Nonetheless, the energy test will almost always show that the person's central meridian remains strong, even as I tell the audience, "Give it your best shot." Zipping up can help you be present with another person in a conflictual situation with less chance of the other's negative attitude dragging down your energies. Many people have told me how zipping up has enabled them to speak with a boss, a difficult parent, an angry child, or an ex-lover and stay centered in their own truth and self-validation.

Since the central meridian is closely attuned to your thoughts and feelings, it is also quite responsive to hypnosis and self-hypnosis. A powerful way to psychologically implant affirmations such as "I am clear, centered, and well-organized" is to state them while you are tracing the central meridian, imagining you are zipping them into every cell in your body. In fact, after you have zipped up the central meridian, imagining that you are *locking* the zipper in place and hiding the key tends to prolong the technique's effects.

Testing the Zip Up. Thinking a positive thought strengthens the flow of your energies. Thinking a negative thought weakens their flow.

1. Think a positive thought and have someone energy test you.
2. Retest after thinking a negative thought.
3. Have your partner hold a negative thought and then retest *you*. Your partner's thoughts probably weakened your central meridian.
4. Have your partner continue to hold the negative thought, but this time Zip Up as your partner thinks it. Test again to see if zipping up protected your energy field from the impact of your partner's negative thought.

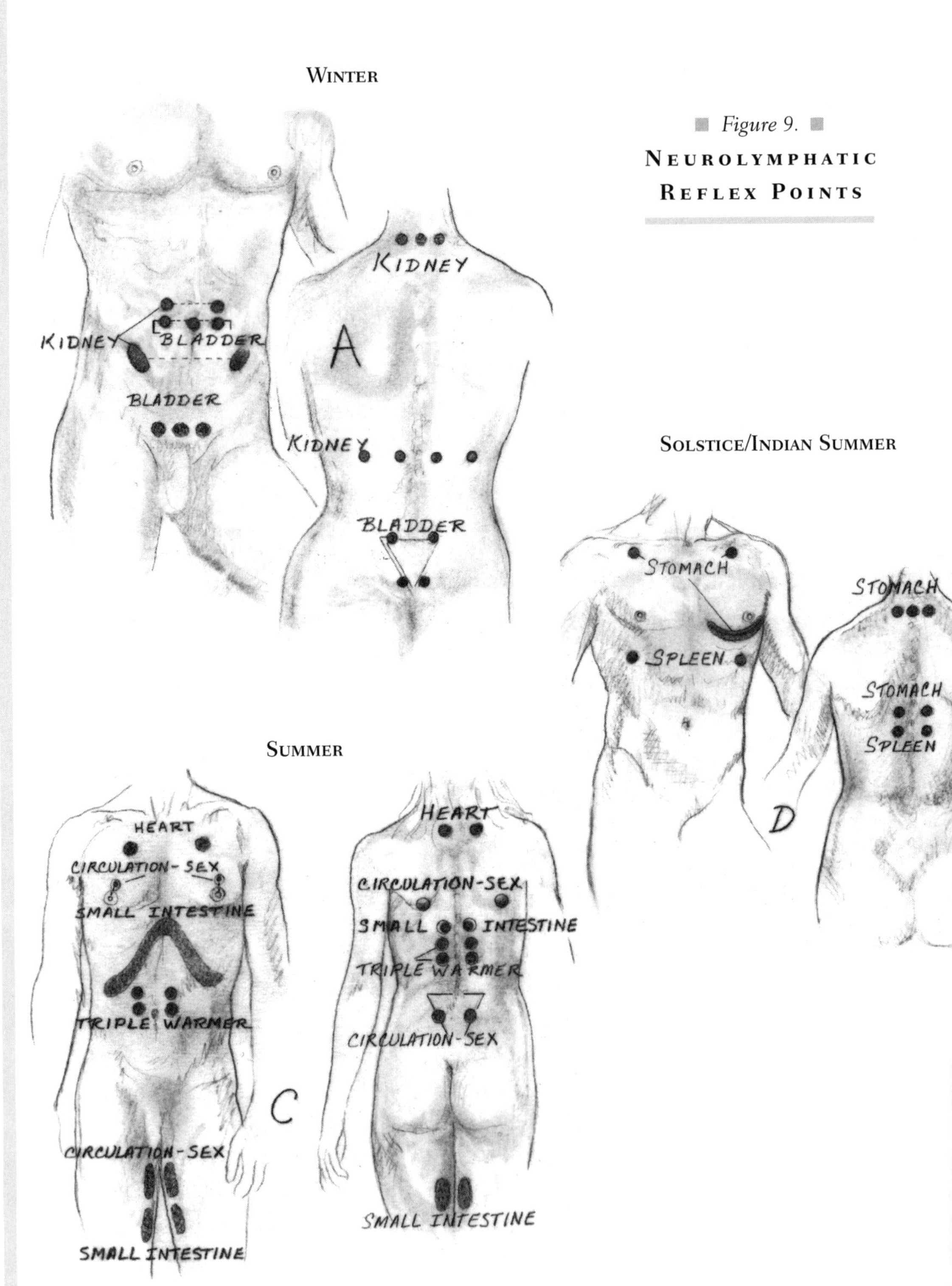

■ *Figure 9.* ■

NEUROLYMPHATIC REFLEX POINTS

Spring

Autumn

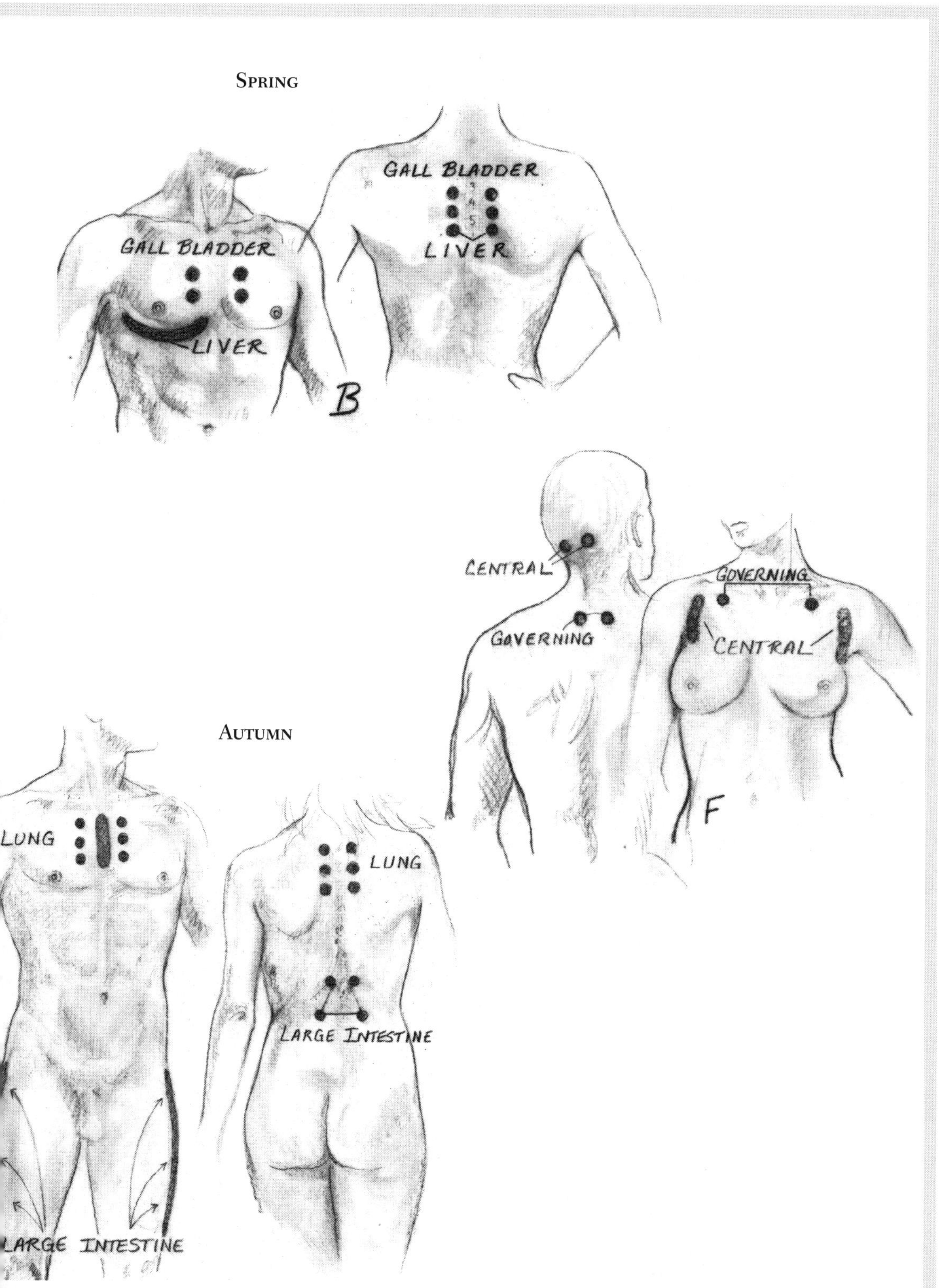

5. End by clearing the negative thought field, with both you and your partner zipping up two or three times, breathing deeply.

A DAILY ENERGY ROUTINE

Just as you may take vitamins each morning to supplement your diet, these six "energy vitamins" can help you become healthier, more vibrant, and more resilient against illness and stress. Combine these techniques into a daily five-minute energy routine (it is not necessary to energy test them each time), and you will be doing much toward keeping your energies unblocked and flowing. Remember to breathe deeply and deliberately with each activity, always in through your nose and out through your mouth.

The Daily Energy Routine

1. The Three Thumps
2. The Cross Crawl
3. The Wayne Cook posture
4. The Crown Pull
5. Neurolymphatics
6. Zip Up

Some of the techniques will have greater impact for you than others, but in combination they will strengthen and balance your energy field. You are likely to notice gradual improvements in your stamina, vitality, and overall health. As you make your way through this book, you will be adding additional techniques. Over time you may also realize, based on how you feel or how you energy test, that one or more of these techniques isn't as necessary in your personal routine, but for now, I suggest that you experiment with the entire routine.

What is the evidence that the Daily Energy Routine will help you? Beyond my own experiences with thousands of clients and class participants, a growing body of research suggests that these techniques can improve performance and reduce anxiety. Educational kineseology is a specialized approach that has introduced many applied kineseology and Touch for Health techniques into educational settings, and its practitioners have frequently tracked the outcomes of their interventions. A series of both small-scale and large-scale studies in school districts from a variety of countries, including the United States, Canada, Israel, and Australia, have demonstrated improvements, sometimes impressive improvements, on such measures as concentration, organization, productivity, reading skills, spelling skills, math skills, writing skills, self-awareness, self-expression, and confidence.[7] For instance, a group of first-year nursing students learned a six-minute daily routine using techniques similar or identical to the Cross Crawl, K-27 massage, and the Wayne Cook posture. After several weeks of practice, they showed a 69.5 percent reduction in self-reported anxiety and an 18.7 percent increase in performance on skill tests, as contrasted with control subjects.[8]

How To Get Your Kids Out of Bed in the Morning

Among the greatest contests that nature has engineered is getting your kids ready for school. Step one is to get them out of bed. The following can help. It can also help you get yourself out of bed. Try it for yourself and then show your child. While still lying in bed (time—about a minute):

1. *Tug around your ears, gently pulling along the lobes. This stimulates acupuncture points, tiny electromagnetic reservoirs on the skin, that will open you to the energies of the new day.*

2. *Stretch your legs and arms with three deep breaths.*

3. *Do the Three Thumps* *(page 63)**.*

REPROGRAMMING YOUR RESPONSE TO STRESS

Stress has an immediate impact on every system in your body. Its effects are cumulative, and learning to better manage your body's responses to stress is one of the most important steps you can take for maintaining vibrant health. A simple but invaluable energy technique for interrupting the stress response while it is happening can also be used to reprogram the way your body responds to stress. Because our primal behaviors were wired into us millions of years ago, our bodies often react to the pressures of civilized life as if they were the perils of life in the jungle.

Try a simple experiment. Have a friend energy test you. If you show weak, tap or massage K-27 and cross crawl to restore your energy. Then think of a stress. Energy test again while you are holding this stress in your mind. You will find that the effects of stress on your body's energies are instant and automatic. Based on having had thousands of people do this experiment in classes, I can predict with confidence that by simply thinking about a significant stress, the energy leaves your muscles. Place the pads of your fingers on your forehead, take a deep breath, and your energy will be restored.

The Emergency Response Loop. Many of us get caught in the following loop: the daily stresses of life trigger the primitive brain centers into an emergency response condition, up to 80 percent of the blood leaves our forebrain, stress chemicals pour into our bloodstream, primitive stress response emotions sweep over us, and we proceed through another day in the modern civilized world with the biochemistry of an early ancestor in mortal danger. We wind up trying to adapt to the complex surroundings that caused the stress with the most primitive parts of our brain. Our more recently developed cognitive abilities are annihilated. Our perceptions become distorted. Our capacity to respond creatively is in meltdown.

Posttraumatic stress disorder exemplifies this loop in extreme. A harmless sight or sound or smell or impression activates a concentrated stress response, and your body relives a situation of overwhelming threat. But in a milder form, we are all dealing with stresses and pressures that unnaturally trip our fight-or-flight response, cage us in the limited reality of our primitive brain centers, pulse stress response hormones through our bodies, and leave us feeling more fearful or anxious or angry or aggressive than the situation warrants.

When we are experiencing intense stress, we are not biologically programmed to sit around thinking about our problems. Your forebrain is not even well designed for saving you from immediate danger; your reptilian brain structure is far better organized for producing quick and effective defensive responses. But it doesn't distinguish whether the alarm was set off by a physical threat, a relationship spat, pressure on the job, or a myriad of daily irritations. When the crisis response is needlessly engaged, not only is it useless in helping you survive, it plays havoc with your health and tranquility.

What You Can Do. You can reprogram your autonomic nervous system to no longer set off a crisis response in the face of daily stresses. By training your nervous system to keep the blood in your forebrain, you will be more able to think clearly and cope effectively, even amid life's ongoing pressures.

Holding a stressful memory in your mind while touching specific spots on your head, called neurovascular holding points, conditions primitive brain centers to have a composed rather than an emergency response to the memory. By resetting your nervous system in this way, the stress response cycle is not activated when the memory occurs. If you carry this out with a number of memories, the effect begins to generalize, both to other memories and to current stressors. This simple technique can not only bring greater peace of mind by keeping you from unnecessarily triggering the stress response cycle, it will help maintain and even improve your health.

Your stress reactions are physical responses. When you fall apart or emotionally "lose it," it has more to do with physiology than psychology. This fact alone can afford you some compassion for yourself or for another. And a wonderful thing about the neurovasculars is that you don't need to try to be positive. It's better, in fact, to sink into the full unpleasantness of the negative feeling while holding the points. As the blood returns to your forebrain, you will feel yourself lifting out of stress. You are at the same time reconditioning your responses. Holding your neurovascular points also provides a cranial adjustment, so symptoms such as chronic headaches, neck pain, or jaw tension may spontaneously diminish.

The neurovascular holding points are situated at various spots on your head as well as three other places on your body. They affect blood circulation. By softly holding specific neurovascular points for between three to five minutes, you can increase the blood circulation to the part of the body the points affect. Two neurovascular points, called the frontal eminences—the bumps on your forehead directly above your eyes—impact blood circulation throughout the entire body (see center drawing of Figure 36). They are particularly valuable because by holding these points when under stress, the energy from your fingertips keeps the blood from leaving your forebrain. More important, this can draw blood back into your forebrain even after you've begun to "lose it." If you cannot locate bumps on your forehead, just find the points about an inch above your eyebrows.

We actually know these points instinctively. When shocked, the hand may naturally find its way to the forehead, often with the exclamation "Oh my God." That is why I call these the "Oh my God" points. Because they are slightly raised in most people, they are also called the frontal eminences. The next time you are hit by stress and feel overwhelmed or highly emotional (time—3 to 5 minutes):

1. Lightly place your fingertips on your forehead, covering the frontal eminences, the "Oh my God" points.
2. Put your thumbs on your temples next to your eyes, breathing deeply.
3. As the blood returns to your forebrain over the next few minutes, you will find yourself beginning to think more clearly. It is that simple!

Years ago I was doing volunteer work at an elementary school, assigned to help a teacher who had the seemingly impossible job of handling about thirty hyperactive kids, mostly boys. It was a wild scene to walk in on. Erasers were flying, kids were climbing on cabinets, the sounds were deafening. It felt hopeless. I was trying to give the teacher some tools and techniques that might help, but, to say the least, it was a challenging mission. Sometimes by mustering the full force of my personality I could engage the class in the exercises, but chaos tended to rule.

One day as I was coming to the class, the teacher noticed me through a window and alerted the kids: "Class, she's coming." When I walked around the corner and into the classroom, what I saw was amazing. Each student was holding another student's neurovascular points. They made a long chain around the room, and they were silent! You could have heard a pin drop. The teacher was able to get them to do it because they liked affecting one another this way, and they also liked how the "calm" felt.

Reprogramming the Emergency Response Loop. Bring to mind a situation where you had a hard time coping. Perhaps you were terrified, panicked, or furious. Perhaps you were overwhelmed with jealousy or grief. Perhaps you "lost it."

Place your fingerpads over the "Oh my God" points, thumbs on your temples next to your eyes, breathing deeply. Keep the scene in your mind over the next few minutes. As you relax, you will be freeing yourself from the memory's emotional grip.

Experiment with a single memory, using the technique daily until you can hold the memory in your mind without feeling a stress response in your body. Then go on to another memory. Not only can you use this technique systematically for working with the accumulated stresses from your past, you can hold the points whenever you are feeling stressed or overly emotional. It will be one of the simplest but most valuable tools you get from this book.

How to Release Built-up Stress

Think in terms of opening a valve that lets off steam. When you are feeling like a pressure cooker, you need to release congested energy from your body. Here is a series of exercises for opening the valve, releasing the stress buildup, unscrambling your energies, and taking yourself out of biochemical overwhelm. Do any or all of them (time—about 5 minutes):

1. *"Expel the venom" (page 219).*

2. *Do the Three Thumps (page 63) and the Crown Pull (page 77).*

3. *Hold the Wayne Cook Posture (page 73).*

4. *Do a stretching exercise such as Separating Heaven and Earth (page 249), the spinal suspension (page 119), a favorite yoga posture, or a power walk.*

5. *Smooth behind the ears (page 235) and then tug at your earlobes, circling from the bottom to the top.*

6. *Rub a stainless steel spoon over the bottom of your feet. The metal draws excess energies down and off the foot. "Spooning" has a subtle centering effect on your nervous system, and it is grounding.*

7. *End with a vigorous Cross Crawl (page 69).*

ON TO THE BODY'S EIGHT ENERGY SYSTEMS

I typically begin a healing session by using some variation of the Daily Energy Routine or the neurovascular holding points to ensure that I will not have to be fighting a scrambled force field. This doesn't take a great deal of time, and the techniques are a good tool kit for keeping your energies humming as well as for correcting energies that are stressed, scrambled, or running backward. In fact, before applying other self-healing techniques on yourself or having a healer work on your body, it is a good idea to go through the Daily Energy Routine and, if you are feeling stressed, to hold your neurovasculars as well. Your body will be more receptive to the treatment, and your health is more likely to build from one session to the next if you are not having to use the session to undo the stresses of the week. When people schedule a second session with me, I often give these techniques as homework so that they develop and reinforce healthy habit patterns in their energy fields.

Once a client's energies have become as calmed and as balanced as simple interventions will allow, I turn my attention to specific energy systems. Eight interrelated energies are continually influencing the state of your body, mind, and spirit. In Part II you will have an opportunity to explore each of these energy systems in some depth.

Part II

The Anatomy of Your Energy Body

Eight Energy Systems

BY STUDYING ENERGY ANATOMY . . .
YOU WILL BE ABLE TO READ YOUR OWN BODY
LIKE A SCRIPTURE [AND] TAKE THE EDGE OFF
THE SENSATION THAT YOU ARE
LOOKING BLINDLY INTO EMPTY AIR
FOR INFORMATION.

— CAROLINE MYSS
Anatomy of the Spirit

Chapter 4

The Meridians

Your Energy Transportation System

THE MERIDIANS NOT ONLY FEED VITAL ENERGIES TO THEIR RELATED ORGANS, THEY ALSO REFLECT ANY PATHOLOGICAL DISTURBANCE IN THOSE ORGANS, THUS PROVIDING PHYSICIANS WITH A CONVENIENT AND HIGHLY ACCURATE TOOL FOR DIAGNOSIS AS WELL AS THERAPY.

—DANIEL REID
Guarding the Three Treasures

We are each a constellation of energy systems, just as the body with its immune, endocrine, cardiovascular, and other systems is a constellation of physical systems. The eight energy systems introduced in this second section of the book work together naturally, usually below the threshold of your awareness. They include: the meridians, the chakras, the aura, the Celtic weave, the basic grid, the five rhythms, the triple warmer, and the "strange flows."

When a new client walks though the door, I am confronted with a medley of impressions. I might experience the person's energies as being chaotic or congested. Some of them may be flowing in the opposite of their natural direction. The blending of the energies may seem harmonious or jangled. One of the energy systems may jump out at me as needing particular attention. Meridians may be running backward. Chakras might be pulled too tightly around the organs. The auric field may be collapsed or have holes that create vulnerabilities. Energies from the left and right hemispheres of the brain may not be crossing over properly to the opposite sides of the body. Over the years

I have learned to identify from within this jumble the eight systems presented here.

Each is a distinct form of energy that corresponds with ancient descriptions. Various cultures and their healing approaches have tended to emphasize some of these energy systems over others. For instance, the *meridians* and the *five rhythms* have guided Chinese acupuncturists for thousands of years. The *chakra* system of India predates acupuncture. Depictions of the *aura,* frequently seen in traditional religious symbolism as the halo, extend back at least as far. The *strange flows* are well known in the East, and descriptions of this energy system can be found in the shamanic lore of Scandinavian countries and in Native American tribes as well. The *Celtic weave* is familiar in the East as the Tibetan energy ring. The *basic grid,* which is related to the renegade psychoanalyst Wilhelm Reich's orgone energy, was most recently identified in the 1960s by scientists breeding cows. They refer to it as the cloaccal system.

YOUR ENERGY TRANSPORTATION SYSTEM

The design of the meridian network is awesome. Rather than thinking of meridians as an obscure or foreign concept, you can know them as fourteen tangible pathways that carry energy into, through, and out of your body. A profound intelligence resides in your meridians. They serve you well without your even being aware that they exist, but ask in their language for an energy boost, and they will do as you bid.

Meridians are energy pathways that "connect the dots," hundreds of tiny reservoirs of heat, electromagnetic, and more subtle energies along the surface of the skin. Known in Chinese medicine as acupuncture points, these energy dots or "hot spots" can be stimulated with needles or physical pressure to release or redistribute energy. Like a river that rises and falls, the flow of the meridians is ever changing, and its fluctuations can be detected by sensitive individuals as well as by mechanical instruments. Acupuncture points are on the surface of the skin, but the meridians they open into travel deep into the body and through each of the organs and muscle groups. Your meridians are your body's energy bloodstream.

Each of twelve of your meridians is actually a segment of a single energy pathway that runs throughout the body, surfacing twelve times and appearing as twelve segments. Each segment is named for the primary organ or system that it services. Two additional energy pathways, called central and governing, are also thought of as meridians. Because they share properties with the strange flows as well (discussed in Chapter 8), central and governing are considered both meridians and strange flows. The other twelve meridians form a chain with one meridian linked to the next. The central and governing meridians open more directly to the environment. The energies that surround you can enter and exit through them.

The ancient Chinese maps of the meridian system correspond with what I see when I look at a body, but they were at first discounted in the West because they had no known anatomical correlate. Radioactive isotopes injected into the acupuncture points have, however, now demonstrated the existence of a system of fine ductlike tubules approximately 0.5–1.5 microns in diameter that follows the ancient descriptions of the meridian pathways.[1] In subsequent studies, pathways of light revealed by infrared photography also show that the maps described in the ancient texts were accurate.[2] The acupuncture points situated along the meridians have been likened to the amplifiers that are found along a telephone cable, boosting the signal so that it can continue to the next amplifier. The orthopedic surgeon Robert Becker found preliminary evidence that "the acupuncture points were just such booster amplifiers, spaced along the course of the meridian transmission lines."[3]

Meridians affect every organ and every physiological system, including the immune, nervous, endocrine, circulatory, respiratory, digestive, skeletal, muscular, and lymphatic systems. Each system is fed by at least one meridian. In the way an artery carries blood, a meridian carries energy. As the body's "energy bloodstream," the meridians bring vitality and balance, remove blockages, adjust metabolism, even determine the speed and form of cellular change. Their flow is as critical as the flow of blood; your life and health depend on both. If a meridian's energy is obstructed or unregulated, the system it feeds is jeopardized.

If you think of the meridians as an energy transportation system, a complex traffic network, you have a concrete model of how meridian energies interact. When a freeway becomes congested, it may be necessary to divert some of the traffic onto another highway. An off-ramp may need to be cleared or widened. If a highway or meridian becomes backed up with too much energy, as occurs in the hubbub of daily life, a bottleneck forms. And the resources needed to support the community or the body are also blocked. It becomes difficult to provide food and remove waste products. Likewise, if a highway is damaged in an earthquake, even the critical support services that form the community's "immune system"—like police, fire, and ambulance units—cannot function properly. Because your body is in use twenty-four hours every day, day in and day out, and because it is under continual stress and has periodic "earthquakes," its pathways need regular maintenance and repair as well as an occasional major renovation.

BALANCING YOUR MERIDIANS

Disturbances in meridian energies correlate with ill health. The amount of light emitted by the meridians of laboratory animals decreased when the animals were ill and increased as treatments such as acupuncture began to improve their physical condition.4 Because disrupted meridian energies often precede illness, meridian readings are sometimes used to predict health vulnerabilities and prevent disease. Some government employees in Japan, for instance, are routinely screened during their annual medical examination by a machine that has twenty-eight electrodes attached to the meridian endpoints. Only people with abnormal meridian readouts are required to go through further diagnostic testing.5

To "balance" your meridians is to get them running smoothly and efficiently. This can be accomplished in a number of ways, which you will be learning in this chapter. You can (1) trace your own meridians, (2) "flush" congested meridians, (3) twist and stretch specific points associated with each meridian, called alarm points, (4) massage the meridian's neurolymphatic points, or (5) hold or tap specific acupuncture points that are located toward the ends of the meridian.

TRACING YOUR MERIDIANS

While many of the essential structures of the human body, including the meridians, evolved millions of years ago, these structures evolved to support a body that was adapting to a markedly different world than we encounter today. So it is not surprising that your body's energy transportation system may sometimes get its lines crossed and may deliver too much energy to one organ, or not enough to another. The stressors we face daily tend to send our most vulnerable meridians into a frenzy of overwork or, alternatively, virtual shutdown. When this frenzy or shutdown is recurrent, other meridians, attempting to compensate for the imbalance, become entrenched in crisis mode. In the ensuing chain reaction, your whole energy transportation system can become ineffective, draining your vitality and leaving you frazzled and susceptible to all sorts of maladies.

The strategy of waiting until our bodies evolve to fit our lifestyle, a process that typically requires eons, may be impractical. Human adaptation, however, does not rely on genetic mutation alone. We learn.

Tracing your meridians is one of the simplest and most effective ways you can correct for the mismatch between your genetic programming and the demands of your environment. Regardless of the pressures, stresses, and new circumstances that tend to overwhelm your meridian system, if you can keep the energy highways open, minimize the traffic jams, maintain the import and export systems, remove stagnant energy, and bring in a fresh energy supply, you will be healthier.

Because your meridians, like your hands, carry electromagnetic and more subtle energies, you can influence their flow by tracing them, keeping your hands in direct contact with your body or a few inches away from it. The energy will follow your hand, and the flow of energy along the meridian's circuitry will be strengthened. By tracing your meridians every day, you can direct the traffic in your energy transportation system. You can communicate to it, in a language it understands, that keeping the energies flowing along their natural routes is more effective than a haphazard crisis response. And you can trace your entire meridian system in two minutes, a procedure you can do as a daily exercise or whenever the spirit moves you. In addition, if you know one of your meridians needs special attention, you can trace that meridian several times each day. You can also add meridian tracing to the Daily Energy Routine you learned in Chapter 3. You have already had some experience tracing a meridian. The Zip Up (page 82) traces the central meridian.

To trace any of the meridians, use an open hand, palm facing the body, directly touching it or staying within two inches of it. As you pass your hand over a meridian, you are aligning your hand's energies with the meridian's energies, like the moon pulling the tide. The following instructions show you how to trace all fourteen of your meridians (also see figures 10–23). You'll notice that a two-hour time span is listed beside each of the meridians, except for central and governing. This is the time period when the meridian's energies are at their peak. The practical significance of this will be explained later in this chapter.

Central Meridian: Place both hands on your pubic bone and bring them straight up the front of your body to your bottom lip (see Figure 10).

Governing Meridian: Place one hand at your tailbone and trace straight up your spine as high as you can. Reach over your shoulder and try to touch the hand reaching up. If your hands cannot meet, connect them with your mind. Then with the hand that reached over your shoulder, trace the energy the rest of the way up your spine, over your head, over your nose, and to your top lip (see Figure 11).

Spleen Meridian (9 a.m. to 11 a.m.): Start at the outside corners of each big toe and go straight up the inside of your legs, flaring out at your hips, up the side of your rib cage, and down to the bottom of the rib cage (see Figure 12).

Heart Meridian (11 a.m. to 1 p.m.): Place your open hand underneath the opposite armpit in alignment with your little finger and trace straight down inside the arm and off the little finger (see Figure 13). Do both sides.

Small Intestine Meridian (1 p.m. to 3 p.m.): Turn your hand over and, starting at the little finger, go straight up the outside of the arm to your shoulder, drop back on your scapula, go over to your cheekbone, and back to the opening of your ear (see Figure 14). Do both sides.

Bladder Meridian (3 p.m. to 5 p.m.): Place both hands between your eyebrows, go up over the crown and down the back of your head and neck. Remove your hands from your neck, reach them back underneath and as high as you can stretch onto your spine. Trace your hands down either side of your spine to below your waist, jog in and up toward the waist, and then in and around your gluteus maximus. Leave the meridian there and come up

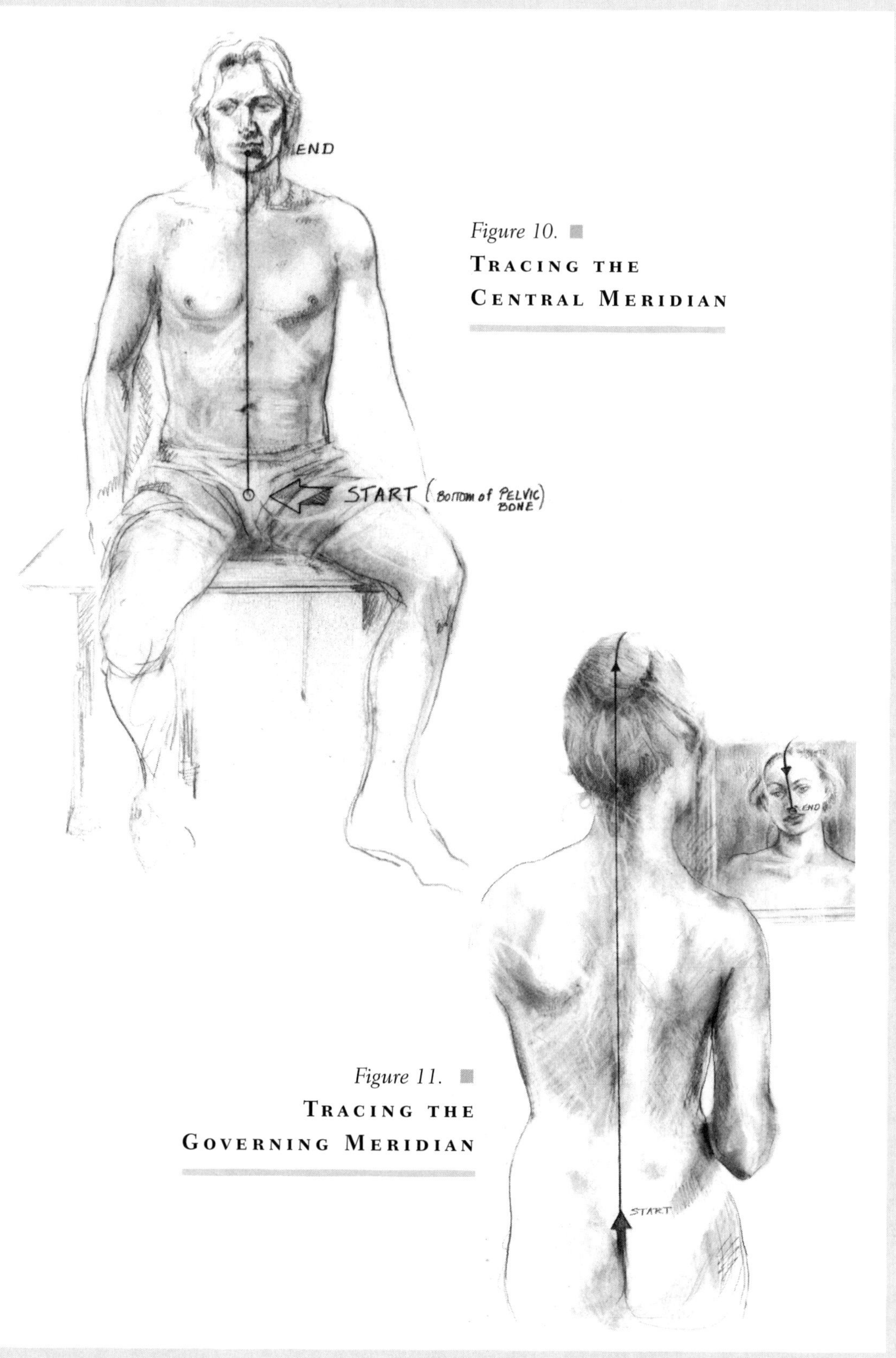

Figure 10.
TRACING THE CENTRAL MERIDIAN

Figure 11.
TRACING THE GOVERNING MERIDIAN

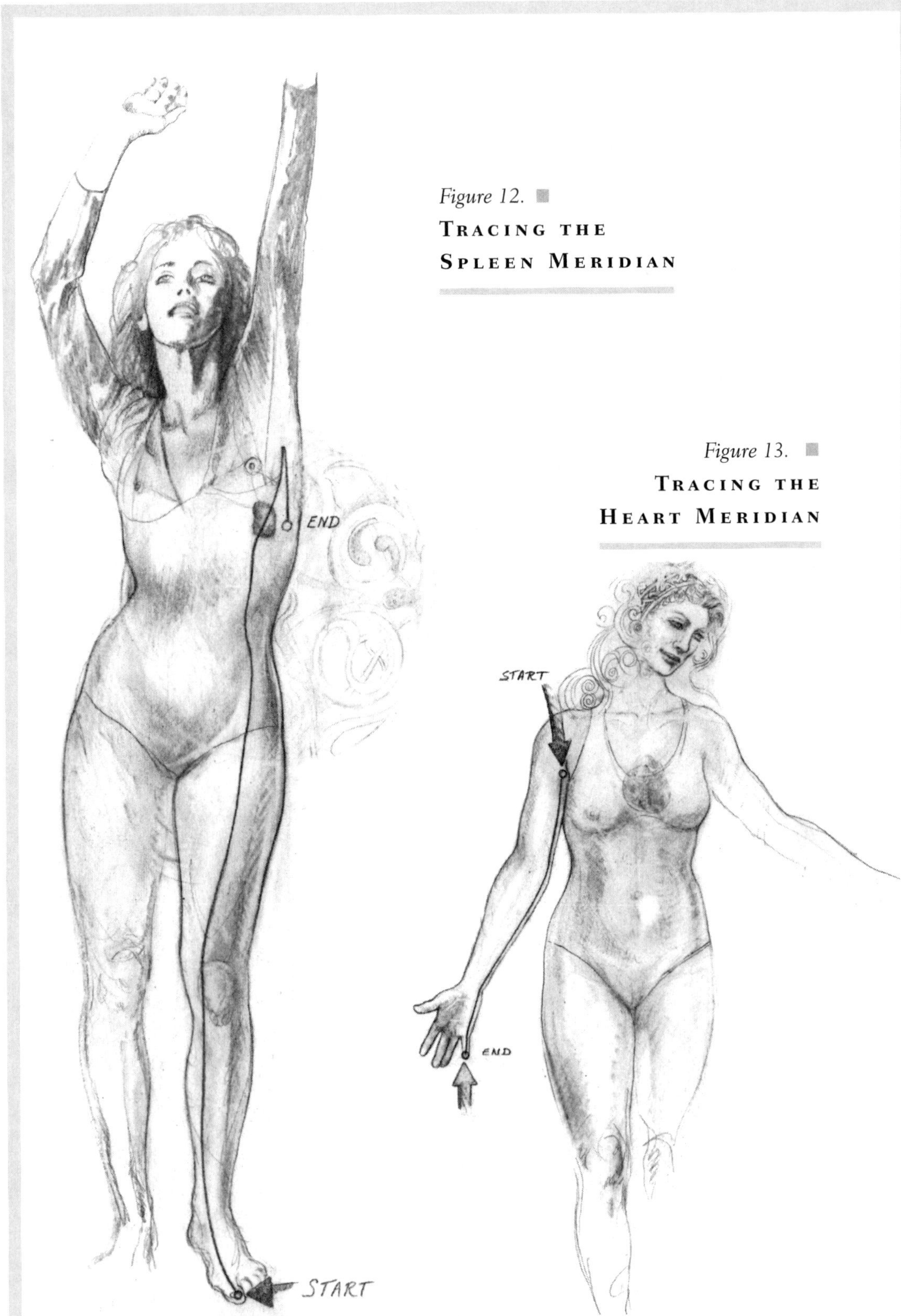

Figure 12.
TRACING THE SPLEEN MERIDIAN

Figure 13.
TRACING THE HEART MERIDIAN

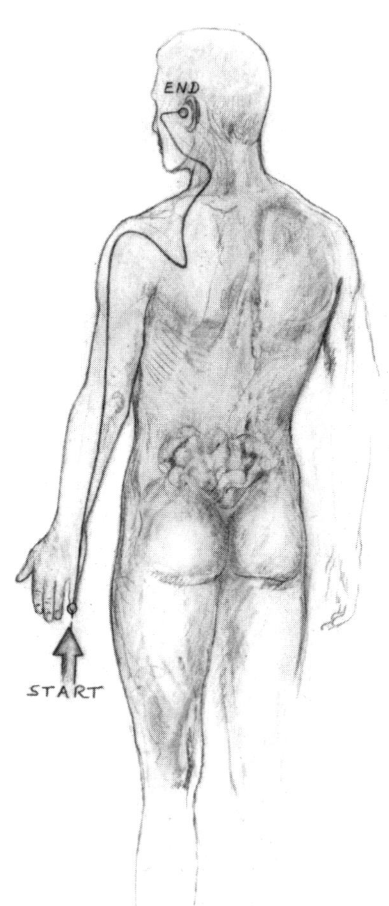

Figure 14. ■
Tracing the Small Intestine Meridian

onto your shoulders, go straight down to the back of your knees, in at the knees, down to the floor, and off your little toes (see Figure 15).

Kidney Meridian (5 p.m. to 7 p.m.): Place your fingers under the ball of each foot, middle finger in line with the space between your first and second toes. Draw your fingers up to the inside of each foot, circle behind the inside of each ankle bone, and go straight up the front of the body to K-27, the points beneath the clavicle at the top of the sternum (see Figure 16). Vigorously massage these points.

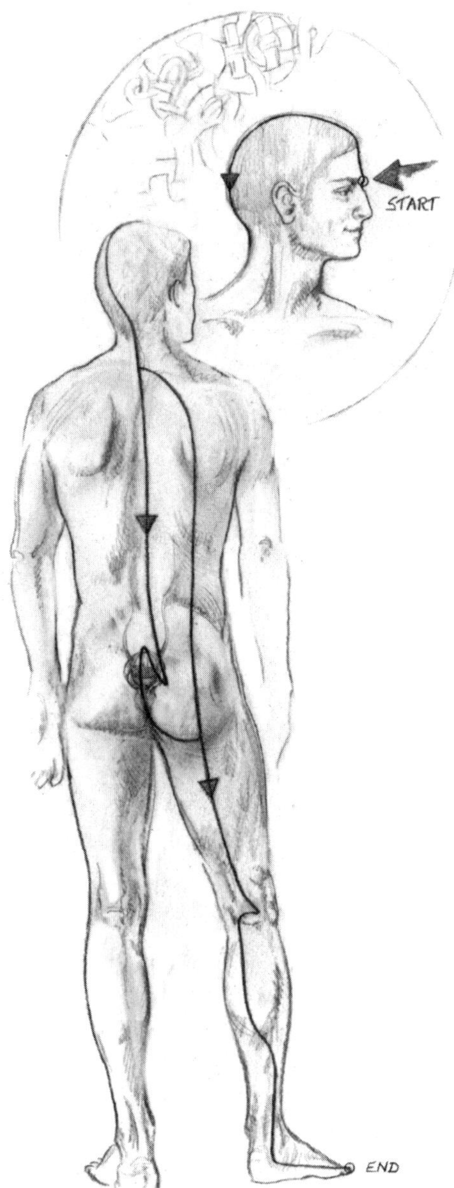

Figure 15.
Tracing the Bladder Meridian

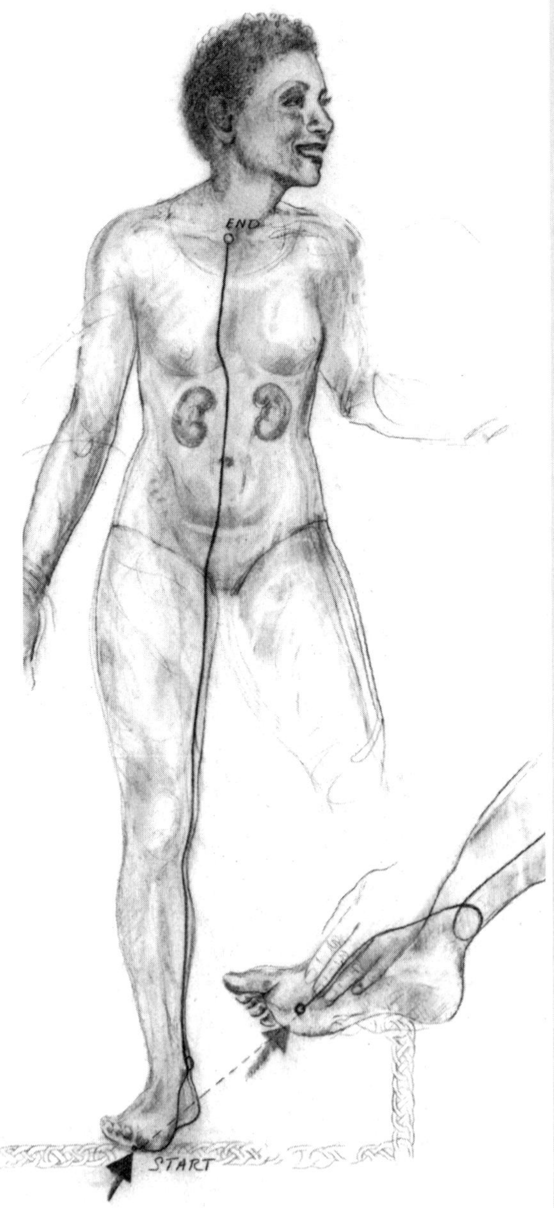

Figure 16.
Tracing the Kidney Meridian

Figure 17.
Tracing the Circulation-Sex Meridian

Figure 18.
Tracing the Triple Warmer Meridian

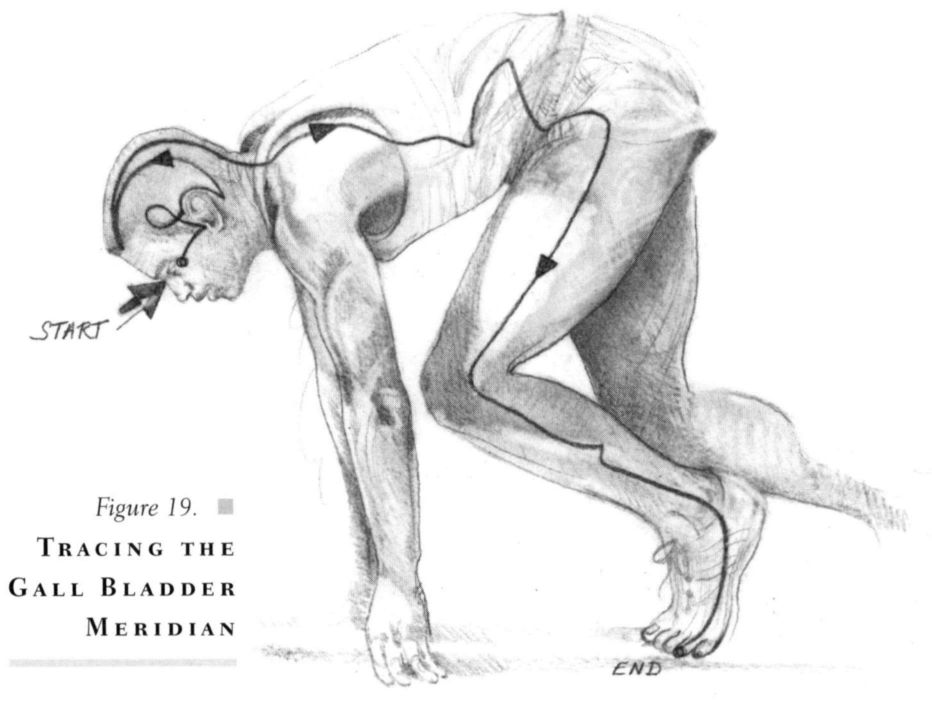

Figure 19. TRACING THE GALL BLADDER MERIDIAN

Circulation-Sex Meridian (7 p.m. to 9 p.m.): Place the fingers of one hand at the outside of the opposite nipple, come up over the shoulder, go down inside the arm and off the middle finger (see Figure 17). Do both sides.

Triple Warmer Meridian (9 p.m. to 11 p.m.): Turn your hand over and, starting at the ring finger, trace straight up the arm to beneath your ear, follow your ear around and behind, ending at your temple (see Figure 18). Do both sides.

Gall Bladder Meridian (11 p.m. to 1 a.m.): Place the fingers of both hands on the outside of your eyebrows, drop to the opening of your ears, take your fingers straight up about two inches, circle forward with your fingers, and drop back behind the ears. Go forward again over to your forehead, back over the crown of your head, and around your shoulders. Leave your shoulders, take your hands to the sides of your rib cage, go forward on the rib cage,

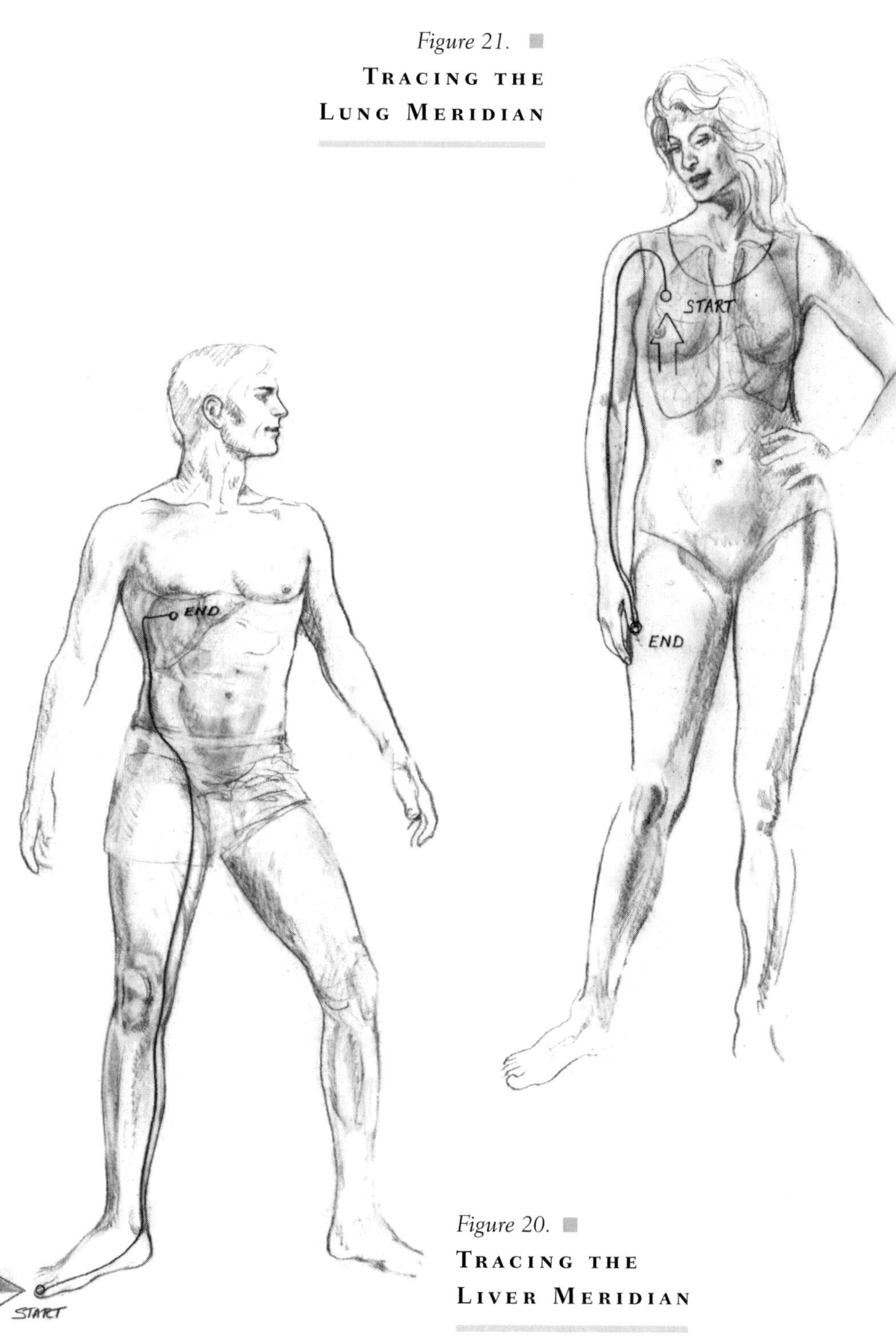

Figure 21.
Tracing the Lung Meridian

Figure 20.
Tracing the Liver Meridian

back on the waist, forward on the hips, straight down the sides of the legs, and off the fourth toes (see Figure 19).

Liver Meridian (1 a.m. to 3 a.m.): Place your fingers on the insides of your big toes and trace straight up inside the legs, flaring out at your hips, up the sides of your rib cage, and back to underneath your ribs, in line with your nipples (see Figure 20).

Lung Meridian (3 a.m. to 5 a.m.): Place one hand over the opposite lung and move it up over your shoulder, straight down your arm, and off your thumb (see Figure 21). Do both sides.

Large Intestine Meridian (5 a.m. to 7 a.m.): Place the open fingers of one hand at the end of the pointer finger of the opposite hand, trace straight up the arm to your shoulder, cross the neck to beneath your nose, and go out to the flair of your nose (see Figure 22). Do both sides.

Stomach Meridian (7 a.m. to 9 a.m.): Place both hands underneath your eyes, drop to your jawbone, circle up the outside of your face to your forehead, come down through your eyes to your collarbone, go out at your collarbone, over your breasts, in at your waist, out at your hips, straight down your legs, and off the second toes (see Figure 23).

Rather than attempting to memorize how to trace each meridian, I suggest that you read the above instructions into a tape and let the tape guide you. Read slowly, and where the text says, "Do both sides," read those instructions a second time. Just as the routine begins by tracing the central, governing, and spleen meridians, it finishes by tracing them as well, to overlap and strengthen the circuitry. So add the instructions for tracing them a second time at the end of the tape, in the order of spleen, central, and governing.

You can actually begin with any meridian and proceed in the order as listed. You get an extra boost when you begin where you are most vulnerable. So, after central and governing, begin with the spleen meridian or any other your intuition suggests. The Chinese typically started with the lung meridian because it brings in the breath. I have found there are many benefits to starting with the spleen meridian. First of all, the spleen meridian provides critical energy for the immune system, and each of the meridians is affected by

Figure 23.
TRACING THE STOMACH MERIDIAN

Figure 22.
TRACING THE LARGE INTESTINE MERIDIAN

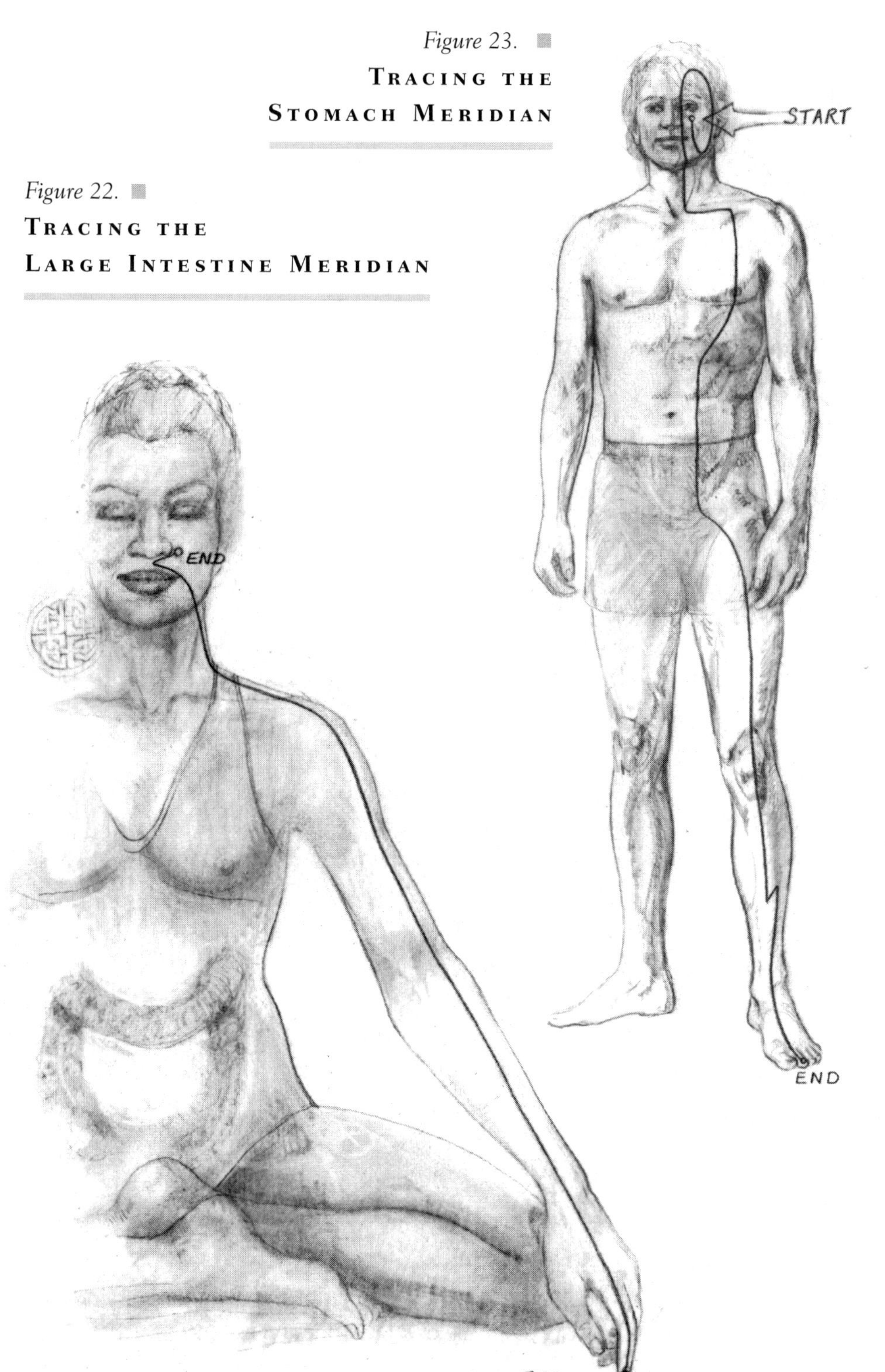

it. It corrects chemical imbalances and problems with blood supply. If you are ill or don't have much energy, the spleen meridian will be weak. Tracing the spleen meridan first sets a strong foundation for tracing the remaining meridians.

Tracing your meridians daily will do a great deal toward keeping you healthy and feeling good. However, a meridian can be so entrenched in heroic efforts to bring balance to the organs it governs, or may be so overwhelmed and shut down, that simply tracing it cannot bring it back to its natural flow. The remainder of this chapter will teach you how to identify and correct such imbalances.

MERIDIANS FUNCTION LIKE THE ORGANS THEY SERVE

Each meridian flows through and serves at least one organ or physiological system. For example, like the kidneys themselves, the function of the kidney meridian is purification. The kidney meridian filters toxic energy, allowing energies that have been obstructed to begin to move. I'll choose to work with the kidney meridian when I want to "flush" a particular system—for example, if the lymph system feels sluggish, the back feels tight, energy is stuck below the waist, or the body is carrying an illness. The kidney meridian filters the body's energies, leaving them fresh, clear, and vital, and the kidney's energies renewed as well. The first points on the kidney meridian, at the bottom of the feet, are called the Wellspring of Life points (Figure 24). The kidney meridian is said to contain the life force of beginnings and renewal.

An emergency room physician took one of my classes because he was trying to be more open to alternative healing. He was disappointed with the class, however, because he was not able to sense energies that others seemed to be sensing. Despite my assurances that this was not necessary and would come in its own time, he was growing so frustrated with all the talk about energies he couldn't feel or see that the class was having the effect of closing rather than opening him to these new ideas. One evening while he was at work, a man was rushed to the hospital in shock. The patient was unresponsive to standard medical procedures, and he was dying. Suddenly a demonstration I had done for the class on the Wellspring of Life points flashed through the physician's mind. He pushed these points with his middle fin-

Figure 24.

WELLSPRING OF LIFE POINTS

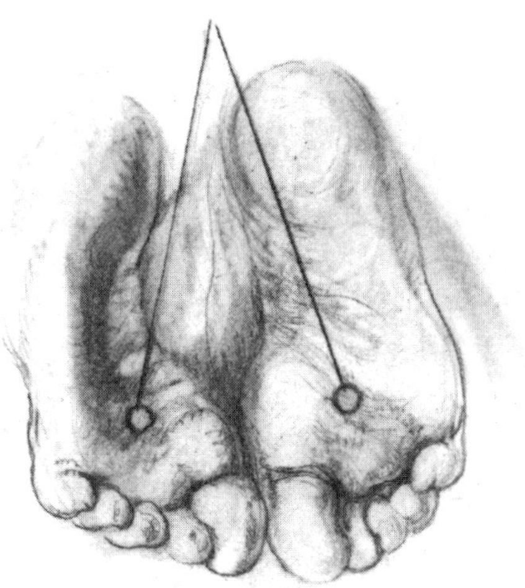

gers, as hard as he could. He said, "It felt as if I had put a plug into the container from which his life force was slipping away." Suddenly all the instruments showed that the patient was coming back. The patient did live, and at the next class the doctor triumphantly described his first experience of *knowing* he was feeling energy.

ENERGY TESTING YOUR MERIDIANS

One of the most empowering tools you can develop for taking charge of your own health is to learn how to test the flow of energy through your meridians. Testing the meridians enables you to make healthy choices based on your body's actual needs. Meridians are not only associated with a primary organ and energy system but also with muscles, which is the basis of energy testing.

An interesting piece of trivia is that a meridian does not necessarily flow through the muscle it affects. So if you study the traditional muscle tests used in applied kineseology or Touch for Health, you will find that the

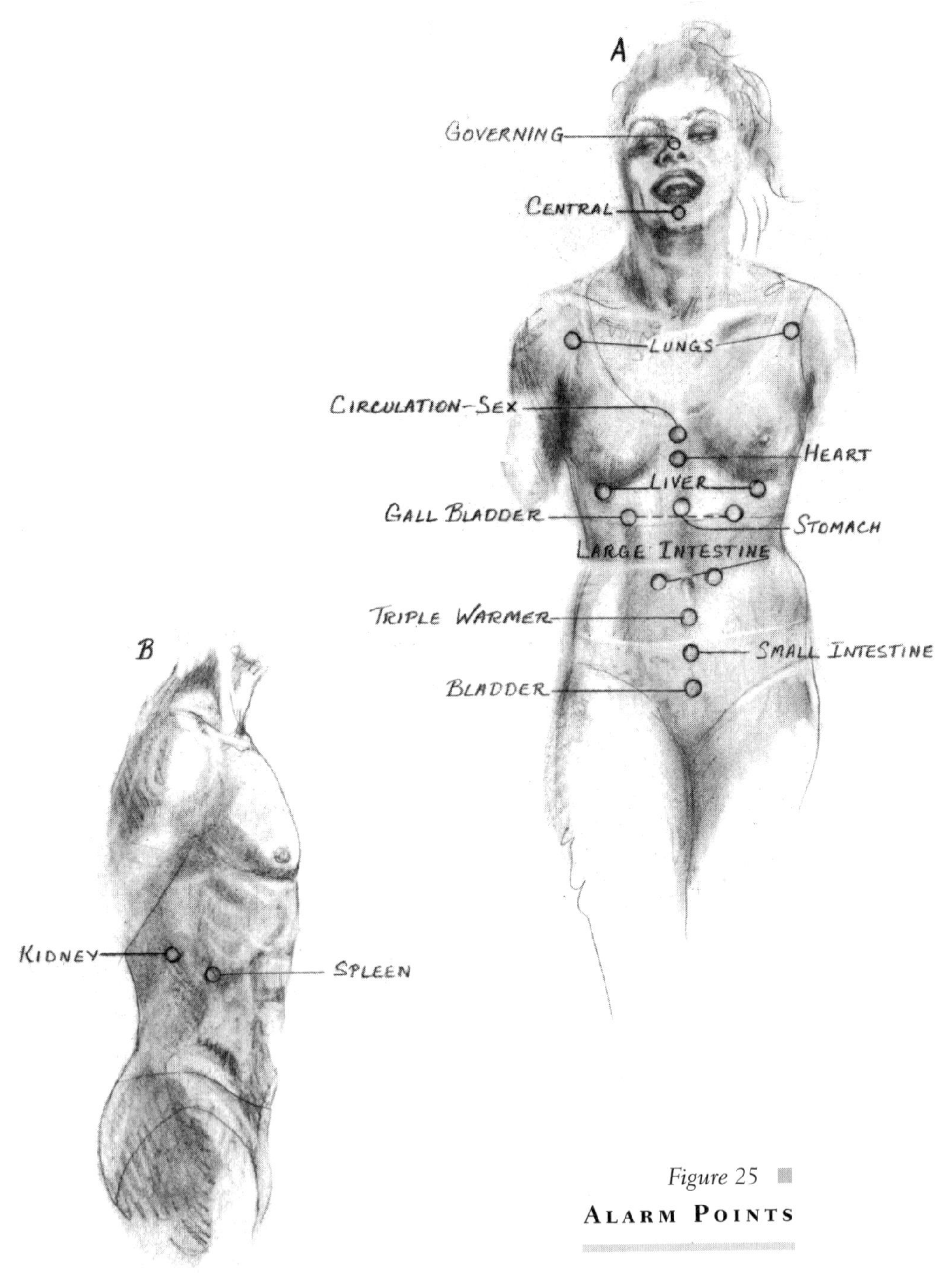

Figure 25
ALARM POINTS

meridian being tested does not always flow through the muscle used in the test. In the developing human embryo, each meridian *does* flow through its associated muscle. While the meridian and the muscle may separate (as the limbs grow out, for instance), the initial relationship between the two remains imprinted. The energy bridge jumps the physical gap, and energy testing takes advantage of this enduring relationship.

While energy tests have been worked out for each meridian and are necessary to master for more advanced work,[6] an easier way to test the meridians is by using a single test—the general indicator test—and specific "alarm points" (see Figure 25). Alarm points are acupuncture points that act as circuit breakers on the meridian system. An alarm goes off when attention is needed. There is an alarm point, or a pair of alarm points, on each meridian. When too much energy is flowing through the meridian, it "pops" the alarm point, interrupting the energy. When you find an alarm point that has been "popped," there is a disturbance of the energy somewhere on the meridian or on an organ the meridian serves. If such disturbances become chronic, they are precursors to illness.

Testing its alarm point can tell you if a meridian's energies are overloaded, deficient, scrambled, or sluggish. When touching an alarm point weakens the indicator muscle, the alarm point is saying, "Pay attention," and the circuit breaker needs to be reset once the crisis has passed. Long ago, when people lived closer to nature, I believe we intuited the imbalances caused by the body's alarm response, and we instinctively knew how to correct them through touch and massage. Today, we are exposed to many more stressors, which are continually activating our alarm system, and we have lost touch with our natural instinct to rebalance ourselves.

You can identify which alarm points are activated by simply noticing if there is tenderness when you massage the alarm point:

1. Locate each alarm point on Figure 25 (see also the verbal descriptions that follow).
2. Place your thumb and first two fingers on the point and, with pressure, twist it about a quarter of a turn.
3. Write down or remember which alarm points were tender.

If you have access to a partner, I suggest that you energy test each alarm point. The procedure, basically, is to touch an alarm point with your thumb, second, and middle fingers bunched together and have your partner do the general indicator test you learned in Chapter 3 on your other arm. If touching the alarm point reveals a weakness in the flow of energy, the weakness is on the meridian associated with that alarm point. Begin by energy testing the governing meridian:

1. Locate the alarm point for the governing meridian in Figure 25.
2. With either hand, hold the tips of your thumb, index finger, and middle finger together and place them on the alarm point.
3. Use the general indicator test: Place either arm straight out in front of you, parallel with the ground, and then move it outward 45 degrees (left arm to your left, right arm to your right). Your elbow should be straight and your hand open, palm facing the floor (see Figure 4). Have your partner place the fingers of an open hand behind your wrist. As you resist, your partner pushes down firmly and smoothly, for a maximum of two seconds, to determine if the muscle holds firm.
4. If you are working with a meridian that has paired alarm points, as indicated in Figure 25, test each alarm point separately.
5. There are three possible results:

    a. If the muscle stayed locked when testing the alarm point, the energy on the meridian is flowing smoothly.
    b. If the muscle became weak, there is a disturbance somewhere on the meridian: overenergy, underenergy, or a combination, as when some parts of a highway are congested and others are empty.
    c. If the meridian had paired alarm points and the test for one showed weakness and the other held strong, the disturbance in the meridian is on the side of the body that tested weak.

You can quickly energy test all fourteen meridians using the alarm points. Test from governing, then central, and continue through the remaining twelve meridians, locating the alarm points as shown in Figure 25 and described here.

The first eight alarm points are on a line that goes straight down the front of your body.

Governing: Tip of the nose.
Central: Center of the chin.
Circulation-Sex: Centered between the nipples.
Heart: Bottom tip of the sternum.
Stomach: Halfway between the bottom of the sternum and the belly button.
Triple Warmer: An inch below the belly button.
Small Intestine: Drop down another inch.
Bladder: Again, drop down an inch.

The remaining alarm points are paired, and each side must be tested. Return to the top of the body.

Lungs: Swing your thumbs up to where they naturally land on the upper outer edge of the chest, near where the arms are attached to the torso.
Liver: Make a plumbline from the nipple straight down to one rib beneath the breast.
Gall Bladder: Drop about another inch to the edge of the rib cage.
Spleen: With your hands on your rib cage at the sides of your body, take your hands straight down until you reach the last rib. This is the body's only "floating rib."
Kidney: Follow the rib cage about another inch toward your back, to the top of the bottom rib.
Large intestine: Move your fingers to the belly button and come out an inch to either side of it.

How to Overcome Nervousness About an Upcoming Presentation, Job Interview, or Confrontation

The Three Thumps (page 63), the Wayne Cook posture (page 73), and a vigorous Cross Crawl (page 69) are a potent one-two-three routine for unscrambling your energy field, crossing the energies in the right and left brain hemispheres, and moving stress hormones out of your system. Use it while preparing for the event, and just prior to it, zip yourself up with an affirmation (page 82).

CORRECTING IMBALANCES IN YOUR MERIDIANS

If the arm becomes weak during the alarm point energy test, it means there is a disturbance in the energies that are flowing through that meridian. This might have been caused by a reaction to something in your environment, a temporary physiological condition such as indigestion or an infection, an emotional upheaval, or, if the imbalance is chronic, it could indicate an illness. But because many of the body's needs are served by overlapping systems, a single meridian rarely causes an illness on its own; illness is usually a more extensive breakdown. The meaning of the energy test is also related to how long the problem has existed, which other meridians are out of balance, and your overall health. Nonetheless, it is comforting to know that you can correct existing imbalances before you even look into their meaning. The following procedures will build a stronger energy field and strengthen your overall health. Establish a strong energy field, and you will be opposing any disease process.

So far in this chapter, you have traced your meridians to clear and strengthen them, and you have energy tested all fourteen meridians to determine if any of them are still out of balance. If a meridian does not respond to being traced, it usually indicates a chronic pattern rather than a fleeting disturbance. There are numerous additional ways to correct such chronic problems, including the following four techniques:

- Stretch and twist the alarm points
- Flush the meridian
- Massage the meridian's neurolymphatic points
- Hold the meridian's acupuncture strengthening points

Stretch and Twist. A simple but powerful way of correcting a meridian that is out of balance is to stretch and twist the skin directly over the alarm point you held when testing the meridian (time—10 seconds per point):

1. Place the fingers of either hand on the alarm point as you inhale.
2. Exhaling, press the alarm point with your thumb and three fingers and twist the skin about a quarter turn in one direction, and then the other.

Flushing a Meridian. Select one of the meridians that tested weak or whose primary organ or system is problematic for you. If you don't know of a problem, use the lung meridian. The following procedures, like taking a bath, will never hurt your body, and with the numerous pollutants your lungs must filter daily, flushing the lung meridian even when it is balanced will at least be good preventive maintenance. To flush a meridian, do the following (about 30 seconds):

1. Begin by finding the diagram that shows how to trace the meridian you have selected (from Figures 10–23).
2. Use the full sweep of either or both hands to trace the meridian in the *opposite* direction of the natural flow that is shown on the diagram.
3. As you pass your hands over the meridian, inhale. You may want to imagine that your hand is a magnet, pulling stagnant energies out of the meridian.
4. Exhale and shake these energies off your hands.
5. After flushing the meridian, trace it in its normal direction three times, slowly and deliberately.

Flushing a meridian backwashes its energy, so the meridian pathway will be clear when you then move your hand in harmony with its flow. Flushing may produce either a sedating or an energizing effect, depending on whether the meridian was overcharged or undercharged. Flushing the energy out of the meridian is like backwashing a filter. It is often useful to flush a meridian before you strengthen it because this clears it of energetic debris, leaving more space for fresh energy to enter.

Neurolymphatic Massage. When you trace a meridian, or flush it, the energies of your hands move the meridian's energy. When you massage the neurolymphatic points, you move toxins out of the meridians. The spinal flush (pages 79–80) cleanses the lymph system as well as each of your meridians. Another habit worth developing is to become familiar with the neurolymphatic reflex points you can reach with your hands (Figure 9) and to make it a daily practice to check if any of them are tender. Each meridian is associated with specific points, but rather than having to energy test or even learn which points go with which meridian, you can simply massage any points that hurt. You will get immediate feedback from your body as blocked energies begin to open. If the same point is consistently tender, identifying the meridian it is on provides valuable information as you track your own health patterns. The neurolymphatic system's job is to keep the lymph, blood, and meridian energies flowing, and by massaging the points you move toxins out of your muscles while shooting fresh energy through your whole body.

Holding Acupuncture Strengthening Points. Acupuncture strengthening points are sites on the meridian that, when held for two or three minutes, can break up energy blockages so harmony can be restored. Each meridian contains between nine and sixty-seven acupuncture points. Certain points—called strengthening points and sedating points—are particularly potent. *Sedating* points are used for pain control and for sedating overcharged meridians later in this book. For now, you will find that by holding the *strengthening* points for a particular meridian, your hands become a jumper cable that creates a channel, drawing energy from another meridian to it. These points can bring a burst of energy into the area, and holding them is one of the primary procedures used in the healing art known as acupressure. Figure 26 shows the acupuncture strengthening and sedating points for twelve of the meridians, central and governing excluded. To strengthen a meridian (time—about 5 minutes per meridian):

1. Find the figure(s) that correspond with any meridians that remained unbalanced after being traced, flushed, and massaged. The abbreviations followed by a number represent the name of the acupuncture point. SI 5, for instance, means the fifth acupuncture point on the small intestine meridian.
2. Lightly but firmly hold the strengthening points labeled "first" for 2 to 3 minutes.
3. Hold the strengthening points labeled "second" for about 2 minutes. Continue with any additional meridians that were disturbed.

If the energy test showed an imbalance in either the central or governing meridian, the following procedure, the Hook Up, will bring both into balance. The Hook Up is great to do any time your energies are feeling a bit "off" (time—about 2 minutes):

1. Place one thumb or middle finger on your forehead between your eyebrows and the other in your belly button.
2. With a slight pull of the skin upward on both points, close your eyes, take a deep breath, and relax.

People often ask how they can know when they've held someone else's hook-up points long enough. I ask them to watch until they see the person spontaneously sigh or take a deep breath. This indicates that the central and governing meridians have hooked up, and an energy test will confirm it. The Hook Up can also strengthen the auric field, connect energies that flow from the front to the back of the body, and bridge energies between the head and the body, and we will return to it later.

How to Loosen Up When You Are Feeling Stiff

The Spinal Suspension affects all the meridians and chakras by elongating and stretching the spine and opening the shoulder blades. It can rejuvenate you, and it will ward off fatigue (time—1 to 2 minutes):

Figure 26.

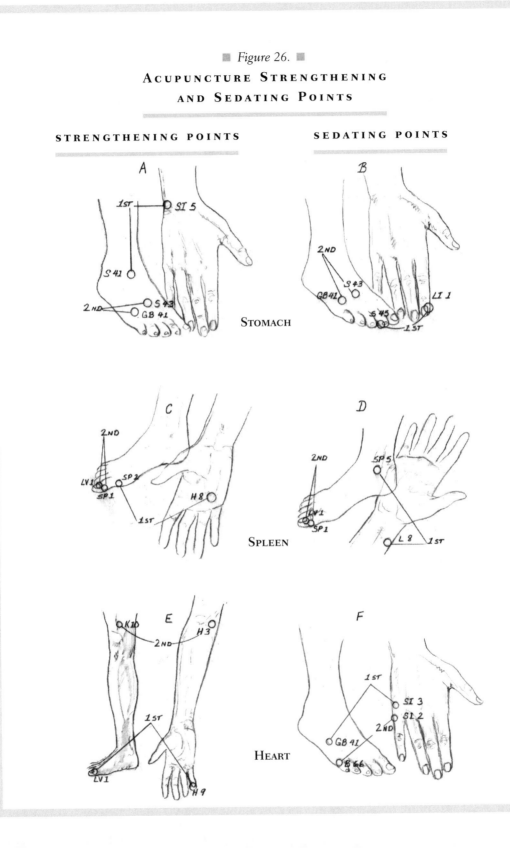

Acupuncture Strengthening and Sedating Points

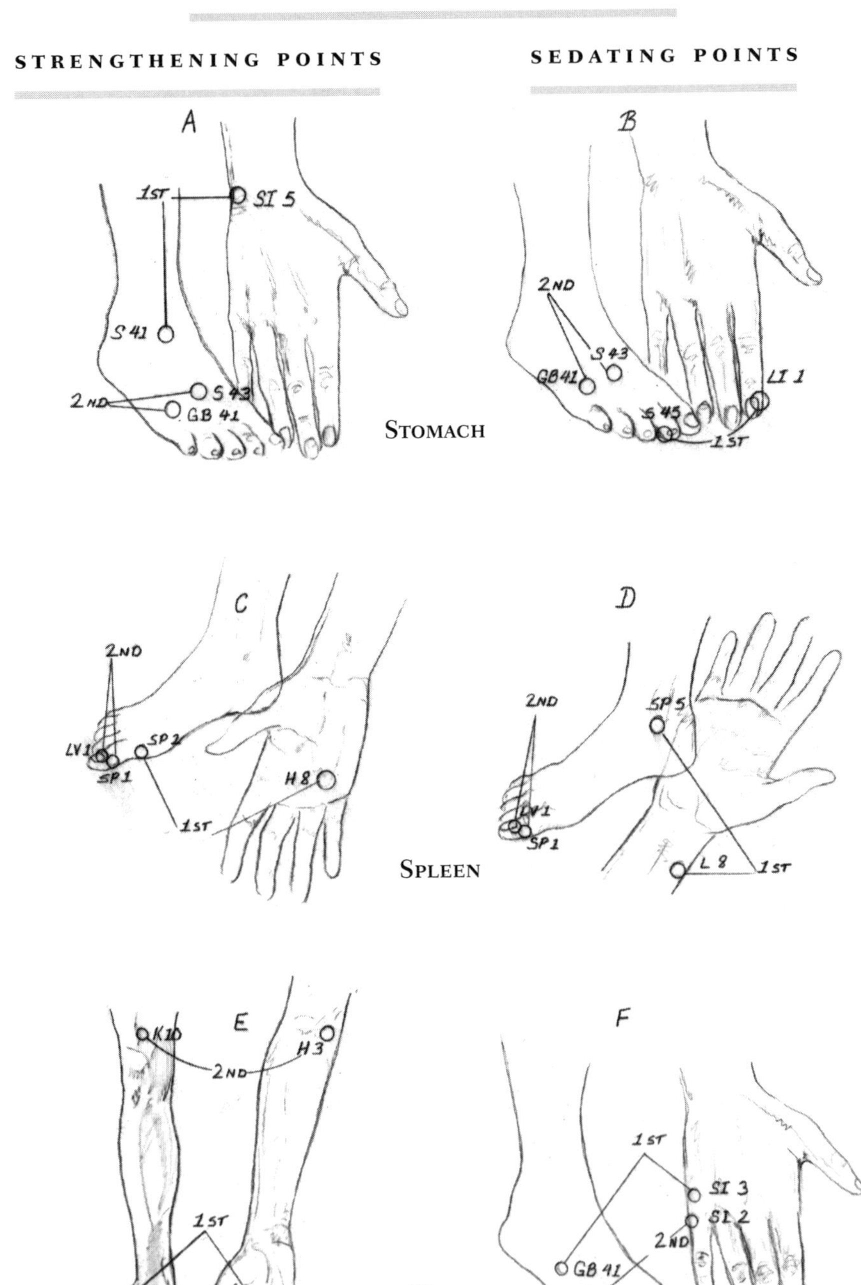

STRENGTHENING POINTS | **SEDATING POINTS**

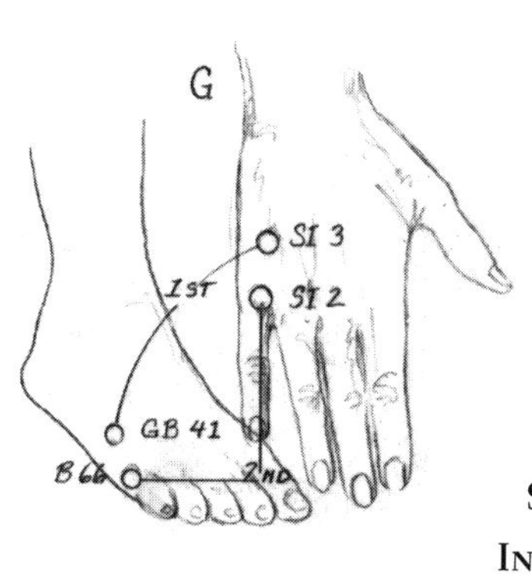

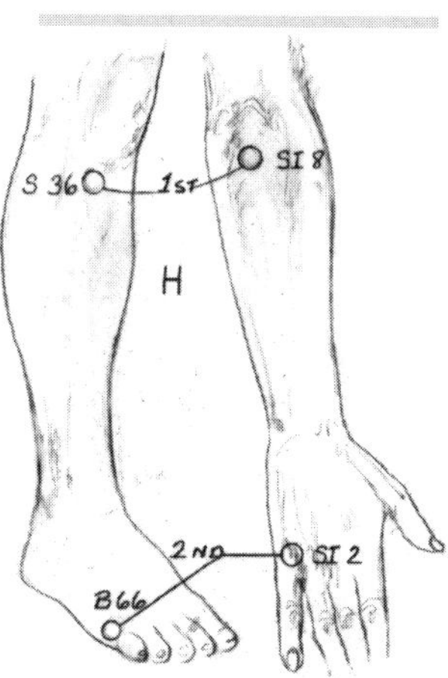

SMALL INTESTINE

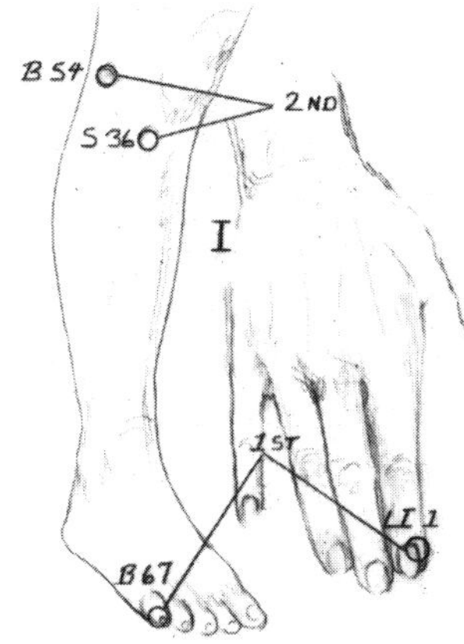

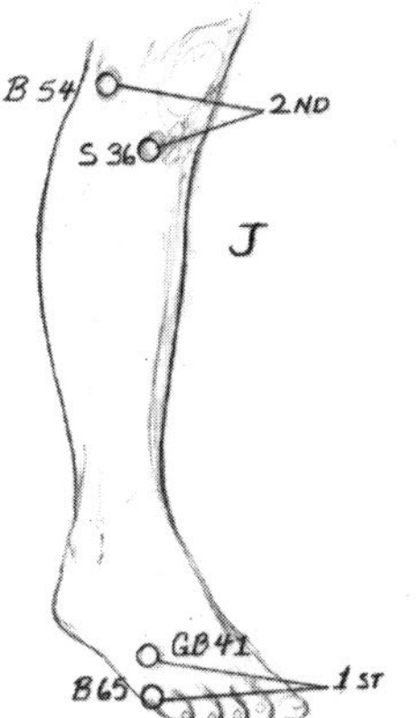

BLADDER

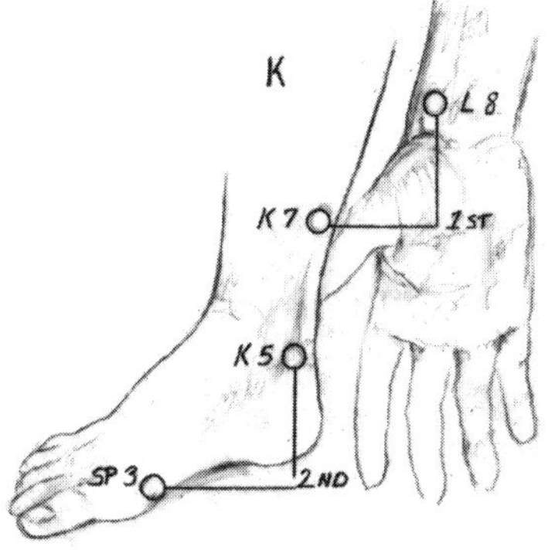

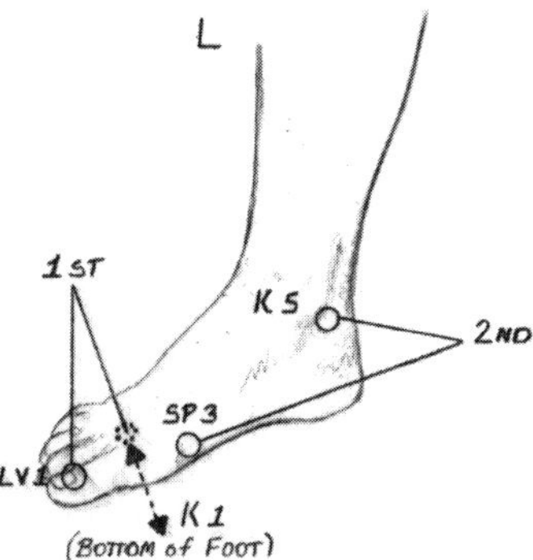

KIDNEY

Figure 26.

Acupuncture Strengthening and Sedating Points continued

STRENGTHENING POINTS | **SEDATING POINTS**

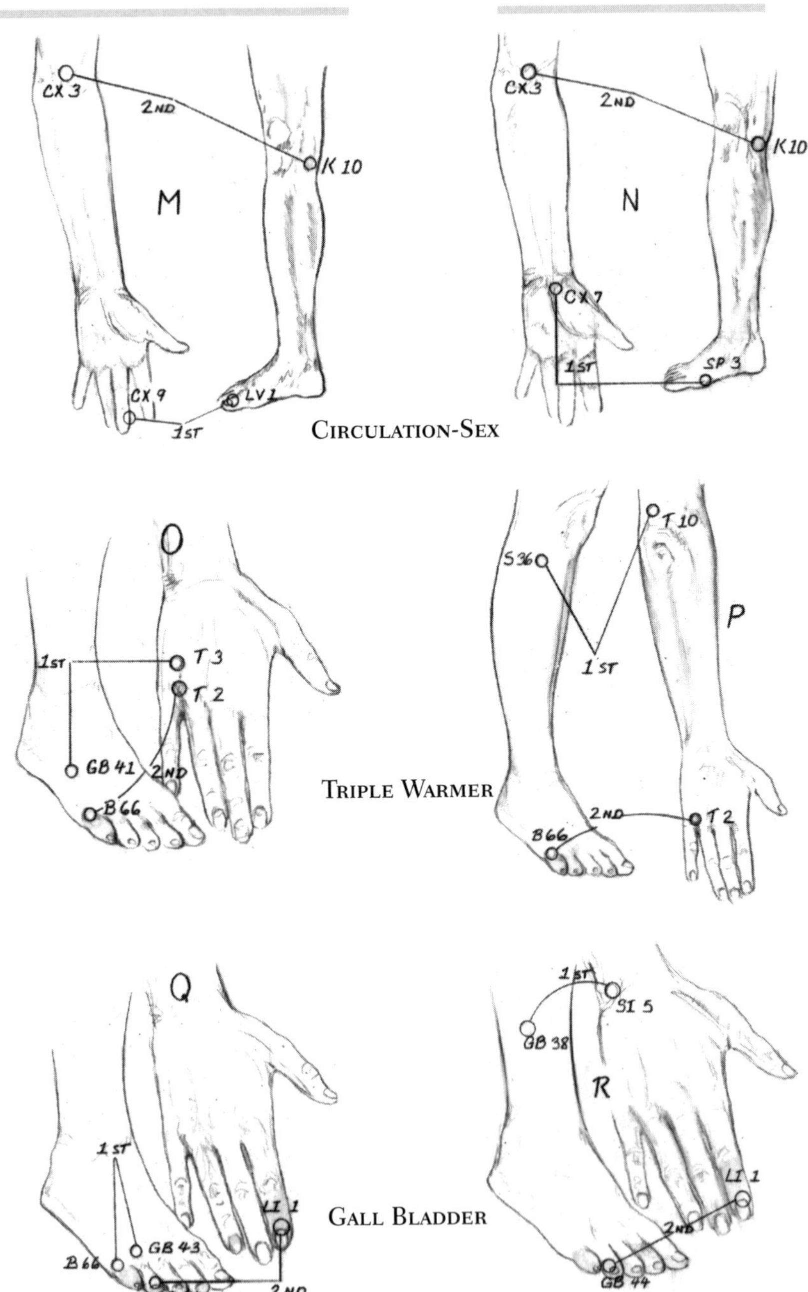

STRENGTHENING POINTS | **SEDATING POINTS**

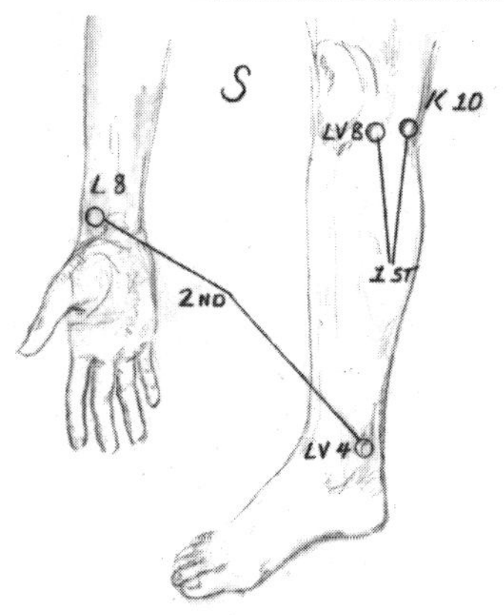

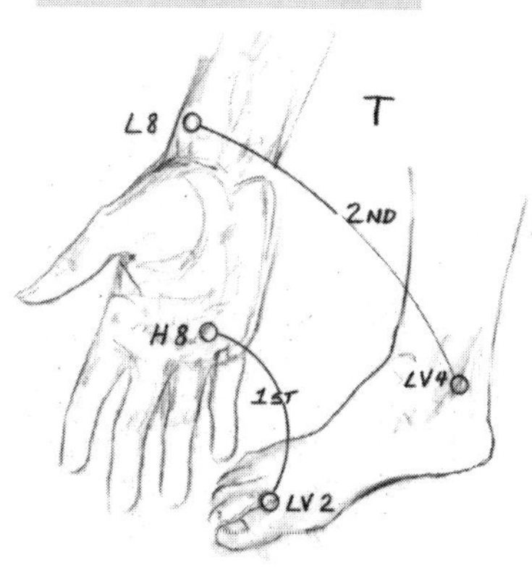

Liver

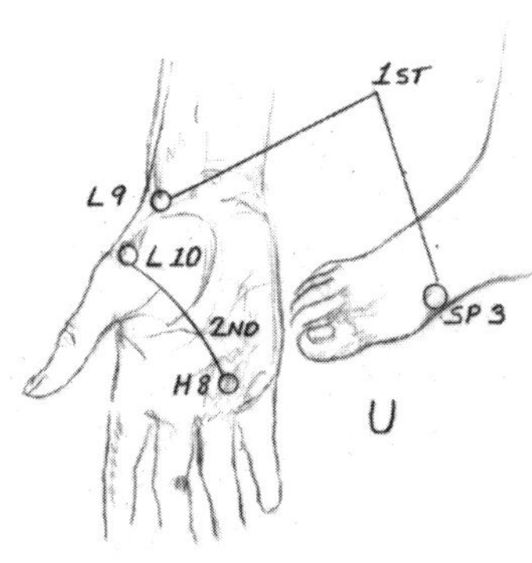

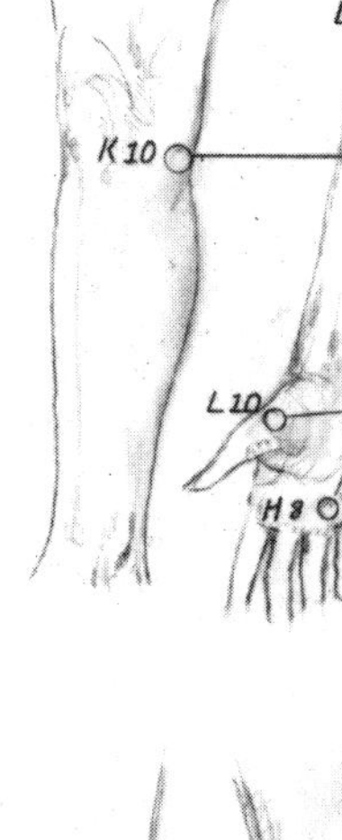

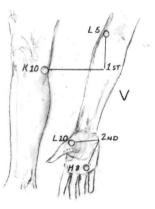

Lung

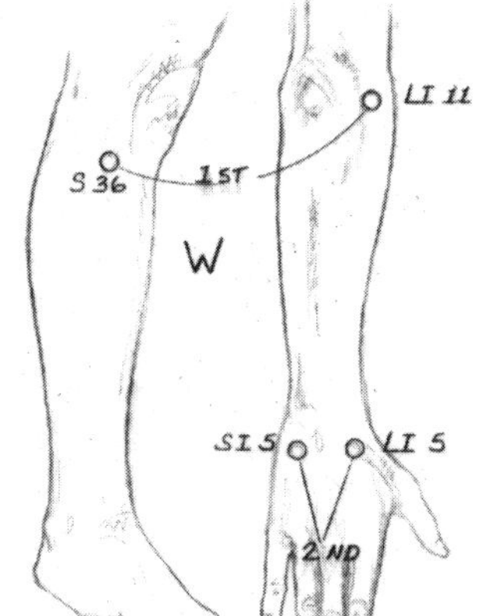

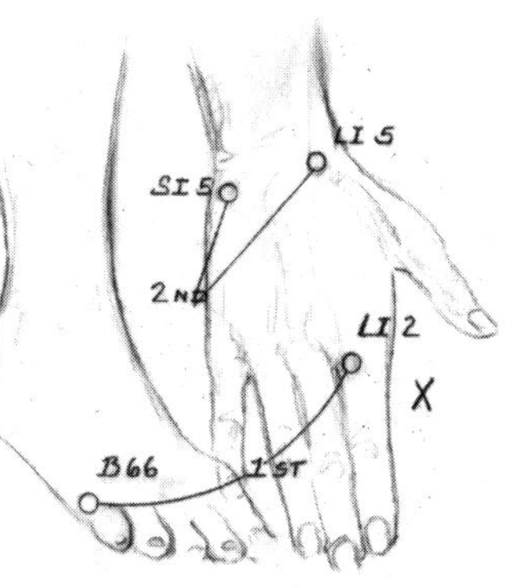

Large Intestine

1. *Stand with your feet wider apart than the width of your shoulders.*

2. *Place your hands on your thighs above your bent knees and straighten your arms. Take several deep breaths. The position feels like sitting on an invisible chair (see Figure 27).*

3. *With your head forward and your bottom back, adjust your feet so your knees are directly above your ankles and your arms and back are straight. You are creating a kind of suspension bridge.*

4. *Slowly stretch one shoulder across and down toward the opposite knee. Repeat with the other shoulder. It is a crossover exercise. You will feel the stretch across your back. You can do this stretch more than once.*

5. *Very slowly raise yourself with your arms hanging down, until you are in a standing position.*

ALIGNING YOUR MERIDIANS WITH THE EARTH'S MERIDIANS

The earth's daily rhythms, time zones, and seasonal changes affect the flow of energy in your meridians. Just as the tides go through a daily cycle, each of your meridians has its own twenty-four-hour cycle. With twelve major meridian segments, each has a two-hour "high tide" period when its energy is strongest, its pulses are sharpest, and its absorption of energy is the most active. Twelve hours later it is in its two-hour "low tide" period when it is at rest. Each meridian comes into its high tide or "power time" like clockwork, and these times are shown on the Meridian Flow Wheel (Figure 28). Notice also the relationship between each meridian and the meridian sitting opposite it on the wheel. When a yang meridian is at the peak of its cycle, the corresponding yin meridian is at its nadir. Yang is the active, outward, and expansive principle in Chinese cosmology; yin is the receptive, inward, contracting principle.

You can become familiar with how the cycles of your body correspond with the Meridian Flow Wheel by keeping, for a week, a careful log of the

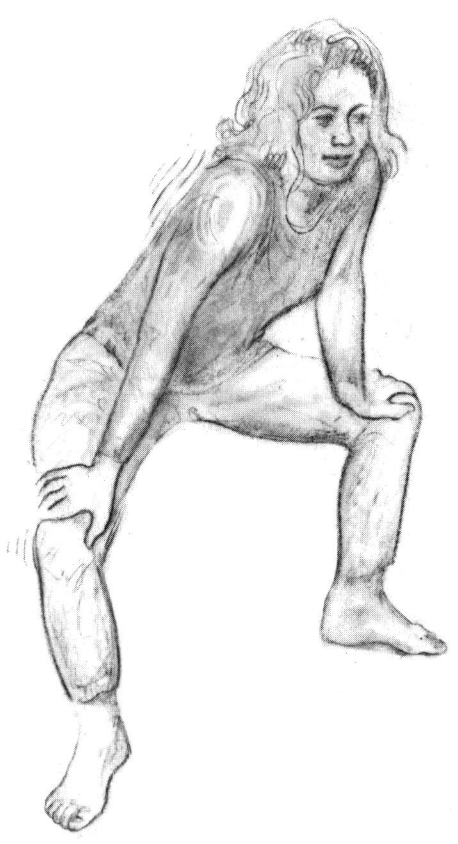

Figure 27. ■
Spinal Suspension

times of day that you notice changes in the way you feel. Do you lose your energy and get drowsy at a particular time of day? Is there a time of day that you are prone to headaches? Does your heart seem to race at the same time every day? When do you feel the most enthusiastic? The most grumpy? When are you most susceptible to cravings? When is your resolve the weakest? Record even subtle changes. After about a week, look for patterns. Note any condition that seems to occur or increase at a particular time of day and consider whether the meridian that is at its power time during that period might be involved.

Strengthening a Weak Link in Your Energies. For instance, people often get tired between 3 p.m. and 5 p.m. That is the period when the energy in the bladder meridian should be the strongest. The bladder meridian governs the nervous system because the meridian houses all the nerves along the

■ *Figure 28.* ■

MERIDIAN FLOW WHEEL

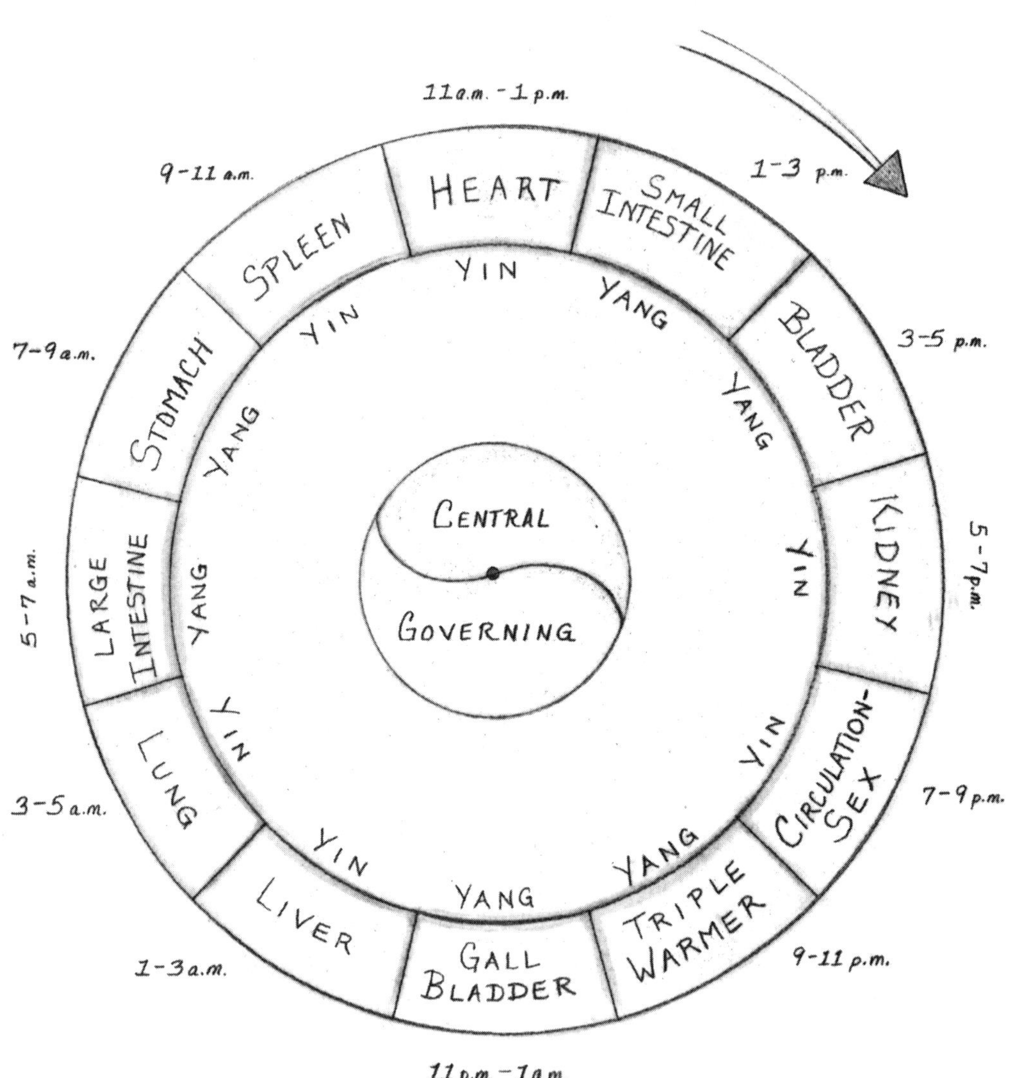

spine. If you rest during this period, the energy is available to restore your nervous system. Perhaps nature set it up this way in order to rejuvenate you for the evening. You've been active all day, you've just made a huge expenditure of energy for eating and digesting (the small intestine meridian was strongest from 1 to 3 p.m.), and it is not just a coincidence that in many cultures that are closer to the rhythms of the earth, people take a siesta during this time. Looking at the meridian that is at its low tide may complete the picture. The lung meridian is bladder's polarity (see the wheel), and it is at its low point during this period. Your body is receiving the least amount of oxygen at this time of day, which also makes you want to nap. It is a natural time to let the digestive and nervous systems restore, and there is not necessarily something wrong with you if you get tired in the afternoon.

But you may be required to get your rhythms in line with your realities in order to perform well at an eight-to-five job or during another period where the demands are high but your energies are low. One of the most frequent complaints I hear about the 3 p.m. to 5 p.m. time period is from mothers who say, "It's the time of day I want to be the most alert because it's when my kids come home from school. They need me to be present, but I'm falling asleep on them. I don't even look like I'm interested in anything they're saying."

How to Be at Your Best When It's Time to Greet Your Kids After School or Your Spouse After Work

Ten minutes before they usually walk through the door, stop everything, and (time—10 minutes):

1. *If you are exhausted, your energies are probably running backward. Begin with the Three Thumps (page 63) and the Crown Pull (page 77).*

2. *If this does not revive you, your energies are probably scrambled. Do the Wayne Cook posture (page 73).*

3. *If you are agitated, release any stored anger or frustration with the exercise for turning off your anger (page 208).*

4. If this does not calm you completely, do the Taking Down the Flame exercise (page 220).

5. Whether you were exhausted or agitated, end your ritual with Separating Heaven and Earth (page 249) and zip up with an affirmation (page 82) that you are fully present for your children (or spouse) as you prepare yourself for the onslaught.

You can also do this with *your children to realign their energies after school or with your partner to ensure a better homecoming.*

By tracing your meridians each day in a manner that adjusts for your low energy time, not only can you help your body adjust to the schedules that are imposed upon it, but you also may be preventing a future illness. If you regularly trace your meridians during your low energy time according to the following instructions, you boost your energies while strengthening a vulnerable system in your body.

The times shown on the Meridian Flow Wheel indicate each meridian's power time, the high tide period in which that meridian's pulses are strongest. Another meridian comes into its power time every two hours. When you trace your meridians, you always begin with central and governing. You can trace your meridians exactly as instructed earlier in this chapter, following central and governing with spleen, but you get an extra shot of energy if you "enter the wheel" on the meridian that is in its power time.

1. Match the time of the day with the high tide meridian shown on the wheel. For instance, if you lose your vitality, at around 3:30 each afternoon, notice that the bladder meridian is in its power time. So, following central and governing, you would begin with the bladder meridian.
2. Continue tracing your meridians clockwise around the wheel.
3. Finish by again tracing the bladder, central, and governing meridians, in that order.

Jet Lag Breaks Your Alignment with the Earth's Meridians. Jet lag occurs when your body is unnaturally yanked from the electromagnetic environment to which its meridians have acclimated and into a time zone that is not in phase with them. The body continues to operate according to the time zone in which it began. If you fly from Los Angeles to London, leaving at noon, you begin with your body in heart time (see Figure 29). During the ten-hour trip, you pass through five two-hour time segments, so your body has moved into triple warmer time. It is triple warmer time in the city where you started, and it is triple warmer time in your body, but it is large intestine time in London. Because your meridians are no longer in synch with the Earth's meridians, you experience jet lag.

Any energy technique for jet lag will work better if during the flight you also get up every couple of hours and stretch. I travel a great deal in my work, and I have found over the years that what works best for me is to:

1. Do the three thumps every couple of hours while breathing deeply. The first thump, K-27, ensures that your meridians are not running backward; the second thump, the thymus, fortifies your immune system, protecting you against any germs on the plane; the third thump, the spleen, also assists in immune strength, and it balances blood chemistry as well.
2. Do the Separating Heaven from Earth exercise (page 249) every couple of hours. Also stretch the calves and any other parts that want to be stretched.

An easy method that stabilizes your body as you go through time zones involves the use of magnets. Get two round magnets and determine which is the true north and which is the south side of each (see pages 300–301). Tape the magnets to your belt so one magnet is on the left side of your body and the other is on the right side. The magnet on the left should have its north polarity facing your body. The magnet on the right should have its south polarity facing your body. Remove the magnets or the belt shortly after reaching your destination city. Never wear magnets for more than twenty-four hours. The benefits of this technique are related to the fact that when you fly, you are not as connected to the Earth's electromagnetic field, and the magnets on your body act to counterbalance the deficit.

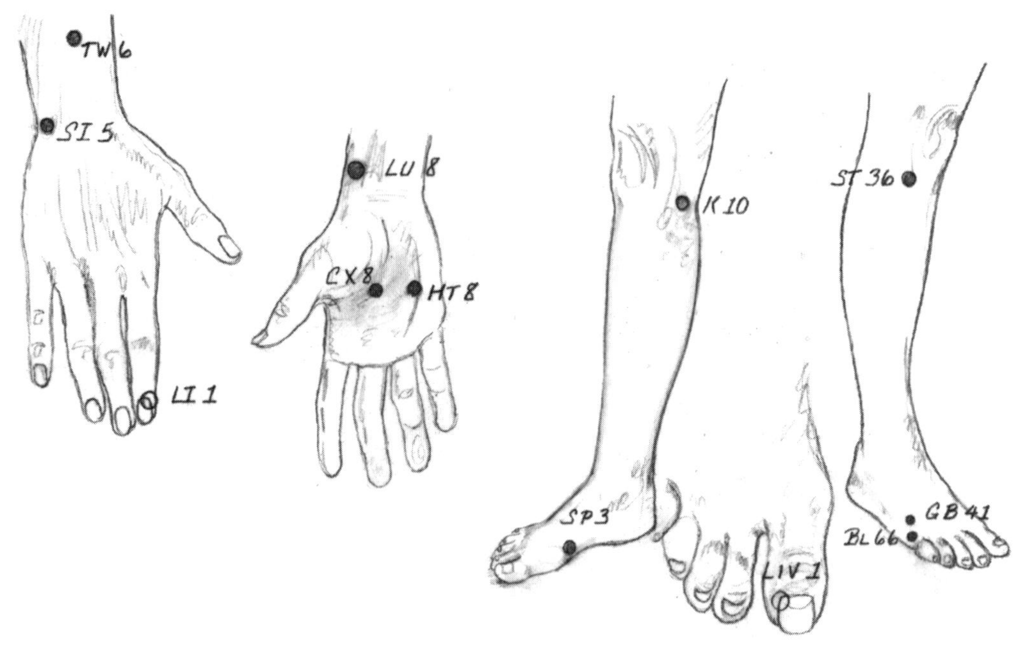

■ *Figure 29.* ■

JET LAG PRESSURE POINTS

TIME OF DAY	MERIDIAN	ACUPUNCTURE POINT
5 a.m.–7 a.m.	Large Intestine	LI 1
7 a.m.–9 a.m.	Stomach	St 36
9 a.m.–11 a.m.	Spleen	Sp 3
11 a.m.–1 p.m.	Heart	Ht 8
1 p.m.–3 p.m.	Small Intestine	SI 5
3 p.m.–5 p.m.	Bladder	BI 66
5 p.m.–7 p.m.	Kidney	K 10
7 p.m.–9 p.m.	Circulation-Sex	Cx 8
9 p.m.–11 p.m.	Triple Warmer	TW 6
11 p.m.–1 a.m.	Gall Bladder	GB 41
1 a.m.–3 a.m.	Liver	Liv 1
3 a.m.–5 a.m.	Lungs	Lu 8

Two additional methods for countering jet lag actually reset the low and high tide periods of your meridians according to the time zone in the destination city. The first technique involves tapping acupuncture points. Photocopy Figure 29, take it with you on the plane, and:

1. As soon as you are seated on the plane, identify the current time and the current time at your destination city.
2. Look at Figure 29. Next to the two-hour time segment that includes the local time is the name of the meridian that is in its power time and an acupuncture point. Find this acupuncture point in the drawing.
3. Next to the two-hour time segment that includes the current time at the destination city is the name of another meridian and acupuncture point that are in their power time. Find this acupuncture point in the drawing.
4. With some force, tap both acupuncture points, simultaneously or in sequence, for about a minute. First tap the points for the current time and the destination time on one side of your body, then the other.
5. Repeat every two hours on the next pair of meridians. That is, move down the list to the next meridian below the original local time and to the meridian below the original time at the destination city. If you sleep through the specified time, simply tap the points that bring you up to date after you awake.
6. If the flight will be eight hours or longer, find the acupuncture point associated with the anticipated arrival time. For the last four hours of your flight, tap this point, simultaneously on the left and right sides of the body, for a minute every hour.
7. This technique can be augmented by a round magnet with a hole in the center so you can hang it from a string or from dental floss. After tapping each acupuncture point, spin the magnet over the point, three or four spins in each direction.

A second way of countering jet lag involves tracing the meridians rather than tapping the acupuncture points. Although it is more complex, if you already know how to trace your meridians, you might prefer this technique.

1. Tap K-27 (page 64), Cross Crawl (page 69), and trace your meridians (Figures 10–23) before you get on the plane. After tracing central and governing, trace the meridian whose power time on the Meridian Flow Wheel (Figure 28) matches the time of day. If you are tracing your meridians at 6 p.m., for instance, you would find that the kidney meridian is in its power time from 5 p.m. to 7 p.m.
2. Continue clockwise on the wheel through the other eleven meridians. End with the meridian you traced first (e.g., kidney if you began at 6 p.m.) and then central and governing.
3. About halfway through the flight, reset your watch according to the location over which you are flying and, in a place where you can stand, trace your meridians according to the current time, entering the wheel at the meridian that is in its power time. Always begin with K-27, Cross Crawl, central, and governing.
4. After arriving at your destination, trace your meridians once more according to the new local time.

This chapter has provided an introduction to the meridian system. We will return to the meridians when we consider illness in Chapter 9. The focus here turns to your chakras, your body's energy reservoirs.

Chapter 5

The Chakras

Your Body's Energy Stations

MAJOR CENTERS OF BOTH ELECTROMAGNETIC ACTIVITY AND VITAL ENERGY ARE RECOGNIZED IN INDIGENOUS CULTURES THE WORLD OVER. IN THE HUNA TRADITION OF HAWAII, THEY ARE CALLED *AUW* CENTERS; AND IN THE CABALA, THEY ARE THE "TREE OF LIFE" CENTERS. IN THE TAOIST CHINESE TRADITION THE TERM IS *DANTIEN,* AND IN YOGIC THEORY THEY ARE CALLED "CHAKRAS."

—WILLIAM COLLINGE
Subtle Energy

The word *chakra* means disk, vortex, or wheel. Whereas meridians are an energy transportation system, the chakras are energy stations. Each major chakra in the human body is a center of swirling energy positioned at one of seven points, from the base of your spine to the top of your head. Memory is energetically coded in your chakras just as it is chemically coded in your neurons. An imprint of every important or emotionally significant event you have experienced is recorded in your chakra energy. If I know your chakras, I know your history, the obstacles to your growth, your vulnerabilities to illness, and your soul's longings.

A sensitive practitioner's hand held over a chakra may resonate with pain in a related organ, congestion in a lymph node, subtle abnormalities in heat or pulsing, or areas of emotional turmoil. I have learned that these sensations have specific meanings. The hand can vibrate so intimately with a chakra's energies that the practitioner has an experience that mirrors the client's or reveals the condition of one of the client's organs. If I think I am going to have only one session with a person and want to provide the best information I can, I will focus the session on the chakras. I know that as I delve deeper and experience the successive levels of each of the chakras, information will emerge that I could not initially access.

Each chakra regulates distinct aspects of your personality, has specific tasks, and dedicates its energies to those tasks. Each encodes a slice of your story. When two chakras are energetically cut off from one another, the aspects of the personality governed by each may become warring subpersonalities, as is seen for instance in the classic conflict between head and heart. While your chakras tell your story, they can also hold your story back. Once a chakra that has been energetically blocked is released, your life is able to again unfold.

The organs that are in the proximity of each of the primary chakras are related to that chakra's functions, and each chakra is named after the part of the body over which its energy spins (see Figure 30). They are called, from the bottom up: the root chakra, womb chakra, solar plexus chakra, heart chakra, throat chakra, pituitary (third eye) chakra, and pineal (crown) chakra.

The chakras reveal how your history plays into current symptoms. By clearing chakras that are still locked into the energies of your past, you can give your entire energy system a boost, and, depending on the chakra, this will also enhance: your life force (root chakra), creativity (womb chakra), assertiveness (solar plexus chakra), love and compassion (heart chakra), expressiveness (throat chakra), ability to transcend your own story (third eye chakra), or flow in your connectedness with the universe (crown chakra). This chapter introduces you to the chakra energies and shows you how to clear, balance, and strengthen your own chakras and those of others.

A CHAKRA WAKENING

A man whose wife had been deeply hurt by his periodic love affairs was bewildered by her pain, since it was she that he clearly loved. He was also puzzled about why the other women seemed hurt. He looked at these liaisons as a primal biological function, and he couldn't grasp their responses to them. He reasoned that because these fleeting encounters were so passionate that the women must have had a great experience and that the termination of each affair was so complete that his wife had no reason to feel

■ *Figure 30.* ■

THE SEVEN MAJOR CHAKRAS

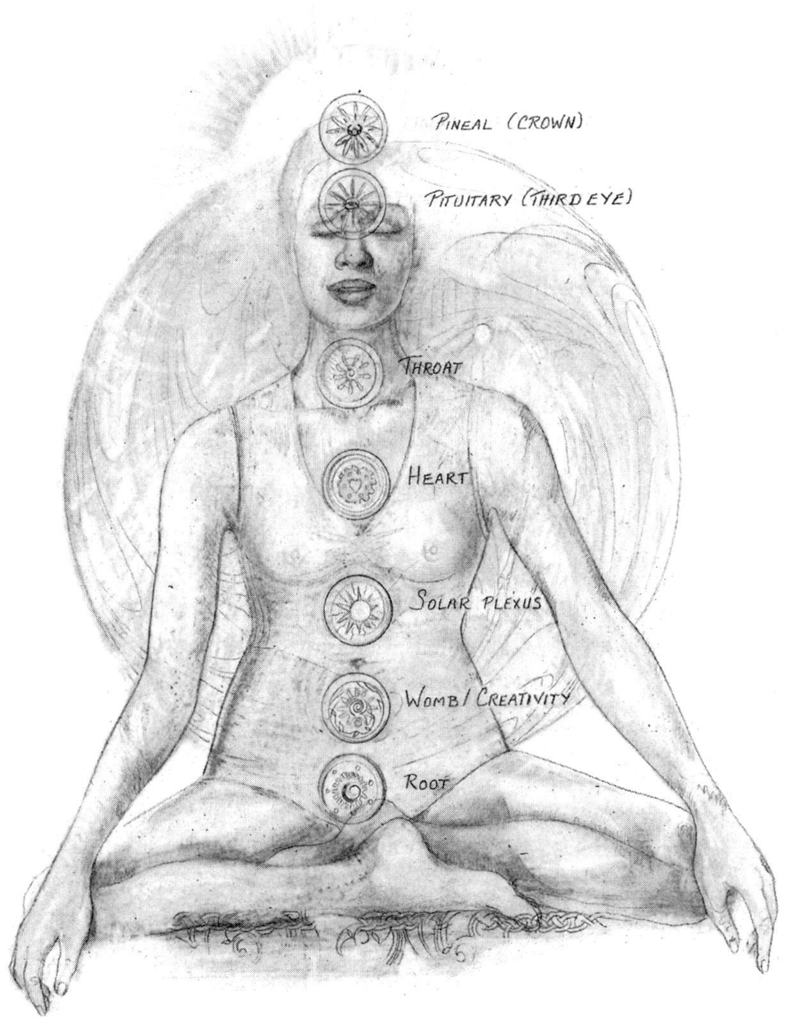

threatened. He was genuinely baffled that these women made such a fuss about it all. Attempting to explain what he considered obvious, he would repeatedly avow, "You can't fight biology!" His wife finally decided to stop trying and announced that she was leaving.

They came to me for a session, not to save the marriage, but rather for damage control, so that they might end it more civilly, less harmfully. He sincerely didn't want her to be hurt, and he didn't want his children hurt, but he just could not comprehend her feelings. She believed he was a scarred human being who turned around and scarred others. She did not want him to scar their children any further or pass his values on to their son or his devaluation of women on to their daughter.

Early in the session I decided to work with the husband alone. I asked the wife to return in an hour. I wanted to see his energies separate from hers, and I wanted to determine just how deeply scarred he was. I focused on his chakras. There was an obvious collapse of the energy in his heart chakra, but his root chakra told the most poignant story. As I sensed into its deeper layers, I began to have feelings and images about his childhood. More specifically, it was as if he hadn't had a childhood, as if there never had been for him a time of innocence. I sensed that he had been in survival mode from early on. I had images of severe and unpredictable beatings, which he later confirmed, and it seemed that in order to survive he had had to disconnect from his heart. The only pure spontaneous joy he knew was having sex with someone to whom he had no responsibility.

It was clear that I wasn't going to convince this guy of anything, but even if he couldn't let the concepts through, I thought he might make some headway if he could feel all the energies involved in his relationships. This was indeed what happened. Sensing into each of his chakras, it seemed to me that the best thing I could do was to hook his fourth chakra, his heart chakra, to his root chakra, the chakra of his sexuality. His was actually a surprisingly sweet, very loving heart chakra, but it was essentially cut off from both his awareness and from his root chakra.

With repressed energy moving from deep within his heart chakra, he felt acute sorrow about losing his wife. He began to have many other feelings as well. He was able to feel his wife's pain and sadness, as well as the pain of the other women. But his heart chakra was so separated from his root chakra that he still couldn't grasp why what he had done had caused them this pain. He still believed they thought his sexual casualness was bad because they had been so fully indoctrinated into an outdated morality. He contended that it was their stuffy sexual values that caused them their pain, not him.

As we worked, the circuitry of his heart, his root chakras began to connect with one another. I made wider and wider circles with my hands to expand the energy of his heart chakra until it was reaching so far beyond the area of his heart that it extended into the field of his root chakra.

He felt a buzzing force as the areas between his heart and his pelvis began to bridge. He began to comprehend what his wife had said about how, for her, making love always relates to the heart. While this was not an instant conversion experience, it did leave him willing to entertain her point of view and to look at the consequences of his having been cut off, so early, from his heart. Now, four years later, they have remained in their marriage, and they both report that it has been a monogamous journey.

AN OVERVIEW OF THE CHAKRA SYSTEM

While the term *chakra* has its origins in India, many cultures have identified and worked with these spiraling energy centers, and any healer who is sensitive to the body's energies will eventually stumble upon them. Electrical oscillations in the skin above the chakras are in the frequency of 100 to 1,600 cycles per second, as contrasted with 0 to 100 in the brain, 225 in the muscles, and 250 in the heart.[1] When advanced meditators consciously project energy through a chakra, the strength of the electrical field emanating from that chakra multiplies.[2]

Seven major chakras are usually distinguished, although I've heard of systems that identify only five major chakras and others that set the number as high as ten. Part of this discrepancy may be because there are also many minichakras throughout the body. The hands and eyes act as chakras, and new spiraling vortexes may appear anywhere that fresh energy is required, for example, after an injury or if an area needs to be cleared of toxic energies.

Each chakra influences the organs, muscles, ligaments, veins, and all other systems within its energy field. Chakras also influence the endocrine system and so are strongly involved with your moods, personality, and overall health. Your physical and psychological evolution, as well as your spiritual journey, are all reflected in your chakras.

Each chakra is a conduit for a particular form of energy within the universe. The seven chakras resonate, respectively, with the principles of survival (root), creativity (womb), power (solar plexus), love (heart), expression (throat), transcendence (third eye), and unity (crown). Each chakra's energy is the microcosm of one of these universal principles, expressed within your body. The chakras also feed off both the positive and negative energies of the cosmos—archetypal forces such as those for bonding *and* for separation; for growth *and* for atrophy. They spiral these energetic prototypes into your body, feeding your entire physical structure and linking your body to the subtle energies that surround you.

The scale from the first to the seventh chakra follows a natural developmental line—from the urge for life, to life's unfolding, to its empowerment, to its capacities for love, expression, and comprehension, to the longing for connection with the cosmos. I disagree, however, with the commonplace notion that the lower the chakra, the less spiritual it is. This hierarchical idea strikes me as something like saying that a forty-year-old man is spiritually superior to his four-year-old son. To each his season. Each developmental stage is as sacred as the next. Each chakra connects with a universal force, has a soul-deep task, and holds a transcendent beauty. For many people, in fact, personal evolution is more about *dropping* deeper and deeper into their "lower" chakras than about cultivating the "higher" ones.

How to Rid Yourself of Chronic Headaches

The following Headache Isometric Press helps overcome and prevent headaches (time—about 2 minutes):

1. *Sitting, relax your shoulders. Tilt your head to the right, bringing your ear close to your right shoulder.*

2. *Place your right palm against the right side of your head (see Figure 31). Take a deep breath in as you press your hand and your head against one another. Push hard while holding your breath.*

3. *Slowly but completely releasing your breath, drop your hand and stretch your head closer to your right shoulder. Repeat two more times.*

Figure 31.

HEADACHE ISOMETRIC PRESS

4. *Repeat on the left side three times.*

5. *With your head straight, place the pads of your fingers on the bones at the bottom of your head. Take a deep breath in and isometrically push your fingers and your head together.*

6. *Drop your hands into your lap and open your jaw as you release your breath through your mouth. With a deep inhalation through your mouth, jut your jaw out, pulling your lower teeth up to your top jaw. Release your breath, open your mouth, and let your jaw relax.*

7. *Tighten your jaw one more time. As you release your breath, let your head drop forward toward your chest.*

8. *Take a deep breath in, with the pads of your fingers pushing your head upward as your head continues to push down. Release your fingers with your outbreath and allow your head to drop further downward. Repeat twice. On the last time, as you let your breath out, lock your fingers behind your head and allow your elbows to pull your head downward.*

SEVEN CHAKRAS, SEVEN LAYERS

Each chakra spirals down seven layers into the body. (At least I've never been able to reach deeper than a seventh layer.) The top layer, which sometimes feels more like debris than a coherent energy, seems related to recent events that have not yet been integrated. As you go deeper, the energies get clearer and more consistent, and the chakra's distinct influence becomes more obvious. The fourth level begins to pick up incidents from earlier in the person's life. If I move into the field deeply enough, and reach the fourth, fifth, and sixth levels, I get images and stories. When I tell the stories, the person usually responds with a surprised confirmation. Working with the heart chakra of a morose thirty-six-year-old-woman, I related, "I feel I am looking out at the world from the age of about seven, and I have just lost someone I love dearly. It is not a parent, perhaps a sibling? My grief is too much to bear. My heart is closing down." Her startled and tearful reply: "That's when Robert, my older brother, was accidentally shot by a neighbor boy who was playing with his father's gun. He died two days later." I do not think of this as a supernatural ability, but rather an example of the natural but often atrophied capacity to read the story imprinted in a chakra's energies.

The seventh layer down carries forth what the person has learned and embodied in previous lifetimes about the issues governed by this particular chakra. This knowledge or strength has been earned. The energy in this layer will often be a totally different color and density from the chakra's other layers. It is more like light emanating from a deep source, filtering up through each of the other layers. As we mature, the layers become better aligned and the light more accessible, bringing forth wisdom and peace. When the light at the chakra's source is blocked, we may feel estranged, cut off from deeper sources of who we are and why we are here. Clearing and aligning a chakra's seven layers begin to correct this alienation from self.

Working with the seventh layer of a chakra can be confusing because the information it holds often seems to have nothing to do with the person on my treatment table. I have, however, come to understand that the energy of a person's unfulfilled potential stay here, and it seems to be nature's design that this potential stays dormant until the timing is right for it to blossom. Sometimes I know I am touching into greatness that the person is still unaware of possessing. At other times, even if I am well connected with the person, I can't get down to the seventh layer. It is as if a higher wisdom does not yet deem it fitting to reveal to the person the potentials stored in this deepest layer, perhaps abilities cultivated in other lifetimes. But when the seventh layer does appear to me, it seems as if the soul's wisdom is saying, "Okay, it is time for this to be known."

A woman was having her eighty-sixth birthday and someone had given her a session with me as a birthday present. I eventually began to sense that I was picking up something about a past lifetime. I said, "Excuse me, do you believe in past lives?" With annoyance in her voice, she said, "I most certainly do not!" So I tried to ignore what I was sensing, but it continued to plague me. Finally, I said, "Images that are not of this time or this place are presenting themselves to me very strongly. I know you said you don't believe in past lives, so we can just take this as a metaphor, but I feel I won't do you justice unless I tell you about it."

She finally agreed, and I told her that I was having a vision of a man in his forties who lived in England. He was a real gentleman. He was sort of short and squat, and he was a wonderful piano player. Along with his family, the piano was his great joy. He had two daughters and a wife, and they were all accustomed to life's finer things. But he wanted them to have more than just an ordinary, unadventurous life, and he found the courage to take them to the New World.

The only piece of furniture they had sent over was his piano. When they got to what is now Virginia, it was not at all a genteel world. There were fears and concerns of the unknown, of the Indians, and of the hardships. The town had a fort around it. Because the man was short and squat and did not look at all refined, they wanted him to be a guard outside the fort. He protested that he was a musician, but there was no piano to play, as his had not yet arrived. He was ridiculed for trying to get out of the work as a guard by claiming to be a piano player, and he was never recognized for having taken the enormously courageous step that brought him to the New World. He was not at all cut out to be a guard, and he was very unhappy with his life. His daughters hated living in the New World, they blamed him for their misfortune, and they were quite cruel to him. They were also ashamed and embarrassed by his ineptness, and they soon became estranged from him.

He had no one to talk with where he stood guard, except little children from the Indian communities who would come by. These playful boys and girls became the light of his life. His higher-ups told him he must make them go away so he not create problems with the Indians. But every time the children came around, he enjoyed them so much that he allowed them to stay.

His piano finally arrived, after having been lost for years in a foreign port. When it showed up, he was much older. His daughters were married and gone. No one took him seriously about the piano, and he couldn't find the courage to play it after so many years.

I also saw that he died without ever realizing the gift he had provided his community. He had a great deal to do with maintaining trust with the Indians because they saw that their children loved him so much. Just as he had shown tremendous courage in coming to the New World, he had also been an unsung hero in ignoring his superiors and not sending away the children.

After relating this story, I looked at my client, and she was choked with emotion. She told me that she was a pianist, and that she had two daughters, and her daughters felt that she did not have courage because she was a much better pianist than anyone knew. "I am very very good, but I cannot play in front of people." She looked me in the eye and said, "That was me." She kissed me goodbye when she left. That night was her birthday party. The next day she called to tell me that she had played her piano at the party. Everybody was stunned by her abilities, and her daughters spoke eloquently of their pride in her.

THE MECHANICS OF THE CHAKRAS

Your chakras pull their energy in from your environment (including, for better or worse, the energies of other human beings) and distribute them in your body. Your chakras also send energy outward. Chakra energy that spins in a clockwise direction tends to move outward and envelop other energies; chakra energy that spins in a counterclockwise direction tends to move inward and draw in other energies. It may seem odd that a chakra's energy can simultaneously be spinning in both directions. It is hard, for instance, to imagine a whirlpool that spins clockwise and counterclockwise at the same time. Understanding that chakras are layered helps solve this puzzle. Each layer can spin in either direction, and depending on its energetic state at a given moment—its need to take in or to release energies—the direction may change.

Just to make it more complex, counterclockwise energy can also move outward. Because counterclockwise energy draws other energies to itself, this suction force can reach out into the environment and pull energy into itself. This is what is sometimes happening with people who "drain your energy." They are unconsciously sending out, with a counterclockwise spin, an energy vortex that can suck on your own energies like a straw in a glass of milk.

Light and color also move through each chakra, spinning up and out from the vortex of its deepest layer. We see the colors of another's chakras through our own filters. Therefore, there is no standard, no dictionary that can tell you, absolutely, the meaning of a particular color on a given chakra. The many books that try contradict one another. What I can offer are guidelines. Your filters will be different from mine, but there is enough common ground from one person to the next that the guidelines can be useful. For instance, clear green in any of the chakras seems to signify that a healing in the chakra is under way. Green is in the middle of the color spectrum, and its appearance often means the energies are coming into balance.

The color red, as the first and most dense color on the spectrum, has traditionally been associated with the root chakra; violet, the last color, with the seventh chakra; and the remaining colors are believed to neatly correspond with the other chakras just as they fall along the color spectrum. While this is accurate in a certain pure or theoretical sense, I've never seen a person whose chakras perfectly match this pattern. Rather, each chakra is as unique as a thumbprint, holds many colors within its seven layers, and reflects the energies the person was born into as well as the life that has been lived. So while there is some basis for the correspondences between the chakras and their associated colors—from red at the root to violet at the crown—the vibrational relationships of the chakras and their colors are far more complex.

Thinking of the chakras as corresponding with the color spectrum is, nonetheless, a framework that reflects an organic *tendency.* And since colors are vibrations of energy, you can use them to regulate an energy system. You can restore the energies of a chakra, for instance, by bathing it in a color that would give it greater strength, clarity, or balance. You can intuit or energy test for the color that will serve the development of a particular chakra. Simply lay a piece of colored material on the chakra and do a general indicator test (pages 65–67).

THE SEVEN CHAKRAS

I first know a chakra by the way its energy feels to me and by the colors and shapes I see. As I move my awareness to these sensations and colors, a story often opens.

The Root Chakra. Because it is so widely taught that red *is* the color of the root chakra, students sometimes assume they are imagining it when they correctly see other colors. I almost always see other colors there and sometimes no red at all. When I see gold in the root chakra, it often reflects an essential goodness in the person, like the saying "good as gold." Purple or silver bands spinning around the root chakra tell me that the person experiences spiritual guidance or protection, which the person confirms when I comment.

I have seen light shows in the root chakra, rich with colors so humbling and beautiful that I feel I am looking directly into the realm of spirit. Most stunning, I have been present at births where I was overwhelmed by the amount of energy and light around the mother's root chakra, as if it were proclaiming the miracle of birth. I have seen at the furthest layers of the root chakra a color that is like the night sky, unfathomably deep. Tapping into that energy, the room often fills with mystery, and I begin to understand the present lifetime within a larger context. Wisdom from past lifetimes seems to be concentrated most strongly in the seventh layer of the root chakra. It then moves up to the seventh layers of the other chakras, drawn according to that chakra's theme. Past life lessons related to power, for instance, are stored in the seventh layers of the solar plexus or third chakra, and past life lessons related to loving are stored deep in the heart or fourth chakra.

The root chakra sits at the base of the spine, and its energies move to the front of the body, spinning over the pelvis and sexual organs. It exerts its influence both up the body, carrying the life force, and down the legs, providing support. This chakra is a channel for the primordial energies of the Earth. All life, as we know it, evolved from the face of the Earth and her fluids. Earth, impregnated by the energy of the sun, is the mother to us all. The root chakra is a receiver for the Earth's subtle energies. Through this connection, the term "grounded" takes on a literal meaning.

A person's earliest life experiences are imprinted on the first chakra, and they become lodged there before the brain has developed the myelin sheaths that are necessary to physically code such memories. These early stories may still be affecting the person fifty years later. Not only is childhood and past life material coded here, so are ancestral memories—perhaps a trauma that goes back many generations. Even if traumatic experiences faced by your parents or grandparents aren't directly coded in your DNA, their emotional imprint may be coded in your energies. I have unmistakenly seen an ancestor's pain or courage in a client's energies, and it can be freeing for the person to realize that an energy that feels alien traces back to previous generations.

I recall, for instance, strongly sensing that I was going to find disease in the body of one woman as soon as I touched into her root chakra. To my surprise, as I explored the energy of each chakra, looking for the one that carried the illness, she was beginning to appear thoroughly healthy. Returning to her root chakra, and going deeper and deeper through its layers, however, I became absorbed in a terrifying sense that death was imminent. While I could find no diseased energy, this terror about dying was embedded in every layer of her root chakra.

I began tuning into painful stories rising up from what felt like previous generations, mostly concerning the women of her family: dying during childbirth, starving on the prairie, natural deaths for the circumstances of the times. A feeling spiraled up from her root chakra that seemed to say, "No one can save me; nothing can be done; I'm not going to survive!" I related this and explained my sense that the feeling she wasn't going to survive belonged, I believed, to her ancestors, not to her.

"All my life," she said, "I've believed death lurked around the next corner. I have never understood it. I've been in therapy trying to find out if some childhood trauma had caused my death fears. I've often run to doctors, just to make sure I'm really well. I can't even commit to a relationship because I'm afraid he may be the reason I won't survive. I know it makes no sense, but it is an overwhelming force that comes up in me, and this feels like the key I've been looking for."

The more "civilized" a culture, the more muddled the root chakra energy of its members tends to be. A direct way to cultivate this energy, and to feel more sexual and sexier, is to take up a form of dance where the primary impulse of the movement comes from your pelvis. The root chakra resonates with the Earth's primordial life force. I love free-form African dance done to a primal beat. To dance from your root chakra is to dance with the Earth.

The base of the root chakra is like a pilot light: Its fire is sometimes called psychic heat. It directs energy both internally and externally, up and out. When it sends its force upward, it fuels the other chakras, travelling all the way up through the crown chakra. Visions in the crown chakra are just as likely to have originated in the root chakra as in the heavens. When a person's root chakra is closed, you get the sense that the person isn't present or grounded. Something substantial is missing. Not only is the power of the root chakra unavailable, its power to fuel the other chakras is also lacking.

When the root chakra sends its force outward, it connects with the environment and with other people. While you connect with people through other chakras as well, the root chakra has a particular quality. It can exude a sense of safety and groundedness, making others feel physically secure. It can be a bridge from one person's passion to another's. Or it can convey threat, hostility, and aggression. Its connection to others is primal and basic.

The root chakra is the foundation of our most visceral drives, as contrasted with more calculated motives. Thus the drives to eat, acquire, hoard, copulate, stay safe, or protect a family member are all governed by the root chakra. Even the impulses toward honor, loyalty, religious fervor, and ethnic conflict are governed more at this level than by rational thought.

The root chakra is, in fact, associated with our tribal nature. It is an expression of the force in the universe that says "We are one." Through the root chakra, we know our identity with the Earth, our identity with our offspring, ultimately our identity with all who live. Not just self-survival, but also survival of the lineage have their energetic foundations in the root chakra. The mother or father leaps in front of the child to take the bullet. The lineage must go on. When I witnessed my daughter accidentally cutting herself on some broken glass, the feeling was like having a knife plunged through my root chakra and up my body. Historically, this force ensured our allegiance to the survival of the tribe and all the members of our community. In its most evolved sense, it is the primal urgency—which increasing numbers of people are feeling today—that we all have to make it together on this planet.

Many people are divided against themselves at the level of the root chakra, so they carry conflicting primal beliefs and responses. Perhaps the person's parents were so fundamentally different from one another in their outlooks and values that they provided the child with incompatible messages. Perhaps the parents were united, but united in opposition to the culture or to the child's basic nature. Perhaps the impossible paradox of being abused by an otherwise loving parent created the split. Perhaps a man is set against his female side, so he does not have access to his own receptive qualities. Perhaps a woman is so terrified of the male energy that is out in the world that she doesn't allow the male part of herself to protect her. All these splits show up in the energy of the root chakra. When a person is deeply divided in these ways, the top layers of the root chakra seem more influenced by life experience while its deeper layers seem more aligned with the person's intrinsic nature.

My understanding of the chakras departs from the accepted wisdom in three basic ways. I believe that the "lower" chakras are as spiritually relevant as the "higher" chakras. I believe that chakra colors are more complex than those shown on the charts. And while sexuality has for thousands of years been associated with the second chakra, I believe it is most basically a root chakra affair. I think the men who articulated the system wanted to see their desire for sex as a divine rather than animalistic drive, and they disassociated it from the primal impulses of the root chakra and exalted it to the chakra of creation. This has rarely been challenged, but it seems impossible to imagine that sexuality originates anywhere but the root chakra. When sexuality arises, when the juices begin to flow, you feel it in the first chakra, and that is where I energetically see the origins of sexual energy. In a woman, it is the receiving chakra, where the male's energy enters her, where sexual union is consummated. When a person has been molested or is cut off from his or her sexuality, I see the scars in the first chakra.

Root chakra energy seeks out other root chakra energy, like the attraction of two magnets or, in some instances, more like a heat-seeking missile. That is the nature of sexual attraction, and it is nature's arrangement to ensure the continuation of the species. Gender enters the story because yin (feminine) energy is attracted to yang (masculine) energy. But since we all have both yin and yang energy in each of our chakras, no man is all yang and no woman is all yin, we are each a unique combination, and the more extreme variations baffle sexual stereotypes.

While sexual feelings begin with a root chakra wanting to connect with another root chakra, sexuality may embody all seven chakras, which is why it is so confusing and why it is so wonderful. A sexual union where all seven chakras meet is a merging of the urge for life, the urge to create, the urge for individuality, love, expression, expanded consciousness, and spiritual union. And with seven chakras, each having seven levels, and two people, that makes a total of 2,401 possible kinds of sex between any pair of lovers, if my math is correct, based on chakra energies alone.

Second Chakra. Regardless of the circumstances of a person's life, regardless of the pain a person is carrying, I often unintentionally smile when I get to the second chakra. It is a sacred vessel, a womblike container of imagination and creative impulse. It is the energy that makes a baby so lovable, and it is in fact within this energy that the embryo was nurtured. It holds a world of joy, freedom, and laughter, removed from the traumas and pain of life. It is a protected domain, where the creative force can flourish. Within the second chakra, babies grow, imaginative projects germinate, and the boundless creativity of the universe pours into each of us.

The second chakra extends from the top of the pelvic bone to the belly button. It encompasses the small and large intestines—organs that take from food the nourishment that can create new cells and eliminate what cannot be used. This chakra also encompasses the organs involved in creating a baby—the womb, ovaries, and fallopian tubes. The womb is the second chakra's emblem; sometimes oval, sometimes round, it actually looks like a womb, even in a man, and it is a safe womb, a world unto itself within which your core energies are free and unfettered.

The energy of the second chakra resembles the aura that surrounds the body. There are seven layers in the aura. The seven layers of the second chakra are also like concentric spheres, unlike the tiered layers I experience in the other six chakras. At its center is the protected place, the essential "me" that existed before the wounding—still pure, glorious, beautiful. I feel honored when I am able to step through this doorway to a person's inner sanctum. When I enter, I often find myself blubbering with awe about what I see and feel there and about what it tells me about the person. More often than not, people feel "seen" and validated, and they might respond with something like, "I always wanted somebody to know this side of me." Even more than this, when I begin to describe what I see, people will often remember a part of themselves they have forgotten, a particular quality of playfulness, innocence, goodness, or generosity.

Whereas the first chakra is basic to the survival of a body, it is at the second chakra that the soul embraces that body. The second chakra is my nominee for the least understood of the seven. Its role has sometimes been reduced to the "bathroom functions" of the bladder and intestines and the "bedroom functions" of the sexual organs. In my experience, this is only the basement of a very large mansion. Your navel is at the top of your second chakra, and just as it was once the physical cord that connected you to your mother, it is the energetic cord that still connects you to the pure creative energies that nourish your being.

The second chakra reveals the sweetness of a person's soul, the innocent self as it was before being hardened by life's tribulations. When people meditate and contact the true, precious essence within, they are often making contact with the energy contained at the deepest layer of the second chakra. People who are strong in the second chakra have a tendency to exude an atmosphere that makes other people feel cared for and comfortable. Like a womb, this energy provides a home for others to dwell in.

The relationship between the second and third chakras is often problematic in our culture. Laura was a mess when she traveled several hundred miles for a session with me. She was trying to keep her local community from knowing how much trouble she was in. She was a chronic alcoholic, and her life was starting to unravel. She had spent twenty years drinking. Her doctor was telling her that she had nearly destroyed her liver, had damaged her heart, and had done injury to her brain as well. She had proven herself to be a good businesswoman despite her addictions, but in the year prior to seeing me, her business was near bankruptcy. Most of her friends had given up on her. Her daughter never wanted to see her again.

Her mother had been drunk during much of the pregnancy. Laura was an alcohol baby, and her addiction was deeply rooted in her biochemistry. Her parents had a volatile relationship, saturated with alcohol, and Laura's childhood had little joy. As an adult, her self-esteem was very low as she continually beat up on herself for her many errors of judgment and her incessant drinking.

She hoped I could help bring health back to her organs, particularly her liver, and then she might have a chance to reclaim her life. She had been in many facilities and had dried out many times, but she didn't have the internal support she needed to stay sober. She often sabotaged herself as she was so ashamed of what she had made of her life.

While I knew there was a great deal of work to do in her third chakra, which is where her diseased liver sat, I felt her salvation lay with her second chakra. Her second chakra took my breath away. It moved me so deeply that I cried. Its color was one of the purest, deepest indigos I'd ever seen, and I knew this meant she was a natural healer. Like people who have a green thumb and can make plants grow, it was plain to me that Laura had an energy in her that could heal people simply by her touching them. But this fact was utterly cut off from her awareness, partially because her third chakra was screaming so adamantly and loudly, telling her how bad a person she was and how poorly she was doing with her life. All her organs were sluggish, and none of her energies were moving well. Her liver was in serious trouble, her kidneys were being overworked, and her pancreas was swollen, leaving her severely hypoglycemic.

Yet Laura embodied the "wounded healer." How I wanted her to know what I was seeing in her second chakra. I felt she might break through a barrier if she realized that she had extraordinary natural healing abilities. Her field of energy was very wide and filled with a force that others would need only to be near to be healed. If she knew what she had, her longing to get beyond her powerful addictions would receive a great boost.

Even in our first session, she was able to sense the beauty of her second chakra, and the next time we worked together, her experience of it was even stronger. This experience was a source of great comfort, and it helped motivate her to want to affirm life and come into sobriety.

I could see in her second chakra that the childhood she never experienced was still an energetic potential, waiting to be aroused. It was very sad to witness. Beyond the chemistry that had engulfed her while she was still in the womb, she was also drawn to alcohol because it could loosen her inhibitions, giving her access to a sense of this lost childhood. Having a way to enter that lost realm, in itself, made the attraction to drugs and alcohol overpowering. Having her second chakra begin to open naturally is now giving her a similar access. While she still has a long road ahead of her if she is to regain her physical and emotional health, she has begun that journey with a fierce determination that makes me optimistic for her. Laura is actually a great deal of fun to work with. She has a wondrous, creative, free, alive second chakra, and laughter spills up and out of her as the barriers break down between her second chakra and her sober awareness.

Musicians and artists often have an extraordinary second chakra, expansive and rich in colors and textures and complexity, but it may at the same time be extremely delicate. Exquisite creativity is exquisitely sensitive to judgment and manipulation, particularly in childhood. Many people with talent are plagued by their parents' need to push them to greater heights, which often is then echoed in the way their own ego, their third chakra energy, marshals their abilities for the ego's purposes. Rather than being able to savor the glory of their creativity, the experience of these individuals is fouled by an unrelenting internalized parental voice badgering them, telling them, "You're not good enough!"

One of the most important things to understand about the second chakra is that you don't make sense of things with it. You move up to the third chakra when you want to make sense of things. The second chakra is governed by faith and trust in the larger picture. It is artistic in its logic rather than linear, childlike rather than sophisticated, organic rather than doctrinaire, flowing rather than time-bound, innocent rather than cunning, trusting rather than suspicious, and free rather than responsible.

The first three chakras play themselves out very differently in extreme behaviors, such as the act of murder. People do not kill with second chakra energy, regardless of how much pain or anger or deep frustration may be coded in its outer layers. A person may get hysterical or throw tantrums, but impulsive murder is moved by first chakra energy, and planned murder, such as warfare and revenge, is catalyzed by third chakra energy.

The color of the second chakra is usually considered to be orange, but I have seen many other colors here as well, including all the other autumn colors. When you penetrate into the contained place in the center of the chakra, you may see a field of color. I have seen purples and blues and sapphires and indigo. I have also seen the person's *life color*[3]—the part of the auric field that does not change from birth to death—in the second chakra. The life color carries a thread of continuity throughout the person's life. Sometimes the life color I see on the second chakra turns out not to be the person's life color, but rather the life color of the child the person will later conceive. I have seen this occur in both men and women.

The energies of the second chakra reflect a person's natural healing abilities. I can see an expanded energy in people with a strong capacity for healing, a protective womblike vibration extending outward that people can almost sit in, as if sitting in a large lap. A blue or indigo color will be strong in the second chakra of such individuals. Indigo is traditionally the color of the sixth chakra, which has to do with psychic ability. Having indigo in the second chakra is like being psychic at the body level. It is more than a gut knowing. It is as if the energy of the umbilical cord is still active and somehow receives information from the cosmos. The mind or ego isn't particularly involved, and carrying this openness at the body level produces a healing force.

Third Chakra. The third or solar plexus chakra is the force that maintains your individual identity. Your personal ego—your sense of who you are and who you are not—is forged within these energies: "This is who I am. This is who I want to be. This is how I want to be seen."

The third chakra holds an energy of discrimination and assertion, which is why it is also called the power chakra. As usual, though, the truth is more complex than a simple name can convey. For instance, the third chakra, encompassing the area between the belly button and the rib cage, governs more organs than any other chakra. On its right side are the liver and gall bladder. On the left side are the spleen, stomach, and pancreas. In the upper center of the chakra is the diaphragm. The kidneys and adrenals are toward the back. Every one of these organs carries a rhythm and acts as a force field within the larger field of the chakra. Each contributes to the complex task of carving a distinct individual out of the infinite possibilities of your inheritance.

Each organ's function in your body parallels its role in your emotional life. Consider the organs within the third chakra. As the filtering system that detects toxins in your bloodstream, the kidneys are your prototype for fear and caution, for detecting and eliminating that which is dangerous. As a factory that breaks down whatever is harmful for your system, the liver is your prototype for self-protective anger. As the alarm system that triggers great rushes of energy for emergencies, the adrenals are the prototype for the panic response that mobilizes you in a crisis. As the body's producer of metabolic juices, the pancreas is the prototype for assimilating what you can embrace. As the organ that sends stale air out of your body, the diaphragm is the prototype for grieving and finding closure with whatever is passing out of your life.

Beyond the diverse influences of the organs that sit within your third chakra, your identity is also shaped by your family and culture. Your essential nature—your genetic and soul level inheritances—provides only a part of your identity. The world around you profoundly impacts your third chakra. It is here that parental messages are imprinted. It is here that social expectations are encoded. It is here that the tensions among "who I am," "who others wants me to be," and "who I should be" are continually slugging it out. These conflicts begin with the earliest stages of learning, and they hit you directly in the center of your body. The pure and innocent energy of the second chakra moves up into the crucible of the third chakra, where it meets multiple contradictory forces, where it is pushed and pulled, stretched and constricted, deconstructed and reconstructed.

The third chakra is the only chakra that usually does have the color that is traditionally assigned to it. Its color is yellow, and I have seen every shade of yellow imaginable in the third chakra. Its shade tells me a great deal. A mellow, wheatlike yellow, for instance, generally indicates that the person isn't particularly distorted in the third chakra. When someone has a bright, sharp yellow, however, it is a blistering energy to move my hand through, like lightning or a nuclear power plant.

The qualities that typify the third chakra are almost the reverse of those of the second chakra. Third chakra energy is logical rather than artistic, sophisticated rather than childlike, doctrinaire rather than organic, cunning rather than innocent, suspicious rather than trusting, and responsibility-bound rather than flowing. If the second chakra holds the natural child, the third chakra holds the controlling parent. Where the second chakra is more like the right brain side of identity, the third chakra "thinks" more like the brain's left hemisphere. Third chakra thought also differs from sixth chakra thought. Sixth chakra thought, though it may be profound, is more detached and cerebral. Third chakra thought is entwined with your identity, your fears, and the needs of your ego.

The energy of the second chakra can be cut off from that of the third chakra, as well as the others above it. There is an energy system between the second and third chakras called the belt flow. It is one of the strange flows that will be described in Chapter 8. The belt flow, which goes around the waist, can become like a huge wall above the second chakra that cuts off the second chakra's natural spontaneity, trust, and faith. When this happens, as it frequently does in modern Western cultures, the person may be tormented by merciless self-hatred, a desecration that often originates in the third chakra's attempts to shape an identity that differs vastly from the natural child of the second chakra. Once the connection between the second and third chakra is opened; energy rises; knowledge about a truer, kinder, gentler self can filter up; and self-condemnation and self-hatred begin to dissolve in that knowldge.

I am not intending to negate the value of the third chakra's function, only to challenge our overemphasis on it. In terms of humanity's evolution, Western cultures have taken the lead in advancing the third chakra to new levels of development, allowing an unprecedented order of individuality. Never before in history has individual consciousness and expression been so independent from the mind of the group, the traditions of the past, or the ways of nature. It is a great and sacred experiment on which rests humanity's future, along with the fate of the planet. If we succeed, Earth could be a paradise that few can even imagine.

But as the culture has made it a collective obsession to glorify the third chakra, the ego, we have lost access to resources the other chakras provide. Our flow with nature and with our primal instincts, our appreciation of our creative innocence, our love connection with nature and with others, the expression of this love, our ability to journey into expanded states of awareness, and our knowledge of our unity with the cosmos have all been relegated to the cultural back burners. As a society, we may have bet on the wrong chakra. But we have also taken our individuality into dimensions unimaginable at any time in the past, and that may be exactly what is called for within a larger plan. Culturally, this situation is not likely to shift overnight. But personally, you have the capacity to establish a healthy balance among your chakras, and you can initiate that process with the exercises offered later in this chapter.

Fourth Chakra. The fourth or heart chakra lives up to its reputation. People whose heart chakra is well developed "see," "hear," and "feel" other people through the filter of the heart chakra, making judgments according to its loving nature. Their energies move out toward others and seek a heart connection. They evaluate actions according to the probable increase or decrease of love in the world that the action is likely to cause. A person whose heart chakra is dominant is guided more by "heart" than "head," feeling than thought, love than logic.

In addition to the heart, the pericardium, thymus, and lungs sit within the heart chakra. The heart chakra is the chalice of both joy and sorrow, and this chalice needs protection, which the pericardium provides. The pericardium is a sac that surrounds the heart and protects it. The heart is, in fact, the only organ in the body that cannot protect itself. Every other organ has its own defense system, but the heart, as the unconditional lover of the body, happily pumping its life-giving fluid to wherever it is needed, is truly made for love and not war. The pericardium is its bodyguard, like a knight guarding the queen. It is also a good personal secretary, buffering the heart from the fluctuating demands of the other organs. The triple warmer meridian, for instance, can draw energy from any of the organs when the immune response has been triggered, but the pericardium does not allow its conscriptions to reach the heart. Your heart needs this total protection because if it is interrupted for even a few minutes, you die.

One of the problems in modern cultures is that the heart chakra is underdeveloped in too many individuals and its principles are underrepresented in too many of our institutions. Having a large heart chakra that is out of balance with your other chakras is, however, also detrimental. Many people do "love too much." They may overidentify with other people's pain, suffering so much for the other that they become emotionally disabled, codependent, and ultimately no more capable of a successful relationship than a person whose heart chakra is shriveled. By loving so large that reason has little influence, by so not wanting the other to suffer, their choices do not give the other person room to develop, and both become disabled. Their growth lies not with silencing the heart chakra, but in cultivating greater balance with the other chakras, allowing each to have its power and its voice.

The heart chakra is the middle chakra, and green, the middle color of the spectrum, is traditionally thought of as its color. Green depicts a lush and balanced energy. Emerald green in the heart chakra shows that the heart's loving energy goes out not only to others, it is also able to nurture the self. However, again, any number of colors may be present. Gold and soft pink indicate an effervescent quality in the person's love. Gold reflects a universal love that people flock to like the warmth of the sun, although such "universal lovers" do not always find personal love to be as easy. Soft pink is evident in those who are soft, kind, loving, and have a special ability to see goodness in others. Occasionally I will see a deep maroon, with a velvetlike texture, directly over the heart. It is a joyful sight, as it reveals a person who has learned about love over many lifetimes and, through the struggles of the heart, has developed the ability for both universal love and intimate love.

The heart itself is an underrated organ in modern cultures. While we are impressed with its resilience as a pump, most indigenous cultures believe that thought originates in the heart, not the brain. Recently, to our collective surprise, we are learning that the heart thinks as well. Researchers at the University of Melbourne have shown that the heart has a nervous system and that it makes decisions independent of the brain, such as how fast to beat. Moreover, the electrical wave produced by the beating heart has about fifty times the amplitude and about a thousand times the strength of a brain wave. As a result of this vastly greater magnitude, the heart tends to pull the brain and other organs into synchronization or "entrainment" with its rhythm in a similar manner to the way, in a room full of pendulum clocks, all the clocks will synchronize their ticking with the clock that has the largest pendulum.[4] We may be run by our hearts far more than we realize. The heart, as Pascal said, "has its reasons which reason knows nothing of."

Fifth Chakra. If love is the product of the heart chakra, expression is the product of the throat chakra. The throat chakra has been referred to as the Holy Grail of the chakras because it holds information from all the chakras. Energies from the sixth and seventh chakras move down through the throat on their way to the trunk of the body; energies from the root, womb, solar plexus, and heart chakras move up through the throat on their way to the head. Within the sacred container of the throat chakra, all of this energy and information is "metabolized"—broken down and put back together into a form that becomes your unique expression in the world.

The thyroid gland, whose secretions govern metabolism, sits within the throat chakra. Metabolism involves two processes, catabolism and anabolism. Catabolism is the breakdown of certain substances in order to release energy. Anabolism builds new tissue from less complex substances. Just as the thyroid gland breaks down food and synthesizes it to build and maintain the physical body, the throat chakra breaks down the energies that travel through it and synthesizes them to build and maintain the energy body. It is also through the throat chakra that we speak, the most characteristically human form of expression.

The most common colors in the throat chakra are aqua or turquoise. Like the other chakras, the throat chakra is comprised of spiraling energies, but unlike the others, when I look deeply into this chakra's spiraling energy, I see chambers of energy as well. These chambers extend up and down the throat like bridges that connect the brain to the torso. Seven columns carry the energy of the other chakras through the throat, facilitating communication in both directions. The energy may flow up or down any one of them, and it may simultaneously be moving upward in some columns and downward in others. The three chambers on the left side of the throat are catabolic. Their function is to break down energy. The three chambers on the right are anabolic. They build and synthesize energy. The seventh chamber, which is in front, maintains a balance between these catabolic and anabolic processes, which sets your metabolism. The energies that cross the right and left brain hemispheres also cross at the throat chakra.

The two basic behavioral difficulties people have that are connected with the throat chakra are that they can't speak up or they can't shut up. Speaking up is an anabolic process—putting things together and expressing them. Keeping quiet is a catabolic process—receiving and assimilating. If there is too much energy in the catabolic chambers, the person has difficulty speaking up. If there is too much energy in the anabolic chambers, the person has difficulty knowing when to remain quiet.

A feisty, articulate, outspoken young woman came for a session. She had strong opinions, was willing to express them, and was admired for her assertiveness. Her sense of timing and appropriateness, however, left much to be desired. Her body's timing was also speeded up, as if her engine were too revved.

She had been fighting chronic sore throats for years. She had been to both traditional and nontraditional healers, and she was well aware of the paradox that this vulnerability in her body happened to be in the area of her greatest strength. But the apparent paradox had a straightforward explanation. When I looked at her throat chakra, I could see that her anabolic chambers (synthesizing, building up, giving out) were glowing with an exceptional brilliance, but her catabolic chambers (breaking down, receiving) looked weak and lifeless.

She experienced herself as having "the truth," but she couldn't *receive* others' truths. Her anabolic nature was so blatant and intimidating to others that her life was devoid of intimacy. And when she would hurt others with cruel words, she would then judge them for not wanting the "truth."

While I could help her with her sore throat, I felt the symptoms would return and told her so, but she could not hear me. The energies in her throat chakra were speeded and not receptive enough to my interventions. I personally do not have a particularly strong throat chakra, and some people don't hear me. She was no more receptive to my words than her throat chakra was to my interventions. Her sore throat did return. I sensed that the key to helping her was to involve her heart chakra. The heart chakra can feed the throat chakra and infuse it with better timing, appropriateness, and finesse. In the second session, spinning and scooping energies up from a deep place in her heart chakra, I tugged and pulled them toward her throat chakra. In essence, I was hoping that the power of caring might give balance to her compulsion to speak the "truth."

I also had her place both hands on her throat so she could become aware of how there was so much more energy on one side than the other. In the center of the throat chakra is the hyoid bone. It works like a gyroscope, setting the balance between the right and left cerebral hemispheres and between fast and slow metabolism. Her hyoid bone seemed stuck, like a throttle that is always open. I taught her how to bring better balance into her throat chakra by stretching and pulling on her neck and hyoid bone. Finally, her tempo and timing slowed and she began to respond to what I would say to her.

As her timing became more balanced, she was also able to give others time to speak and to give herself time to weigh her words and her tone. People began to like her better. She also received another benefit, an unexpected loss of weight that resulted from a better metabolic balance. The "proof in the pudding," however, was freedom from her chronic sore throat.

Another woman exemplified the opposite and more common problem with the throat chakra, the inability to speak up. Perhaps because of genetics, a reluctance to hurt another, an aversion to making waves, or a fear of repercussions, some people's throat chakra can become weak on the anabolic side, and they may have a great deal of difficulty speaking their truth. This woman was lovely, diplomatic, and kind, and she had a beautiful singing voice. Sympathy for others flourished in her. She also was truthful, but she did not want to step on anyone else's truth, and her timing often lingered beyond the point where she could say what she wanted or needed to say.

She'd come to me, however, because her parathyroid glands were not breaking down calcium. This is a dangerous condition, and an operation was being considered. The deepest layers of her throat chakra told a story of numerous lifetimes spent supporting others to go forth and claim their needs. She had logged many hours as an extraordinarily supportive soul.

Calcium makes muscles contract and gives them fiber. Her lifestyle embodied openness and expansion but lacked "fiber." By not allowing her body to utilize calcium, her parathyroid was keeping her muscles flaccid, mimicking her stance in life. Calcium needs to interact with potassium, sodium, and magnesium to have a balanced give and take across the cell membranes. She had lots of give in her, but little take.

Not surprisingly, all of her major relationships had been one-sided. She knew everything about the lives of her friends and intimates and would sit with them endlessly in their pain, but she didn't let anyone truly see her. Not wanting to burden them, she kept others at a distance. This withholding of herself was of course really a kind of selfishness, but she had never recognized it as that.

We worked on various energy systems, but the source of the problem was her throat chakra. It took time for her parathyroid glands to begin to process calcium and her words to express her needs. The breakthrough was not only in her chemistry but also in her mythology. Her ability to tell her truth, state her "no," and ask for what she needed came into balance. As so often happens after a deep inner shift, she began to attract a different sort of person into her life—people who would give as well as take.

Sixth Chakra. Within your sixth chakra are your eyes, ears, pituitary gland, hypothalamus, and brain (except the top of your brain, which sits in the seventh chakra). While the second and third chakras are energy stations where a *self* is created (second chakra) and an *identity* in the world is forged (third chakra), the sixth and seventh chakras are energy stations where the sense of a separate self and identity is transcended. The sixth chakra accomplishes this in two ways. The capacity of the human cortex to perform abstract mental operations makes it possible for us literally to transcend the life of the body. You can, and frequently do, travel into the past, the future, the imagined, the possible, and a world of symbols, theories, and meaning. We can also, via the energies of the sixth chakra, transcend our usual identity by gaining access to a psychic plane that crosses the dimensions of space and time. This makes possible such space-time benders as telepathic communication and precognition, but, more fundamentally, it tunes our radar to energies more subtle than our ordinary senses can perceive.

The flaw that many of us have at the level of the sixth chakra is that we are so filled with our abstractions, mental constructs, and fantasies that more refined processes get crowded out. Our sixth chakras can be dominated by thought that is not balanced by more subtle ways of knowing the world. Seeing the color of energy, hearing guidance from another plane, and being in telepathic rapport with others are all natural ways the sixth chakra can sense the world. Individuals who are mentally brilliant and very strong on the intellectual side of the sixth chakra often neglect its psychic side. Their minds are such instruments of intellectual magic that they don't recognize or utilize a less familiar though more dazzling magic.

On the other hand, I know a man whose sixth chakra's more subtle abilities are amazingly strong. He supports himself by renting it out—using his psychic powers to serve and impress others—but his second, third, and fourth chakras are not well developed. Still, he has gathered a considerable following, and because he could trade on the gifts of his sixth chakra, he simply did not have to grow in his ability to love or respect others. He remains in control, has no sense of allowing a balance of power with others, and lives in a world of hierarchies where he is always on top. Because his throat chakra and his root chakra are also strong, he has no difficulty in speaking his truth or wielding his power, and he is ruthless. While we live in a world that makes gurus out of people with strong psychic powers, we do them a disservice if we do not insist that they open their hearts and tame their egos as well. And we do a disservice to ourselves if we make them authorities over our own deeper sources of knowledge.

People in Western cultures typically have a different energy constellation in the sixth and seventh chakras from those in other parts of the world. In Westerners, the energy of the sixth chakra is often crowded and dense with thought. The sixth chakra can block the seventh, polluting it with an overflow of hazy mental energy. In less intellect-bound cultures, where more subtle ways of knowing are not overshadowed by such mental turbulence, I usually have not encountered this problem.

The sixth chakra as a mental force is at its worst when it becomes the tool of the jealousies, fears, greed, or power plays that may be unfolding in the third chakra. The person then believes with certainty in the ego's distorted perceptions and understandings, and acts on them with confidence. But the wrong chakra is calling the shots. The ego in many ways grows and learns through trial and error, and the more clear and differentiated the sixth chakra is from the third, the more it can serve as a beacon for the ego's development rather than as a defender of its follies.

The sixth chakra is sometimes called the third eye. It is the point situated between the eyebrows just above the bridge of the nose that numerous cultures associate with psychic development. Many people have a strong third eye that is lying dormant. It is as if the eye is there but the eyelid is closed. When my younger daughter, Dondi, was in junior high school, there was a period when her third eye abruptly "opened," and in the aftermath she could not sleep at night. She would lay in bed "thinking," but when I watched her energy, it didn't look like she was thinking at all. It looked like she was laser-beam focused through her third eye, jumping from one spellbinding scene to another. It looked like a tube of energy that was taking her into scenes both familiar and unfamiliar. While lying in bed, she might find herself in the home of a schoolteacher or watching an event unfold far away that would show up on the news the next morning.

A strong and vivid imagination, coupled with her third eye opening, was very intrusive whenever she would become still. I was far more interested in her getting some sleep than in fostering her psychic development. I told her, "Dondi, you've got to close your third eye or you're never going to get a good night's sleep. Put your finger on its invisible eyelid here [middle of the forehead] and softly pull it down to shut it." When she did, the round tunnel of energy would disappear, and she would sleep.

It is not surprising that the psychic side of the sixth chakra is called the third eye. Not only does it foster a quality of "seeing" through veils of mystery, it in many ways *functions* like an eye. Mine feels like a circle or funnel through which I can see images that are as clear as dreams, and far less cryptic. I may see the scene as someone I care about is getting into trouble. Or a few minutes before someone is about to call on the phone, I may see the person's image. Many people have such premonitions but, amid the torrent of mental stimuli in the sixth chakra, haven't learned to distinguish them. With practice, however, the third eye can approach "20/20 vision."

My own sixth chakra has probably saved my life several times. Around 1978, while I was still in San Diego, I had taken my daughters to the Science of Mind Church to hear Terry Cole Whittaker. The services were being held in an old theater in downtown San Diego. I was parked about a block away. As we headed back toward our car, a man was walking quite close to us. I had no reason to think he hadn't himself just been at the service. Suddenly I heard a crystal clear inner voice warn me of danger, telling me to quickly go to the car, get in, and lock the doors. I felt an urging to hurry the girls, who were ambling. I do not normally feel danger, yet my heart suddenly began to beat very fast, and I knew it had something to do with this man. I sensed that this situation was life-threatening, though the man was doing nothing that should have been alarming. He was good-looking and seemed pleasant enough.

My car was on a side street but still visible to the theater. When I put my key in the car door, the man started to come toward me. The inner voice told me to yell to somebody as if I were responding to a friend. I shouted a greeting in the direction of the theater, though I didn't know the person to whom I was shouting. This caused the man to move away just a bit, as if someone were watching. I got my daughters and myself into the car in an instant and locked the doors. He came toward us again, telling me to open the door. I squealed away. I literally had the feeling we were escaping death, and I was badly shaken. We passed a policeman, and I flagged him down. I described the man to him and my concern that someone might be in danger. The policeman took my phone number and said he would investigate.

On the news that afternoon was a bulletin that a little girl was found in Balboa Park, a few miles from the theater, hysterical. She took the police to where her mother had been brutally murdered. The news also showed a man who had just been arrested. Shortly after I saw the news bulletin, the policeman called and wanted me to come to the station to identify a suspect. I said, if it is the man who was just shown on television, that's him. He asked if I was sure. I was. The man had also been identified by the little girl.

A very simple exercise to begin opening your third eye is to place your middle finger at the bridge of your nose and push up a couple of inches, breathing deeply and imagining you are opening the eyelid. Light moves up, from your root chakra, through the second, third, fourth, and fifth chakras, and if you are open, it bends at the sixth chakra and shoots straight out of your third eye. Some people experience it literally as a light or a laser beam. Others experience it more as a feeling. Even if you simply pretend to sense this light, imagine it moving up your spine from the root chakra to your sixth chakra and feel the light bend and shoot out like a laser from your third eye. Try another experiment. With your breath calm and deep, direct a question from your sixth chakra down to your root chakra. Let the question settle into the energy of your root chakra. Then call up the light so it moves back up to your sixth chakra, and be open for the answer to reveal itself as you pull up on your eyelid. I've watched dozens of people be amazed when they have suddenly *seen, known,* or vividly *felt* an answer appear in front of them. While extended courses are offered in psychic development, experiment with this simple exercise from time to time and remain alert for what you see.

Seventh Chakra. Some people need only meditate or direct their attention toward the heavens and they find themselves feeling their oneness with the cosmos, their connection with the realm of the spirit. Their crown chakra is developed, open, and in harmony with their other chakras. People in Western cultures, whose crown chakras are wide open, however, are more likely to receive psychiatric medication than the arduous training that would help them to become more balanced and grounded.[5] Grounding comes from the bottom three chakras. People with only a strong seventh chakra may have fascinating encounters with other realms, but they also seem out of this world, flighty, or disconnected from consensual reality. They are "Om and bliss" without meat and potatoes.

When I am having a hypoglycemic attack or am hormonally imbalanced, I am vulnerable to such psychic experiences. I was teaching a class in Orange County, California, on April 14, 1986. I was premenstrual and very sensitive. Suddenly, sometime after 4 p.m., I couldn't see the room. I felt an explosion in the top of my head, and I had to hold onto the wall to keep from falling. I suddenly seemed to be in a different place. I could see Muammar Gaddafi. I recognized his face from television and newspaper clips. He was cradling a child, and he was weeping. As calmly as I was able, I tried to tell the class what was happening, that I was seeing Gaddafi in a rubble. Something had happened to the spot where he was standing. Then, swoosh, it was as if I was seeing through his eyes, as if I were a part of him. I felt his overwhelming sadness. I did not feel the evil that I'd been led to believe characterizes this man. After a few minutes I was back to my normal self, still a bit woozy, and having no idea about the meaning of what had just occurred. I finished teaching the class and went back to my cottage on the property. Sometime later, people from the class rushed over to tell me to turn on the television. The United States had bombed Gaddafi's family compound in Libya, killing one of his children and injuring two more, at exactly the time the class had watched me go through and describe my experience of being in Gaddafi's body. Why I picked up this episode while I was teaching a class—I don't know. But somehow the oneness of it all swallowed up my isolated self, which is exactly how the seventh chakra works.

Another time, after returning home from a workshop tour in England, I couldn't regain my equilibrium. That happens to me sometimes when I've traveled extensively. Long plane trips can actually blast open my seventh chakra. This is partly from being so high and long off the ground and partly from being whisked through time zones. After returning home, I was finding it difficult to be inside a building, energetically claustrophobic. I drove to the redwoods near the Oregon coast and camped. All night long I kept seeing a woman from my class in England. This woman had been irritated with me because she had wanted an individual session, but my schedule had not permitted it. Now I kept seeing her coming to me through the ethers, calling me, calling desperately, begging, pleading. What was this?! Suddenly it felt like I was leaving my body, being pulled by her out through my crown chakra. I felt I had to mobilize all of my will to stay in, and I literally held onto a tree to ground myself to planet Earth. When I got home, I found out she had been killed that night in an automobile accident.

Unlike the sixth-chakra experience, where I *heard voices* that probably saved my life, seventh-chakra openings can totally engulf you. Had either of these two incidents been only of the sixth chakra, I might have still seen Gaddafi or heard the woman. But because the opening was of the seventh chakra, I merged with the situation, my soul was drawn into it. Crown chakra experiences can pull you right out of the top of your head. Usually, you can find your way back, but it can be terrifying and even dangerous, and I wouldn't wish a sudden or forceful crown chakra opening on anyone. But usually I love it. I may feel a mystical oneness with everyone and everything in nature. You don't get intellectual answers, but you get an absolute peace and joy about life, and you know there is reason and rhyme in all things under heaven.

Meditation, prayer, and ritual are safer but still potent ways of opening your crown chakra and deepening your awareness of your spiritual connection with the cosmos. So is energy work. Methods for clearing, balancing, and strengthening your chakra system follow.

How to Calm Your Kids (or Yourself)

Sometimes you need to calm your kids so they can focus on a task, go to sleep, or be well-mannered for a special occasion. All children are different, and you may have to experiment with which exercises will work for your children, but here are some of my standards (time—about 16 years):

1. *Children love Freeing the Diaphragm (page 266) because it challenges them. They like to see how long they can hold their breath while pushing into their diaphragm. At the same time, they are slowing their metabolic rate and calming themselves.*

2. *Children often spontaneously do the "scissors." You can see them scissoring while watching TV, and they can do it while facing and talking with you as you do the same exercise. Begin lying on your stomach with your chin cupped in your hands, knees apart, and legs bent upward. Drop your feet outward, and then bring them toward one another, crossing, making a scissors, going back and forth, one foot passing in front and then the other. This crosses the energies from one side of the body to the other while aligning the hips, spine, nervous system, and brain. It is good for the circulation in the legs and for hyperactive legs.*

3. *Taking Down the Flame (page 220) could have easily been named "Taking Down the Hyperactivity." If you do it along with the child to keep the child engaged, it will calm you as well.*

4. *Hold the child's frontal neurovascular points (page 89).*

5. *If you have more than one child, you can get the whole family involved and everyone will feel empowered. Children love to massage K-27 together (page 64), Cross Crawl together (page 69), hold the Wayne Cook posture together (page 73), or do Separating Heaven from Earth (page 249) together.*

6. *Find all the ways your child can make sideways figure 8s (page 184): with the hands, the feet, the head, the hips, hands clasped in front and pretending to be an elephant's trunk, or drawing large figure 8s on a blackboard or butcher paper with chalk or colored pens.*

WORKING WITH YOUR CHAKRAS

I often teach about chakras early in my classes because chakra energy is usually the easiest to feel. Most people can sense that their hand is moving through something palpable even the very first time they work with the chakras, and some can see a vortex of spinning energy. Students sometimes see colors for the first time during a clearing of their own chakras. They usually notice these colors in the back of their eyes, whether their eyes are open or closed. It generally matches what I am seeing. We both are seeing the same color, me from the outside, them from the inside.

If you are feeling sluggish in your body, or in your life, you can be certain that the energies in your chakras are sluggish as well, and it is very valuable to know that you can clear them. Clearing the chakras removes toxic energies, both recent and longstanding, and your body becomes better able to adapt to whatever circumstances you must face. A major difference between meridian work and chakra work is that while the meridians fluctuate with whatever is going on daily in your life, only the top layers of the chakra are "blowin' in the wind" in this way. As you begin to work with a chakra's deeper layers, you can make changes and bring about healing at increasingly profound levels.

Chakras govern the endocrine system, so bringing your chakras into balance brings your hormones, and thus your emotions, into balance as well. For many people, working with the chakras leads to an altered state of consciousness or a sense of being in contact with higher powers. You can never predict where a chakra session will take you, because chakra work involves so much clearing of old issues and memories. The more you work with a chakra's energies, however, the deeper you go, and the more complete the alignment with the more profound layers at its core. If you clear and harmonize your chakras regularly, you will begin to notice a cumulative effect.

Generally each chakra is balanced separately. You can work with all seven, or you can focus on those that are in particular need of attention. You can energy test each chakra to get a baseline on its functioning, but again my hope is also to persuade you to check your energies intuitively. If you close your eyes and focus your attention on each chakra, one at a time, you will learn to sense whether it is out of balance or if the organs in the area of the chakra feel sluggish, anxious, or painful. In time you may begin to see the chakra's colors, and you may even sense what these colors mean for you. Because chakra energies are so tangible, as energy goes, bringing your attention to each of your own or another's chakras and noticing what you sense internally are excellent ways to cultivate your abilities to see and feel energy.

Chakra Energy Test. You can clear and harmonize your chakras without doing an energy test, but the energy test is easy to do. It does requires a partner.

1. Lie down. Put the backs of your wrists together, with your arms extending toward the ceiling, elbows straight (see Figure 32a).
2. Your partner lightly taps the chakra being tested (see Figure 30) two times with the middle finger.
3. After tapping, your partner slides both hands between your raised arms, beneath your wrists, hands open and facing outward (see Figure 32b).
4. To test, your partner pushes outward, attempting to separate your hands as you resist.

All seven chakras can be checked in this way. Those where the arms have little resistance to being separated can be cleared, balanced, and strengthened.

Clearing, Balancing, and Strengthening Your Chakras. Chakra work leaves people feeling more centered and grounded, revs up their energies, and aligns them with a deep part of themselves. And it is easy to do. Often a child with a stomachache will instinctively move a hand in counterclockwise circles over the stomach. Moving in this direction pulls out toxic energies, and that principle is the basis of this deceptively simple procedure. Your hands are the only tools you need. Your palms, in fact, are themselves swirling vortexes of chakra-like energy.

You can clear, balance, and strengthen your own chakras, but most people prefer to have a partner do the work while they relax. The partner also serves as a groundwire, and the exchange of energies increases the impact of the technique. Still, I know many people who start the day by clearing their own chakras and report that they love the way it gets their energies moving. Some spin each chakra, exactly as described here. Others sense into their energies

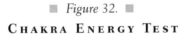

Figure 32.

Chakra Energy Test

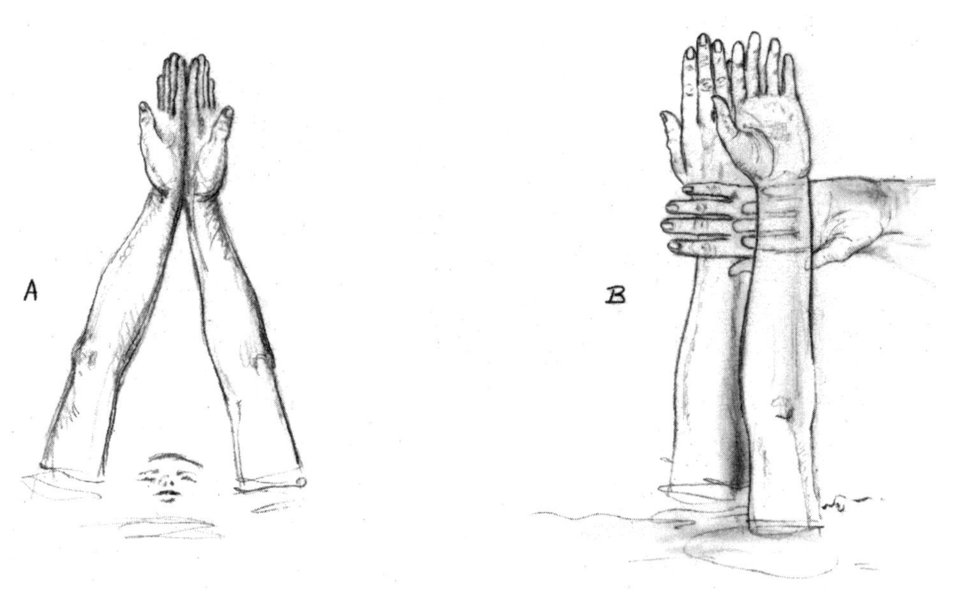

and intuitively choose which chakras need attention. Balancing even one chakra, particularly if it is the weakest link in the chain, strengthens the entire chakra system.

I will describe the technique as it is done with a partner, but the procedure is identical when you are working with your own chakras (time—3 to 9 minutes per chakra):

1. Lie on your back. Your partner clears any excess energy off his or her hands by vigorously shaking them.
2. Begin with the root chakra (see Figure 30). Have your partner place either or both hands, open, palm down, over the chakra and begin to circle it counterclockwise about four inches above the chakra. Imagine that you have placed a clock on the chakra, face up, to determine which way is counterclockwise. Make slow counterclockwise circles that are about the width of the body or a bit less.

3. After completing the counterclockwise motion, your partner shakes off the energy, and then begins to circle over the chakra in a clockwise direction. Continue for at least half as long as the circling in the counterclockwise direction.
4. After completing the clockwise motion, your partner shakes the energy off the hand, and then moves to the next chakra, beginning in a counterclockwise direction, as in step 2. Chakra work generally begins with the root chakra and moves up all seven chakras in the order shown in Figure 30. Do steps 2 and 3 for each chakra.
5. The single exception to the order of circling counterclockwise first and then clockwise is that if you are working with a man, when you get to the crown chakra, the order is reversed. A clockwise motion will clear the crown chakra's energies, and a counterclockwise motion will reharmonize them. A woman's crown chakra, however, spins in the usual direction.

As the chakra's energies follow the hand, all seven layers begin to spin in the same direction, shattering energy blocks and providing the subtle energy version of a Roto-Rooter treatment. The hand can then act like a magnet, pulling stale or toxic energies up and out with its motion. While the amount of time needed for each chakra is highly individual, I would suggest as a guideline that you continue the counterclockwise motion for at least three minutes. If it ever happens that the person feels that the energies are being stirred up too much, simply reverse the direction of the circle, and the person will feel begin to feel fine again.

The counterclockwise circling stirs up toxic energies and moves them out of the chakra. The clockwise motion provides a completion by feeding and reharmonizing the chakra's energies. Most people seem to feel better using their left hand to take out energy (counterclockwise motion) and using their right hand to reharmonize it (clockwise motion). I personally don't experience that difference, and I usually use both hands for each direction.

Should you begin to feel a headache while your chakras are being cleared, it is usually because the toxic energies that are being released have started to move up your spine. If this occurs, it is not a bad thing, and you can help the energy move out by doing the crown pull (Figure 7, page 79). In fact, if you are working with someone who is prone to headaches, you can begin at the crown (seventh) chakra and work your way down.

Holding and Cradling a Chakra. Feeling thankful is good for your health. Having gratitude lifts your energy, and you can single out any or all chakras for channeling in your gratitude and thanksgiving. For instance, hold your hands over your third chakra (Figure 30), bring your elbows to the sides of your body, and hug and cradle the chakra like a baby. Bring your mind's eye into the chakra and breathe deeply. Send thanksgiving and gratitude. Your body responds when you treat it like the holy vessel that it is. Be open to whatever you experience. The solar plexus chakra is so overactive in our culture that it is a good chakra to use in learning this technique. Especially when you are feeling tormented by an inner voice scolding you about what you should do, ought to do, or did wrong, cradle your solar plexus, rocking it like a baby as if rocking it to sleep. I make a "shhh" sound as if to calm the baby. The forces that cause you to judge yourself, or another, helped you survive, and you can sincerely feel gratitude toward your third chakra while encouraging it to evolve. You can cradle and express gratitude toward any chakra in this way, and it will respond.

Connecting Two Chakras. One way to understand personal evolution is that as you evolve, the communication among your chakras becomes more open, clear, and profound. Your heart chakra, for instance, might need more of the protective energies of your solar plexus chakra, or your throat chakra may need more of the creative energies of your second chakra. It is natural for chakra energies to interact, so one chakra benefits from the qualities of another.

If you place your right hand on one chakra and your left hand on another, it takes about three minutes for the energies to begin to connect. People often say this feels like a prayerful experience. It gives them a reverent feeling for their body, evokes their gratitude, and provides a sense that the qualities of the two chakras are feeding one another. Your hands may begin to buzz as the energies, very literally, interact. It is as if your hands are serving as a jumper cable.

Out of the Hands of Babes. I cannot overstate the value of balancing your chakras, the ease with which you can balance them, or how much bang you can get for the buck with this simple technique for clearing and strengthening your chakras. I will, however, close with a story that makes these points. I was working with a distraught woman who was in the middle of a divorce. She had become so dysfunctional that her husband was seriously considering taking her to court to sue for custody of their four-year-old son. The husband didn't want to take on this responsibility, but he felt she was unable to adequately care for their son. I could give her temporary relief in the sessions, but she was stretched so thin, working so many hours now as a single mom, that she wasn't doing any of the energy exercises I was suggesting between the sessions. Whenever she had some time, she just wanted to crawl into bed and cry.

She always brought her son to the sessions. One day I said to him, "How would you like to do something to help your mommy not cry so much anymore?" He smiled a large smile and enthusiastically nodded his head with a big "Yes!" I told him to do just what I was doing, to make large circles over a chakra with his hands. He did, and he said, "I can feel it!" I said, "Whoa, you're really going to help her!" He was very excited. He circled her chakras every day, whenever she was feeling blue. If he saw her looking sad, he would say, "Let me do your chakras, Mommy!" It kept mother, son, and chakras all connected, and it boosted her spirits tremendously to have her little boy cheerfully offering effective support. She credited the daily chakra clearings by her four-year-old son with having helped her begin to cope far more effectively during the divorce. After seeing her changes, the husband, to everyone's relief, particularly his own, dropped the custody suit.

Chapter 6

The Aura, Celtic Weave, and Basic Grid

Protecting, Weaving, Supporting

UNLIKE BODY LANGUAGE . . .
AURAS CAN'T BE FAKED.

—ROSE ROSETREE
Aura Reading Through All Your Senses

Your meridians and chakras bring energy to every organ in your body. The three systems presented in this chapter are more concerned with maintaining your body as an energy system. The aura contains your energies; the Celtic weave knits them together; the basic grid provides their foundation. But no system is more important than another. Whichever might be the weak link in the chain becomes the most important system at that moment.

THE AURA: EMBRACING AND PROTECTING YOU

When you feel happy, charismatic, and spirited, your aura may fill an entire room. When you are sad, despondent, and somber, your aura crashes in on you, forming an energetic shell that isolates you from the world. Sometimes if I feel closed in by too many intruding energies, I will very slowly place my open palms against the inside band of my auric field and push it away. If I go slowly enough, I can feel the pressure of the field as I push on it. Try it sometime when you are having difficulty claiming your space in the world or are feeling sad or small. Imagine yourself surrounded by an eggshell-like energy and exhale slowly as you push it away from you, beginning at about two inches from your body. You may be able to feel an energetic force against your hands, and, in any case, it will give you more breathing room, psychologically and energetically.

The aura is a multilayered sphere of energy that emanates from your body and interacts with the atmosphere of the Earth. It is itself a protective atmosphere that surrounds and embraces you, filtering out many of the energies you encounter and drawing in others that you need. Whereas each chakra is an energy station that is attuned to larger energies in the universe, the aura serves as a two-way antenna that brings energy from the environment into your chakras and sends energy from your chakras outward. Some auras reach out and embrace you. Others keep you out like an electric fence.

Detecting the Aura. The human aura has been scientifically investigated by Valerie Hunt, a neurophysiologist. Over a twenty-year period, she conducted a series of meticulous studies at UCLA's Energy Fields Laboratory that compared aura readings with neurophysiological measures. The colors seen by eight "aura readers" not only corresponded with one another, they correlated *exactly* with electromyograph (EMG) wave patterns picked up by electrodes on the skin at the spot that was being observed.[1]

Aura readers also consistently report shifts in the aura when a person is under stress, has had a healing, or when there are changes in the environment. Psychologist Dorothy Gundling found that aura readers saw changes in the size, shape, action, and color of the "energy fields" of five experimental subjects when the subjects were listening to music. These observations corresponded with changes in blood pressure, pulse rate, respiration rate, and brain wave patterns. Some music produced relaxation in listeners, similar to the physiological response to meditation; other music produced erratic brain waves.[2] While the aura readers tended to agree with one another on the size, shape, and shade of the subjects' auras, there were differences in their perceptions of color. The investigator speculated that this is because the perceptions of aura readers are necessarily filtered through the color of their own aura.[3]

Some studies of the aura do not rely on aura readers at all. The combined use of handheld biofeedback sensors, computing equipment, and special photography devices has demonstrated that the color, shape, and size of an auric field, as captured by these devices on photos or video, shift after the application of acupuncture, hands-on healing, or prayer. The auric field also shifts with changes in thoughts or mood, and there is some evidence that its fluctuations provide reliable diagnostic information as well.[4]

Loosely Knit or Tightly Bound? Traditionally, the aura is said to have seven bands, each corresponding in color and energy with one of the chakras. I don't see it quite that way, and for a long time I didn't know what to make of the discrepancy. While I do experience at least seven bands in the aura, their correspondence with the chakras is not exact, and sometimes it is not even discernible. I have come to realize that while it is generally true that the seven bands of the aura correspond with the seven chakras, they process energies very differently, and they have different energies to process. For instance, the aura is the first energetic filtering system from the environment, but the chakras may then filter out energies the aura has attracted.

Each auric band is attached to the body through the chakra that corresponds with it. For some people, the aura is loosely knit, extending outward, open to the world. For others it is tightly bound, "close to the vest." The density of the aura also fluctuates within a person, reflecting different states of health, excitement, and comfort. When you are ill, your aura can collapse in on you, as if trying to guard your organs, bones, and vital tissue and isolate you from the world. You are more protected, and others are protected from your illness. When people age and begin to decline, the aura begins to fade. Shortly before a natural death, the auric field may have grown so weak that it can no longer hold in the body's energies. If you have ever been with someone who was close to death and had the sense the person was already partially gone, your sensing was energetically accurate.

I was at a conference about nine months after the accident at the nuclear plant in Chernobyl. A woman who had been within one hundred miles of Chernobyl when the accident occurred arranged to have a session with me. She suffered with symptoms that included vertigo, chronic fatigue, and loss of physical strength. Her immune system was also impaired, for she was picking up illness after illness. She looked as if she had been stripped of her aura. She reminded me of an astronaut without a space suit. She had no protection, no buffer zone. I worked with the fields of energy around her, weaving and fluffing the aura. As her auric field began to rebuild, her depleted chakras and meridians could be replenished, and her symptoms began to diminish.

The first band of your aura is situated very close to your body. It is sometimes referred to as the etheric body; it is considered the energetic double of your physical body; and it is critical for survival. My own aura tends to be quite loosely knit. I was taking a class from Sheldon Deal, one of the pioneers in applied kineseology, when he spoke about the thrill skydivers feel as the attachment between the aura and the body literally stretches during a free fall. But then he turned to me and warned, "Donna, you must never try skydiving. You are just too loosely connected. That's why you can connect so easily to other realms. But if you skydive, your body will detach from your auric field and you could die." Apparently, it is well known among skydivers that a small percentage of people die mysteriously during a jump. A strong heart simply stops beating. Your body cannot support life without its aura any more than the Earth could support life without her atmosphere.

How to Develop Abs of Steel

If you cannot get your stomach flattened no matter how much you exercise, the abdominal muscles may not be getting enough energy.

1. *Be sure you are regularly stretching as well as strengthening your abs.*

2. *Massage the small intestine neurolymphatic points on the inside of your legs and under your rib cage (Figure 9c).*

To test if you need these points massaged, stand up and bend over. Note where your fingers hang. Massage either set of points and again see where your fingers hang when you bend over. Your fingers will reach lower after the massage if your abs need to have energy released.

Seeing Auras. I not only believe that everyone can learn to read auras, I believe you already sense them. You know, before a word has been spoken, when you are in the presence of someone who makes you feel energized or drained. When my former mother-in-law walked into a room, every head turned. Everyone felt her presence before even seeing her. We pick up on more than just visual cues. Have you ever had the sense someone is staring at you and turned your head to find eyes gazing your way that had been completely outside your view? Studies of this phenomenon show that not only is it quite commonplace but also your galvanic skin response changes when you are being stared at by someone though they are not within your range of vision.[5] "Good vibes" and "bad vibes" are more than just figures of speech.

While I can "see" auras, I rely mostly on how they feel. Like walking out under a night sky with a full moon overhead, it may be exquisitely beautiful, but it is the feeling, the glorious feeling of a moonlit night's energy, that stirs the soul. Whether or not you are able to see another person's aura, you are already picking up subliminal data. The challenge is to consciously find your way into these perceptions. Again, energy testing is a way of sharpening this ability.

The auric field provides a quick read on a person's general level of health, but preconceived ideas can get in the way. So stay on your intuitive toes. For instance, when a person has been ill, the aura may collapse close to the body, forming a tightly knit protective shield. Early in my work, whenever I encountered someone with a collapsed aura, I would try to pulse the auric field back out to where it was "supposed to be." But if the person was sick, I always felt a resistance, as if the aura were saying, "This danger isn't over yet, and this body requires extra protection and support until it is." The collapsed state is the intelligence of the energy body at work, and I have learned not to override its wisdom.

On the other hand, if the energy imbalances within your body become excessive, the aura can't protect you, and it can become thin and disorganized. It may get in the habit of being chronically collapsed, trapping your energies and preventing your life force from connecting with the world. Like each of your energy systems, the aura learns how to protect you in the increasingly complex situations we encounter today. In the series of compromises called adaptation, it may become chronically locked into a state that is not good for your overall health and well-being, and an intervention can be very useful.

The Hook Up (page 119) is the best single technique for keeping the aura solid. Do it each day. Three disturbances in the aura that can lead to health problems include: (1) slow leaks, lesions, and tears so the body is not protected; (2) an accumulation of more toxic energies than the aura can process; or (3) chronic collapse and stagnation. There is a correction for each of these conditions.

- *Scanning* your aura locates leaks and lesions, and it seals them.
- *Fluffing or massaging* your aura creates movement in your auric field so that toxic energies can be released.
- *Weaving* your aura pulses its energies outward; laces them into a strong, multilayered protective web; and helps maintain them in that pattern.

Your aura is also rejuvenated whenever you clear and strengthen your chakras.

When you scan, fluff, massage, or weave any part of the aura, the vibration reverberates like a spider's web. So it is sufficient to work with only the front of the body. An important exception, however, is that if someone has a particularly longstanding illness, you should do the *weave* on the front and back of the body. I will describe each technique as if you are working with a partner. If you do not have a partner, the moves are identical; you just won't be able to energy test to affirm that they have succeeded.

Scanning Your Aura. Lie on the floor, a bed, or a massage table. To locate leaks, lesions, and tears, the tester "scans" the aura (time—about 30 seconds):

1. As the tester, hold an open hand about three inches above the other person's body.
2. "Scan" the body by moving your hand slowly over it, from head to toe.
3. With your other hand, perform the general indicator test (pages 65–67), exerting an even pressure while slowly continuing to scan the aura.
4. Whenever the arm goes weak, there is a disruption in the aura at the spot being scanned.

5. To awaken your own sensitivities, stay attuned to whether you can feel or otherwise sense where there are breaks in the aura, and see if it matches the energy test.

Any lesions you find may also exist further out than three inches, but it is not necessary to perform a separate test for each band of the aura, as the following correction will permeate through all of the bands (time—up to a minute per lesion):

1. Energetically fill the hole or mend the tear by moving both hands over it in a clockwise circle. Excess energies from other areas will be attracted by the electromagnetic energies of your hands to make the repair.
2. You may find your fingers and hands spontaneously going into a kneading or milking motion as you circle. If so, follow this sensing. You are literally kneading and milking the surrounding energies, redistributing them to heal the lesion.
3. You may also find that your hands spontaneously move out further than three inches from the body, to the aura's outer bands. Again trust that you are organically being led to these areas, and continue to gather and spread the energies to them.
4. If you have no one to test you, moving your hands over your aura with your hands about three inches from your body will also even out your field, filling holes and mending tears. You can cover your entire body, but you also may sense specific areas that need special attention.

Fluffing Your Aura. Like fluffing bedding, to fluff your aura is to bring it fresh air, expand it, and give it space to breathe. If you are feeling flat and lifeless, blue, down in the dumps, or have lost your momentum, fluffing your aura can pump you up. You don't need an energy test to know when you are feeling stagnant and lifeless. Here are two ways to fluff the aura (time—only a few moments for each):

1. Beginning at the bottom of the feet, literally roll the aura's energy forward, with your hands circling one another, as if rolling a beach ball up the body. Breathe deeply. In moving up the center of the

body, this technique, like the Zip Up, has the added benefit of strengthening the central meridian, which in turn feeds all the meridians and chakras.

2. Enter your hand into the aura at any spot that draws you and make a flicking motion that dips down and then scoops the energy upward. Your hands can engage the aura in a dance that is energizing and fun.

Aura Massage. Your energy body can be massaged. This soothes and balances your auric field. It is a great way to close a physical massage or other energy treatment because it seals in that peaceful easy feeling (time—about a minute):

1. Rub your hands together briskly.
2. Place them a few inches above the person's head. Without touching the body, slowly massage the aura's energies downward.
3. With open hands, sweep the energies from the head downward off the toes.
4. Experiment with different movements, slow or brisk, short strokes or long. Cover the entire expanse of the body.

Weaving the Aura. The term "Celtic weave" refers both to an energy system and an exercise. They share a name because the function of the energy system is identical with the goal of the exercise. If the energy system is not weaving your energies properly, the technique will get it going again. The energy test to determine if your aura needs the Celtic weave is also fun to do:

1. The person being tested extends both arms out to the sides.
2. The tester pushes down on both arms simultaneously just above the wrist for no more than two seconds, just enough to see if there is a bounce. If the arms go down, the auric field is not sufficiently supporting the person.
3. If the arms stay strong, test the aura further. It should extend beyond the person's outstretched arms. The tester uses both hands at the same time to slowly trace the energies around the other's head and out to about an inch beyond the other's outstretched fingertips. Then repeat the test.

If the test shows strong, this indicates that the person's aura extends beyond the outstretched arms, as it should. But if the auric field isn't there to protect the person, the arms will test weak, and the following exercise will pump the field out further. If you are working with a partner, adjust the instructions accordingly. To Celtic weave your aura (time—about a minute):

1. Ground yourself by standing for a few moments with your open hands placed on your thighs, fingers spread. Energy will travel down your legs. Tune into it.
2. Rub your hands together. This will generate energy between them. The weave will work whether you feel the energy or not, but you will be exercising your ability to feel energy by seeing if you can feel the energy being generated.
3. Lift your hands beside your ears and about six inches away from them. Hold them there for about ten seconds.
4. As you breathe in deeply, bring your elbows together in front of you. (From here, you will be crossing your arms three times in the following instructions; see Figure 33.)
5. Exhale, crossing your forearms and hands in front of your face, continuing the movement until you are freely swinging your arms down and all the way out to the sides.
6. Inhale, allowing your arms to swing naturally in front of you and to cross one another.
7. Exhale, swinging your arms out again, bending at the waist.
8. Inhale and cross your arms down near your ankles.
9. Exhale and swing your arms out, staying bent over.
10. Swing your arms behind your body, turn your hands toward the front, and with your knees slightly bent, scoop the energy over your head as you stand up. Imagine the energies flowing over your head and down the front, sides, and back of your body.

Your aura protects you from the effects of energy pollutants in the atmosphere such as those caused by high voltage wires and fluorescent lights, as well as the vibrations of people who are stressed, angry, or depressed. Celtic weaving your aura builds this protective surrounding, but the Celtic weave is more than just a technique for strengthening your aura. It organically connects all of your energy systems together.

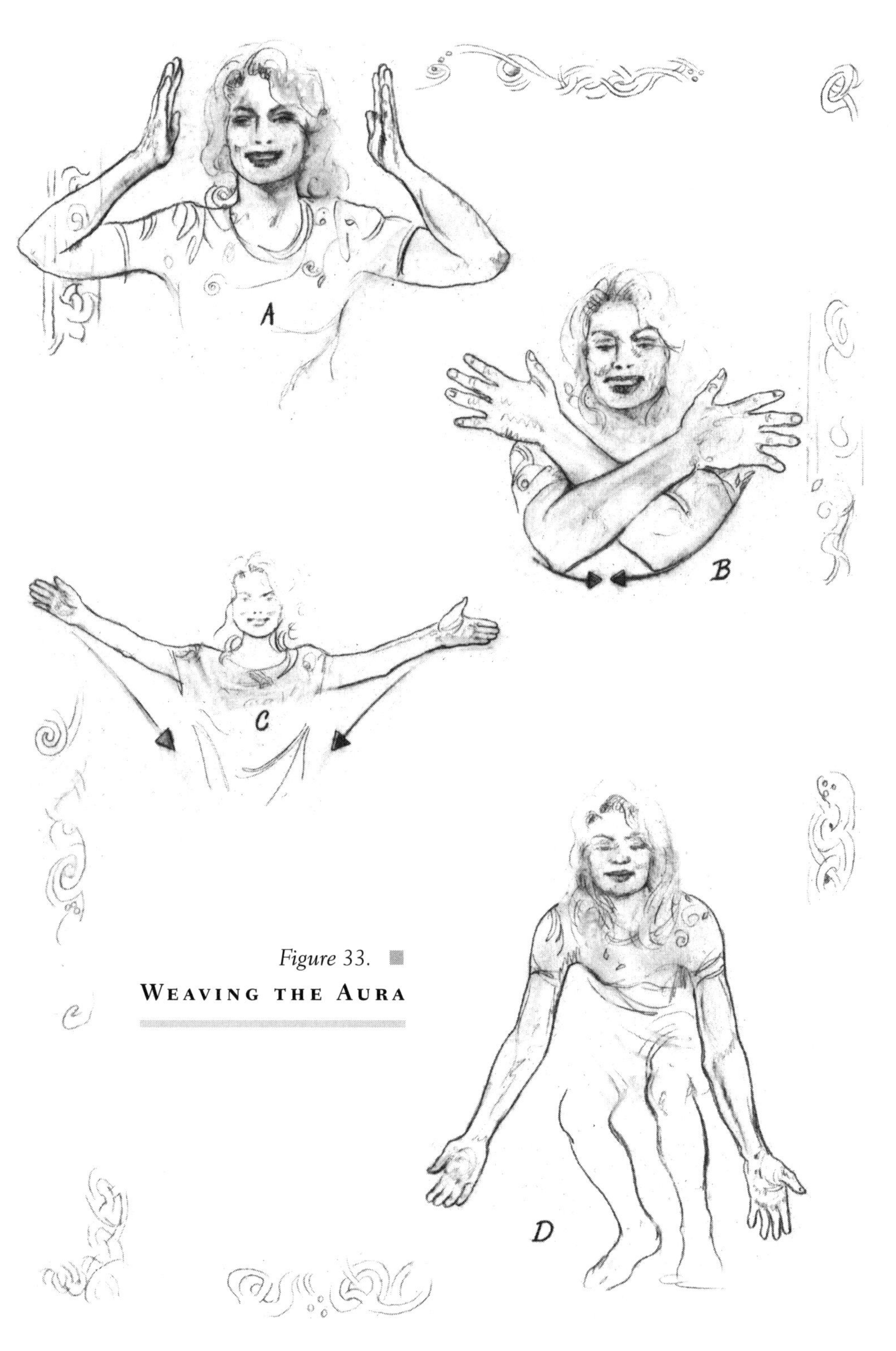

Figure 33.
WEAVING THE AURA

How to Siphon Out Your Child's Pain

Use the pain-siphoning technique (page 283). With your right hand down, out, and away from your body, place your left hand on the area of your child's pain. Explain that you are pulling out the pain, sucking it up and moving it out. Children love this concept, and once they've felt it work, they may want to perform the service the next time you are in pain.

THE CELTIC WEAVE: CONNECTING YOUR ENERGY SYSTEMS

The body's energies spin, spiral, curve, twist, crisscross, and weave themselves into patterns of magnificent beauty. The equilibrium of this kaleidoscope of colors and shapes is maintained by an energy system known by different names to energy healers throughout the world. In the Orient, it is the Tibetan energy ring. In yogic tradition, it is represented by two curved lines that cross seven times, symbolically encasing the seven chakras. In the West, it is seen in the caduceus, the intertwined serpents—also crossing seven times—found on the staff that is the symbol of the medical profession. I use the term *Celtic weave* not only because I have a personal affinity with Celtic healing but also because the pattern *looks* to me like the old Celtic drawings of a spiraling, sideways infinity sign, never beginning and never ending, and sometimes forming a triple spiral. I've had two class participants, both male nurses, one from Brazil and one from Nepal, who immediately recognized this energy system when I described it. Each was expected by the hospital where he worked to regularly "Celtic weave" his patients, one in a burn unit, the other in a general ward, though different names were used for the procedure.

The Celtic weave, *as an energy system,* laces through all your other energy systems and creates a resonance among them. It is the weaver of your force fields. It holds your entire energetic structure together. *As an exercise,* you use the Celtic weave to pulse your aura's energies outward and to strengthen them. The exercise also connects all your energies together so they operate as a single web. Touch one strand, anywhere, and your whole system reverberates in harmony.

People's energies are unique, yet the Celtic weave is universal. It crosses the energies over from each hemisphere of the brain to the opposite side of the body. Its largest formation is a figure 8 or infinity sign that extends from head to toe. In the caduceus, in the yogic figure, and in other systems that focus on the way energy travels up the spine, the Celtic weave is portrayed as a series of figure 8s, looping seven times up the torso. But this pattern can be found throughout the body, over the top of the head, the face, the trunk, the legs, and the feet. These intertwining energies get ever smaller, eventually weaving their primordial pattern at the cellular level. In fact, a single strand of DNA may be the prototype of the Celtic weave as an energy system.

One function of the Celtic weave is to draw all the energy systems together into a tight web of communication so that information can easily travel wherever it needs to go. This enables all the energy systems to work in harmony and cooperation. When the Celtic weave is dynamically engaged, you have a sense of power, a feeling of being charged, and your energies *really* start humming. I've known several energy healers who can *hear* the Celtic weave on their clients, particularly when the person is healthy and the pattern is distinct. They say it sounds like a musical hum or resonance.

Mary Ann was an extremely bright woman who had gone through a series of dreadful crises. Now that she was on the other side of them, she appeared radiant and happy. Her heart was open, and she had clearly grown from her trials. Her body, however, was in significant distress—exhausted, overweight, and in pain, particularly at her joints. It was as if her spirit had managed to race ahead of the vehicle in which it was travelling. While I could have worked with each of the energies that had been so severely taxed by the stresses, doing the Celtic weave was like hitching her body's depleted energy systems to her elevated spirit.

There are many different reasons to use the Celtic weave. It is valuable wherever energies are not crossing over properly, such as in dyslexia and other learning disabilities, in severe immune dysfunction, and in virtually any illness. I often finish a treatment by weaving all the fields together. I may also begin a session with the Celtic weave. Particularly if the person's energies are in such disarray from the start that I can't tell what is happening, Celtic weaving them will allow me to see the underlying patterns. It aligns the energies and clears the static.

Celtic Weaving the Celtic Weave. Review the weaving pattern you used when Celtic weaving your aura (page 180). Each of your energy systems can be woven into a tighter web by using your hand to trace Celtic weave patterns over your body. They can be large or small and over any area. Do as many weaves and crosses as feels right. Your intuition will serve you here. You cannot make a mistake because your body loves the Celtic weave motion. It activates the never-beginning, never-ending energy crossover patterns that are disbursed throughout your body.

Rhythmic 8s. Rhythmic 8s are a variation on the Celtic weave exercise. If you are so caught up in your mind that you can't let go and relax, or if you simply can't think, energetically integrating the right and left hemispheres of your brain will bring balance (time—less than a minute):

1. With your hands hanging, sway your body, shifting your weight from one hip to the other, as if you are moving to music.
2. Let your arms sway with your body (see Figure 34). You will notice that there is a natural figure 8 movement in both your arms and body as you sway from side to side. Let your arms move out further so they are swinging wide.
3. Stretch your hands out in front of you and make a sideways figure 8 with your arms. Go up and over on the right, circle down, then up and over on the left, circle down, and then again up and over to the right. Twist at the waist in the direction you are reaching, allowing your body's rhythm to move with your arms. It is more like a dance than an exercise.

Several Rhythmic 8s will help get the left and right brain hemispheres into better communication. This technique is used in educational kineseology, an application of kineseology that helps children with dyslexia and learning handicaps. A number of studies have shown that making Rhythmic 8s on the blackboard with chalk, while putting their whole body into the motion, helps children with learning disorders.[6]

Figure 34.
Rhythmic 8s

Celtic Weaving with a Partner. There is a simple energy test for determining which Celtic weave patterns are disturbed. Again, the best way is to attend inwardly and let your intuition guide you. In fact, unless you are too well socialized, you automatically and unconsciously tend to cross your arms or ankles when you are in need of the Celtic weave. To find out where your partner needs the Celtic weave and to perform a correction (time—about 3 minutes):

1. Move your hand from either shoulder to the opposite hip in a quick sweeping motion while barely touching the body.
2. Energy test your partner's arm.
3. Go in the other direction, from the hip to the opposite shoulder, and energy test.
4. To correct: With your hands, trace a figure 8 pattern over the torso a few inches above the body. Go in a direction that was strong during the energy test and keep your hand moving in the figure 8 pattern.
5. Move your hand freely, finding your own rhythm, making figure 8s for a minute or longer.
6. If both directions tested weak, which is very unusual, make the figure 8s in both directions.

As well as testing and correcting over the torso, you can test over the head and neck, the legs, or any other segment of the body, large or small.

THE BASIC GRID: THE FOUNDATION OF YOUR BODY'S ENERGIES

When you are standing up, it would appear that your *root chakra* is the foundation of your energy centers. Each of your other chakras sits above it. When you are lying down, it would appear that your *basic grid* is the foundation of your energy centers. Each of your chakras is embedded within this grid, which is the bedrock of all your energy systems. The grid looks very different to me than the chakras. Where the chakras appear as spirals, the basic grid looks like a graph of energy lines. It resembles the steel infrastructure you see at the construction site of a skyscraper. Where chakra energy connects with energies in the environment, basic grid energy is totally contained within your body. And the other energy systems become severely compromised if this foundation is impaired.

I first came to understand the role of the basic grid when it made no sense to me why certain treatments would not hold. I had to look deeper. Because the basic grid is so integral to the body, I'd barely notice it when it was functioning properly. In its collapse, however, all the energy systems are on a shaky foundation.

When I look into the middle layers of a chakra, I get images from the person's past, both pleasant and disturbing, that relate to the chakra's basic theme. If I can peer into the chakra's deepest layers, I usually see the person's latent potential, the inheritance from lifetimes past. But when I look *beneath* the deepest layers of the chakras, I usually see only the barren graph-like pattern of the grid.

When I am working with the basic grid of a person who has suffered severe trauma, however, memories of that trauma may flood into me in a way that can be overwhelming. I have many times been deluged with the person's symptoms as well. For these reasons, I do not teach the most potent techniques for working with the basic grid except to certain students who have both an affinity to grid work and a strong nervous system.

A fifty-two-year-old man, Tom, had been diagnosed with Guillain-Barré syndrome (French polio) over a decade earlier, and while it had abated, he never fully returned to good health. He did not want to tell me about his condition until after the session as he did not want to prejudice my assessment.

I could see that several of Tom's basic grid patterns were disrupted. I held the points that connect to the most seriously affected pattern. A slow yellowish-brown energy began to move up through my arm, and suddenly I was in terrible pain. It was as if this energy were choking the life out of my arm. It was awful, and I had no idea what it was about. Then I began to get a few images, just little pieces of scenes that I couldn't put together. Next I was gripped by an overwhelming feeling of hatred. It wasn't as if I was simply *aware* of his hatred. It was as if I were him, feeling a burning hatred and compelled to act on it. There was no choice. It was huge, and it took over my whole being. The feeling was so intense that I wanted, needed, to take it out on somebody. I'd never experienced hate like this before.

With these feelings, the pictures started to become more coherent. First I sensed that I was walking in the dark, laden with this hatred, which had escalated into a compulsion to kill. Then I was coming to a house, and I realized I had a gun. The pictures were still vague, but I knew that I was going to kill someone inside the house and that I was protecting someone, a woman. A man stood in front of me. I pointed the gun at him, but I couldn't bring myself to fire it. Trauma flooded my body, and I couldn't move. I'd been describing my experience to Tom as it was unfolding, but suddenly I couldn't talk. I was frozen everywhere, paralyzed. Tom saw what was happening to me, and said, "It's Guillain-Barré!" He understood in a flash that I had taken on the residue of this energy that still resided in him.

Twelve years before, one of Tom's closest friends was being beaten by her husband. Through an escalating cycle of abuse and remorse, Tom believed the man would eventually kill the woman. Tom confronted the man with a gun, intending to defend his friend by killing the husband before he killed her. But at the moment of truth, Tom's conscience or his humanity stopped him from pulling the trigger. He ran out of the house and within a short period of time became deathly ill. He woke up the next morning unable to move. He was hospitalized and not expected to live. His diagnosis: Guillain-Barré syndrome, a paralysis far more severe than the paralysis of his trigger finger.

Meanwhile, there I was, paralyzed, standing next to him and thinking, "I'm never getting out of this one. I've taken on something that is too much this time." Energy couldn't move through my body. I felt my life being suffocated out of me. Tom, bless his heart, tried to pulsate his own energy through me. It helped a bit, but I was caught in this horrendous state for some three hours before I again had full control of my body. All that history and energy had been trapped in his basic grid. But once it had been channeled into me and his basic grid reconnected, he was, as he later reported, relieved of years of vague symptoms.

Intense trauma to the body, psyche, or soul is like an earthquake that shakes your energetic foundations. What happens is that the basic grid pulls apart. I believe it acts like a shock absorber—it helps the body withstand the shock of a trauma the body could not otherwise absorb. If the basic grid didn't inhibit the shock, such trauma might have caused the heart to stop or another vulnerable system to fail. In the short term, this softens the blow. But, as when a car's frame has been damaged in a crash, other systems may not function quite correctly until the basic grid has been restored, through time or through an intervention.

A couple was in the midst of a divorce, and their seven-year-old son was traumatized. He was full of rage, acting it out by bullying others at school. He was also inwardly terrified, wetting his bed at night and having frequent nightmares. The mother brought him for a session. While the work did help calm him down for a few days, it didn't hold. In the second session, I looked at his basic grid and could see that a major pattern was disturbed. I'd never worked with the basic grid of anyone so young, and I was concerned how it would affect him.

If the grid pattern restores and the energy it holds is released, the person sometimes, after the session, has to deal with and integrate traumatic memories, pain, or shock. While this is by no means always the case, I didn't know what might ensue, and I wanted his mother to know that they might have a lot to deal with if we proceeded.

We talked about having him see a child psychologist, which I felt could help, but I also had the sense that even if therapy put his life into perspective for him, restoring his basic grid would help him at another level. If the trauma of the divorce became locked into his energetic foundation, it could reverberate for the rest of his life, as so often is seen in people who don't move beyond a trauma that colors their outlook or even causes them to re-create the trauma. His mother decided to let him choose whether to have the treatment.

I was impressed by how tenderly and skillfully she explained the option to him. She said that his spirit had gone through an earthquake, and some of the roads that are necessary for it to be happy had been destroyed. The session would try to mend those roads so he could feel happiness inside himself again. But in mending them, a lot of rubble that could make him uncomfortable for a while might also be unearthed. He decided to take the risk.

I can often experience in my own being the story that caused the break in another's grid pattern. I usually know the exact moment it was disrupted because, in a nonlocal, *Twilight Zone* sort of way, I am there. Sometimes, however, the trauma is at such a deep physical level that I can't make any sense of it, and I never did become sure of the moment the boy's grid pattern collapsed. My primary experience was of a body in shock. I had a quick flash of his father looking distraught, and I then registered deep trauma about the father's leaving—separation, isolation, being cut off from everything that was safe. He had no conscious insight about such issues, but his little body was in shock, and his immune system was on constant alert. It was responding as if he would never be safe again, sending the message to all systems to protect themselves. Perhaps this was the source of his picking fights with other children, the behavioral correlate of an overactivated immune response.

As the grid pathway connected, physical signs of serenity returned to his body. His breathing relaxed. His energy body came out of shock, and with his immune system no longer in a state of constant alert, his nightmares ceased, he quit wetting the bed, and he stopped bullying others.

Children are particularly vulnerable to shock and trauma, which cause disruption in the basic grid, but they can occur at any point in life. There are sixty-four patterns within the basic grid. Eight of them are major branches, and by repairing a major branch, its tributaries are also rebuilt. When too many patterns in a grid are out, it is as if the body has nothing holding it together except the person's raw effort. When people have to use their energy in this manner, they become haggard. They lack vitality, appear stiff, and have little flexibility in their lives. Sometimes a disrupted grid pattern will inhibit natural processes like joy or healing, and in one instance within my experience, it was keeping someone from dying peacefully.

Two sisters were caring for their elderly mother. The mother had for five years been hanging on somewhere between being alive and not being alive. Both sisters were caring for her, which had become a tremendous drain on them. They loved their mother very much, but she didn't want to be alive anymore. She suffered with pain and depression and saw no reason to continue with such a dismal quality of life, but she simply couldn't die. The entire extended family had gathered two or three times to say goodbye. But she didn't die. And now here she was, having outlived a number of the people who had come to be with her on her deathbed. One of the sisters asked me if I would give her mother an energy session. She hoped I might be able to help her be more comfortable and perhaps her energies would tell me something that would be helpful. The mother too seemed eager for answers and release.

It looked to me as if her body could hardly contain her spirit. Two patterns in her grid were totally unstable, and I wondered how she was staying alive with such ruptures in her most fundamental energy system. As I held the points for the first pattern, I was suddenly pulled into her past. I was her. I was four years old. It came like a movie, and I was in the movie. I was experiencing her father sexually abusing her. I described him. I described the scene, and then she began to talk. She'd never wanted to remember. She didn't want her father's memories tainted in any way since she had adored him. But the memories that woke in her were ones she knew to be true, and she began to describe what I was seeing, feeling, and experiencing.

Then another trauma, this one from twenty years later, came into my awareness. She was in a most difficult marriage. Her daughters were young. She was certain that her husband was molesting their daughters. She believed that every shred of their unhappiness was because of this. She felt justified in divorcing her husband because she was so sure he had molested them. While she never spoke to anyone of her suspicions, she told herself that she was taking this step because of the molestation rather than because of her own unhappiness with the marriage.

At some level she had decided not to die until she had told her daughters the truth. She wanted to set them free with the knowledge that they had been molested. But the daughters had become very close to their father, and he was being wonderful to them. She was afraid that if she revealed the molestation to them, they would not believe her and would interpret the revelation as vindictiveness. Still she felt they needed to have the information, and she hung onto life, hoping to find the way to say it.

What came forth in the session, however, was that, having repressed the memory of her own molestation, she had projected her father's transgressions onto her husband. While it had been a difficult and stressful marriage, he was neither a bad man nor an abusive father.

The treatment suddenly became a whole new ballgame. She had me call in her daughters, and she told them the whole story. It was falling into perspective for her that believing their father was an abuser had been her ticket out of her difficult marriage. She now had no need to paint their father as evil, and her relief was enormous.

The other disturbed grid pattern didn't seem to hold another story. It was, rather, as if her inner confusion, fears, and displaced blame kept this second pattern separated. As I held the grid points, I felt a little frightened that the grid pattern was going to reconnect, as this might remove the last obstacle that was preventing her from leaving her body. I thought, "Oh my God, she might let go and die!"

I pulled away from her. It is dangerous to pull away from basic grid points until the energy has completed its flow, and I suddenly felt sick and dizzy. I forced myself to go and get the daughters. I told them I was afraid their mother might die if I continued. One of the daughters was silent, but the other went into the bedroom and said, "Mom, Donna's afraid that in freeing your energies you might release all the way. She's afraid you might die. Do you want her to stop, Mom?"

She answered quickly, "No, but please just tell me what you're doing." I said, "I'm connecting circuitry that because of early emotional traumas may have been separated for much of your life." She looked at me and calmly requested, "Will you do that for me?" I did. With her daughters present, she peacefully died while I was still holding the points.

Because the basic grid absorbs not only the energy of a trauma but also the details of the experience, working with it often evokes powerful memories. I have actually seen evidence for both sides of the "false memory debate" within the basic grid. I have many times while working with root chakra or basic grid energy been confronted with images of an abusive episode that the person did not remember but that in one way or another was later confirmed. I have also worked with people who were certain they had been abused, but their deepest energies told a different story.

A woman came to me bringing memories of cult sexual abuse perpetuated by her father. From her early thirties, when these memories emerged, a significant part of her identity centered around having been ritually abused. She had been involved in therapy, support groups, and numerous organizations that helped her deal with this harsh background. She was in a great deal of misery in both body and soul. Not surprisingly, one of the major patterns in her basic grid was out.

However, when I worked with it, the information it held was not what I was expecting. If a memory surfaces it usually involves the moment the grid pulled apart. Often a series of events led to that moment. The culminating event, the one that ruptured the grid, is often emblematic of the whole series of traumas, and that is the memory that becomes locked in the grid. In this woman, however, three incidents revealed themselves. In the first, she seemed to be about five months old. Her father entered the room screaming at her mother, using abusive language and a threatening tone that went straight into the infant's energy body. Never before had she been in the presence of such brutal energy, and she took from this experience that the world is not a safe place to be. The other two incidents were similar in nature, though escalated, and with the third one her basic grid separated.

In trying to make sense of why she felt so wounded and miserable, her psyche embellished on the essential truth that she was badly traumatized by her father. Media images of cult sexual abuse provided just the right fuel for her psyche to create her elaborate "memories." When I told her that I could find nothing in her most basic energy system that suggested ritualized sexual abuse, but that what I did find provided a plausible alternative explanation, she felt betrayed by me. She had built her identity on the opposite conclusion. In time, however, her own authentic memories began to replace the contorted images her psyche and imagination had created, and she began to heal.

I do not think it is responsible or even fully possible to teach the most potent basic grid techniques in a book. A number of gentler corrections, however, are readily available, and their cumulative effects can be powerful.

Restoring Your Basic Grid by Balancing Your Chakras. Keeping your chakras balanced helps repair your basic grid. Basic grid patterns simply will not reconnect if the chakras are significantly out of balance. The two systems are highly interdependent. Clearing the chakras is a gentle and natural way for healing the basic grid. And the more well-balanced your chakras, the less retraumatized you will be if difficult memories from the past do emerge. So periodic chakra balancings will be restorative for your basic grid. You don't even need to know which grid patterns are out.

Restoring Your Basic Grid by Cultivating Inner Peace. Feeling gratitude is good for you. It strengthens your basic grid. Accepting what you cannot change is good for you. It also strengthens your basic grid. Optimism is good for you. It is better for your health, studies show, if you tend to err by holding positive expectations than by anticipating negative outcomes.[7] I have noticed that people who meditate regularly tend to have an unusual degree of strength and resilience in their basic grid. Meditation and other peaceful practices repair the basic grid, leaving it tight and strong, like a well-exercised body. By finding time each day for an activity that brings you peace, you are fortifying your basic grid.

Restoring Your Basic Grid Using the Celtic Weave. Millions of tiny figure 8 patterns exist throughout your body. Energy weaves through every cell. If these miniature crossover patterns are intact, your energies will be robust, and you will be far less likely to jar your system when working with the basic grid. In over half the configurations within the basic grid itself, energy crosses from one side to another, creating miniature Celtic weave patterns. The basic grid is therefore highly responsive to the Celtic weave technique.

Anyone could have a stronger basic grid. We've all had trauma in our lives that weakened our foundations. In the final section of chapter 3 you learned how to hold your neurovascular points while imagining or remembering a stressful situation. You can engage powerful healing forces that mend your basic grid by doing the neurovascular stress release and then the Celtic weave exercise (time—about 5 minutes):

1. Lie down and do the Crown Pull (Figure 7, page 79).
2. Focus on a scene from the past or present that carries a negative emotional charge.
3. Lightly place your fingerpads over the frontal eminence points on your forehead (page 90).
4. Staying with the stressful scene, and with your fingerpads still on your forehead, put your thumbs on the triple warmer points at your temples next to your eyes and take a deep breath. Blood that had been diverted in the stress response will return to your forebrain over the next few minutes and you will find yourself beginning to think more clearly. Let the scene fade from your mind.
5. Keeping your thumbs on your temples, begin the Celtic weave by moving your fingers in tiny sideways figure 8 patterns at your forehead. When you are finally ready to stand, cement this harmony of mind and body into your aura and your basic grid by doing a full Celtic weave (page 184).

Using this technique on a child who has been traumatized can prevent the trauma from locking into the child's energy foundation. It is most easily done if you can be holding the child in your arms. If the child is old enough to be able to talk about the trauma, lay your fingerpads on the frontal eminence points of the child's forehead as he or she speaks of it. When your child becomes calm, begin the Celtic weave pattern. You can intuitively make many figure 8 patterns all over the body with your free hand two or three inches above the skin. The sense of helplessness and self-blame parents often experience when their child is unhappy can be greatly relieved by the shared bonding and genuine help offered by doing the Celtic weave. You can also weave the 8 while your child is sleeping. Dream patterns may also begin to shift. Even chronic nightmares begin to ease away.

In this chapter, you've learned how three energy systems—the aura, the Celtic weave, and the basic grid—can protect, weave, and support the body's energies. In the following chapter, we turn to the core rhythms that vibrate throughout your energy body.

Chapter 7

The Five Rhythms

The Drummers to Which You March

ENERGY MOVES IN WAVES. WAVES MOVE IN PATTERNS. PATTERNS MOVE IN RHYTHMS. THE RHYTHMS GAVE ME A WAY TO MAP THE MOVEMENTS OF MY OWN PSYCHE AS WELL AS THE EVER-CHANGING STATE OF THE UNIVERSE.

— GABRIELLE ROTH
Sweat Your Prayers: Movement as Spiritual Practice

A person's meridians, chakras, aura, and other essential energies are influenced by a more pervasive energy system. I don't see it as a separate energy but rather as a *rhythm* that runs through all the others, leaving a vibratory imprint on your physical attributes, health patterns, and personality traits.

The rhythms were mapped long ago. Perhaps as early as 3000 B.C., Chinese physicians categorized all of life into "five elements" or "five phases" or "five seasons." This was the basis for understanding how the world works, how societies organize themselves, and what the human body needs to maintain health. It is an elegant framework for sympathetically appreciating human character, temperament, cycles, and illnesses. In this chapter, I offer a brief introduction to this profoundly sophisticated system, provide an overview of each of the five rhythms so you can begin to identify your own primary rhythm and that of others, and present techniques for bringing greater balance among all five rhythms within your own life.

TO EVERY THING THERE IS A SEASON, AND TO EVERY SEASON THERE IS A RHYTHM

We are, Dianne Connelly observes, "a replica of the universe passing from season to season in a natural unending season of life."[1] By carefully observing the Earth's seasons, the Chinese sages gained penetrating insights into the way nature conducts her business. In the seasons of nature they found analogies for understanding the growth and cycles of all things under heaven. In addition to the four seasons of winter, spring, summer, and autumn, the transition times between seasons were collectively treated as a separate season. These periods of transition were originally thought of as occurring for about two weeks four times each year, with one of the solstices or equinoxes at their midpoints. In recent centuries, however, the four transition periods have been abbreviated into a single season, placed between summer and autumn, and compared with Indian summer. Indian summer prolongs the summer, as if trying to hold off the death that inevitably accompanies autumn.

The name of the Chinese system is often translated in the West as "the five elements" because the early pictograms depicted the familiar, concrete, and observable—the five elements of water, wood, fire, earth, and metal. But the system has always concentrated on processes within nature, not her static forms (the literal translation is "the five walks" or "the five moves")—and this dynamic emphasis will be ours as well. Thus, the element of water corresponds with the season of winter, wood with spring, fire with summer, earth with the time of the solstice or equinox, and metal with autumn.

Each person is characterized by one of these elements or seasons, or a specific combination of them. In the human life cycle, we also travel through periods or phases that are analogous to the seasons of nature in tempo, intensity, and function, each potentially lasting for years. I find the language of the seasons to be wonderfully descriptive of the distinct rhythm that vibrates throughout the entire energy body at any moment in time. Each season has its own rhythm. I think of people in terms of the energy of the seasons, and I speak in terms of the *rhythm of summer* or the *rhythm of autumn* when describing a person's "element."

While each of us contains all five seasons, one season or a particular combination of two or three will blend themselves into your personal rhythm. You will vibrate more naturally to people, environments, and activities whose rhythm corresponds with your own. Those that do not will be more challenging for you, but potentially more enriching as their influence expands you.

How to Release Neck and Shoulder Tension

1. *Take a hairbrush that has strong bristles and tap your shoulders in a steady rhythm until you feel more relaxed. The bristles on the hairbrush are like prongs that can break up congested energy.*

2. *Standing, with your arms stiffened and hanging down the sides of your body, make fists, and lift your shoulders high, rolling your head back and to the sides. Repeat this half-circle motion for 15 to 20 seconds a couple of times each day. You are giving your neck and shoulders a natural massage that softens the knots, breaks up calcium deposits, and frees energy to move through the area.*

THE RHYTHMS OF A MAN'S LIFE

While he was apparently dying of heart disease, my father went through a transformation that was both beautiful and profound, and I relate it here as an example of the power of moving through a difficult rhythm within the seasons of one's life.

Daddy was strong in the rhythms of both spring and summer, but he lived the worst, not the best, of each. Spring is the season of absolutes, of taking root, of pushing forth into life. There are no in-betweens: you blossom or you die, and Daddy saw life in black and white, right and wrong. Much of life's richness was lost on him. Spring is also assertive, as the seedling explodes out of the ground, and it is protective, fiercely defending its young. But when stifled, it becomes anger. Daddy's demeanor immediately put others on the defensive. He was a very angry man.

Summer, on the other hand, is carefree, full of passion and feeling. Daddy had a dynamic emotional life, but in the rural South where he grew up, to be so sensitive earned him the label "chicken heart." So he stifled all sentimentality and feeling. Whenever he thought about his shortcomings, he would think of his "chicken heart." His sensitivity, of course, was the part of him we longed to know but rarely saw. He was a bitter and unhappy man, full of emotional contradictions.

Spring is a time when judgments must be made. Where is it safe for seedlings to grow? How shall resources be allocated? To be stuck in the rhythm of spring is to hold on to those judgments, failing to adapt to new information, resisting the coming of other seasons. My father was stuck in his own positions and judgments. But when you do not heed life's fundamental demand that you cycle through the seasons, your body will insist. If all external means for forcing change have been exhausted—disturbances in love, family, work, finances—the body finally gives you a slap to get your attention, to give you one more chance to align your own rhythms with the rhythms of nature and the cycles of your life.

Not only was Daddy stuck in his ways, his heart was stifled. Heart disease is a particular vulnerability for people who are out of balance in summer's rhythm. When he was fifty-five, Daddy had a heart attack. While he was hospitalized, his heart stopped nine times in five days. He was revived each time. The last time, he was pronounced dead, but he came back as the resuscitation team was leaving. During this ninth "death" he had a classic *life after life* experience. He saw a childhood friend who had died long before. The friend greeted him warmly and said:

"You can come with me or go back."

"Why would I ever go back?"

"Because you haven't learned a damned thing, Cecil!"

Daddy started to protest when the friend said, "No, no . . . you haven't learned how to love." With that, Daddy was back in his body. He opened his eyes and began to say "I love you" to each person on the startled resuscitation team, which was already leaving. One doctor became embarrassed and uncomfortably replied, "That won't be necessary!"

While it happened in only moments of clock time, Daddy's near-death experience captures the essence of autumn's rhythm. In the Chinese system, metal is the element that is associated with autumn. The quality that makes metal a good symbol of autumn's rhythm is that metal is enduring. In the autumn of your life, you evaluate what has been of lasting value and what has not. As when searching for gold, you mine eternal truths, discard impurities, and complete the cycle with nuggets of wisdom for the next round.

In that intensified moment of transition to autumn's rhythm, Daddy died to his old prejudices, judgments, and anger. His friend's statement that he hadn't "learned a damned thing," hadn't learned how to love, was like a lens for viewing the impurities in his approach to life. Love became Daddy's gold standard. He died to any aspiration not infused with love, and he was reborn. From autumn's rhythm, he moved for a time into the rhythm of winter, which is deeply reflective. And then back into the rhythm of spring. But this time he found himself on the favorable side of spring, no longer black and white, but rich in its blush of colors. He found truth and beauty in everything. He had the aliveness, vitality, and laughter of a child. He saw wonder everywhere. I remember how he once cried over a rose, he was so taken with its beauty. He became the happiest person I've ever known.

UNDERSTANDING THE RHYTHMS THAT AFFECT YOU

Of the many systems that sort people into one category or another, the five-rhythms approach has the advantage of being grounded in the person's core bioenergies. At the same time, it tells you a great deal about that individual's health challenges, personality, and spiritual journey. When I look at a person's energy field, it is characterized by a distinct vibration that corresponds precisely with at least one of the elements described by the ancient Chinese physicians. I believe I am seeing what they saw. For instance, when I look at someone whom the Chinese would call a *winter* or *water* element person (the derivation of the word "winter" is "to make wet"), the energy literally has a watery, languid, rolling quality, and this manifests in the way the person walks and talks. If the person is well balanced, the rhythm is smooth and flowing, and it runs deep.

You will see that each rhythm has certain strengths and certain vulnerabilities. Many factors determine whether you will manifest the best or the worst qualities of your basic rhythm. For instance, your way of living is related to the way your family and early social environment supported or failed to support that rhythm. Children whose primary rhythm is appropriately recognized and supported grow into adults who express that rhythm in its more positive form. The qualities of a child's primary rhythm may, however, be so prized and reinforced that the child not only learns that rhythm but overdevelops it to the point that little is learned of the other rhythms and there is no balance. If, on the other hand, the qualities inherent in the primary rhythm are punished or bring disapproval, the child may grow up alienated from his or her core rhythm.

While most people embody a combination of two or three rhythms, I will describe each of the types in its unmixed form in order to get you thinking about your own personal rhythm.[2]

The Rhythm of Winter: Embryonic Possibility. Winter's rhythm embodies the seed, the embryo, potential. The time of long nights and little light, winter embodies the promise of the future. While life appears to have ceased, it is growing decisively under the ground, waiting to burst forth.

Winter people, when in their strength, embody a fresh spirit that is infused with childlike enthusiasm because their season is about beginnings. They know how to envision a project and joyfully get it under way. When they feel safe, they utterly trust their surroundings, and they laugh and play with the spontaneity of a baby. Their energies may be limited, since their season has little sun, but like a hibernating polar bear, they are able to retreat into themselves and regenerate. They are deeply reflective about the meaning of life and the direction it should take.

As with each of the rhythms, the winter person's potential weaknesses are the polarity of these strengths. The playful energy of good beginnings is not so well suited for going the full distance of completion. They may have little sense of direction or motivation for the long haul. Just as special care and protection are required to survive in winter, people moved by winter's rhythm often need and demand special attention, so they are particularly vulnerable to narcissism. Rooted in nature's embryonic time, there is a babylike quality to this rhythm. Winter people may be unable to recognize how they are affecting others, focusing only on what others are doing to them. They can have difficulty feeling loved unless love is showered upon them. Needing the mother's succor like the seed needs the unfailing sustenance of the earth, winters who feel unloved tend to retreat within, becoming cold, isolated, and paranoid. Your first cycle of winter's rhythm extends from conception through about eighteen months. But if stress or trauma prevented you from sufficiently garnering its lessons, its issues can become fixated into a lifelong pattern where you behave as if you are the center of the world, for it is possible to become arrested while moving through any of the rhythms.

The talk of a winter person is a slow, flowing kind of groan from deep within. The walk is unhurried and elegant, like a rolling wave, almost a swagger, knees slightly bent so the body seems more aligned with the ground. The sustaining mental state is courage. Under stress, courage may become fear, which is the stress emotion of a winter person. Because the future is hard to see from winter's embryonic shadows, winter people are afraid to move forth, afraid to make a commitment. They reflect deeply, motivated by their fear of what is to come. In the wild, a newborn animal is utterly vulnerable and must quickly learn to distinguish between what is dangerous and what is safe. During your first eighteen months, your first cycle of winter's rhythm, fear alerted you to that which was dangerous. Through fear you learned to establish boundaries. You defined a zone of safety. Dangers, both real and imagined, can tend to paralyze a winter's rhythm, making it even more immobile, more hidden, more pulled toward hibernation. With maturity, however, a winter's fear becomes a wise and discerning caution.

The Rhythm of Spring: New Growth. The energy of a *spring* or *wood* element person is reminiscent of the seedling you might see bursting forth through a rock in the springtime. It is solid within its space. The rhythm is staccato yet insistent, like a marching soldier.

Spring's rhythm embodies the power and insistence of new life. Earth becomes warm, and the hours of light begin to outnumber the hours of darkness. Life bursts forth as the landscape explodes with color and exuberance. Spring is assertive—life pushes onward.

Spring people take a strong stand. They unabashedly claim their space, as if proudly announcing, like a budding rose, "I am a force to be reckoned with." Their strength is that their vision is potent, seeing inequities and assembling forces for justice and truth. Their vision of truth and wholeness inspires others. They see the truth. They see the way. They can marshall their intellect and their energies into a plan. They are sure of themselves and shine in a crisis. Their sense of timing cuts to the quick. Their ability to assert themselves and organize efforts is characterized by sound goals, good judgment, and wise decisions.

The spring person's self-confidence is at risk of becoming arrogance; assertiveness can become inflexible, self-indulgent, and opinionated. They may hold a narrow and rigid vision that causes them to harshly judge those who do not subscribe to their truth or follow their direction. They may righteously hold to this position and become easily and vocally frustrated about the beliefs and actions of others. Or they may lose their vision and be left disorganized, hopeless, and despairing.

The talk of a spring person is choppy and syncopated, almost a shout. The walk is also choppy, hitting the ground decisively, with clear concise movements, like percussion. The sustaining mental state is assertiveness. The stress emotion is anger. In nature, the energy that has been accumulating beneath the ground in winter explodes forth above the earth in spring. Ideas or opinions may take root within a person, growing and expanding until they ferociously burst forth. During the "terrible twos," your first cycle of spring's rhythm, you are exploring, expanding, moving outward, and whoever or whatever blocks this energy will know your fury. If spring is your primary rhythm, your disposition is to push forth. Your roots are firm, your territory is well marked, your purpose is strong. You meet obstacles decisively. If they do not give way, your anger is quick and forceful. With maturity, however, a spring's anger becomes a wise and healthy determination.

The Rhythm of Summer: Fulfillment. The energy of a *summer* or *fire* element person blazes up and out, creating the impression that the person is everywhere at once. Like wildfire, which jumps ravines and spreads in every direction, its rhythm is rapid, random, and rising.

Summer's rhythm embodies fruition. Earth becomes warm and the days long. New light bursts forth in the early morning. The fruit on the tree has matured, ripe and luscious. Summer holds the radiance and joy of youth in all its glory. It gives delight in the richness of the moment.

Summer people move from their heart, open and vulnerable. Their strength is that they are warm, empathic, joyful, and exuberant. With passion and radiance, they are able to draw out the positive and the hopeful in others, communicate with them in their uniqueness, and elicit cooperation. With charisma and a grasp of the whole picture, they ignite the actions of others with insight, compassion, and clarity. In recognizing what is possible, they are the magicians and catalysts who help others believe in themselves, free themselves of self-imposed limitations, and move with confidence to a better future.

Summer people may become junkies for love, for the "high"—whether through parties, drugs, sex, or spirituality. They may go into a panic of frenzied activity, trying to make everyone happy. They often have difficulty with discernment and setting priorities. They may give from their hearts until they have no more to give. Summer people often burn themselves out, overcommitted and exhausted. They are so drawn to the bright side of life that they may not register the dark, the negative, or the dangerous. To those who look to them for leadership, their optimism and enthusiasm may set up expectations that were never meant and are rarely met.

You can hear laughter in the talk of a summer person. The walk is like a skip, with an up and down movement, arms rising and falling like flames. The sustaining mental state is infused with joy and passion, which under stress can escalate into panic or deteriorate into hysteria. In summer, the light is dazzling, the fruit abundant, and the fish are hopping. Excess is all around. During adolescence, your first cycle of summer's rhythm, you lived for thrill and exhilaration. Joys and sorrows were laced with passion, taken to excess. If you are a summer person, you want to enjoy, not strive. The present is all that matters, and as you bask in its warmth, you radiate your excitement. Others may find your Pollyanna optimism either contagious or irritating. With maturity, a summer's nondiscerning enthusiasm, passion, or infatuation become discerning love and involvement.

The Rhythm of Solstice/Equinox or Indian Summer: Transition. The energy of a *solstice/equinox* or *earth* element person, who is oriented toward the *transitions* from one season to the next, has a centering, side-to-side roll. The rhythm sways, as if the person is moving to the rhythm of Earth herself.

The solstices and equinoxes embody the rhythm of transition. As the midpoint between two seasons, the time of transition is governed by a balance between opposing forces, holding both the past and the future in the present moment. Most familiar as Indian summer, its colors are bright and glorious, a last burst of the waning season. This rhythm creates stability amid transition, assimilates change, and coordinates between the season that is ending and the season that is arriving.

Solstice/equinox people know about holding steady. Like the balance scales that are the symbol for justice, they embody fairness. At the center of the cyclone, their strength is to stay stable while nurturing the changes happening around them. Like a midwife or earth mother, they bring support, compassion, and confidence to times of transition. They hold the center, staying in the present moment as they add their tranquil touch to life's changes. Keeping a fresh perspective as the old order passes, they pave the way for stable change, rarely seeming rushed or stressed. Because they exude compassion, people feel safe with them. They bring equilibrium to chaos, peace to the threatened, and shelter to the displaced.

With a compulsion to help others stay in a comfort zone, solstice/equinox people may hinder another's transitions. This aversion to rocking the boat, combined with their characteristic desire to support the other, may also lead to obsessive worry. Or they may involve themselves in a manner that stunts the other's growth, babying and overprotecting. "The helping hand strikes again" is the epitaph of a solstice/equinox person whose life has lost its balance. In their joy at helping others flourish, they may neglect to give enough attention to their own growth. Skilled at helping others integrate lessons and experiences, they may have a harder time integrating their own. Knowing bone-deep that loss is an inevitable part of transition, they may anticipate it and try to prevent it, staying with a bad marriage or an unfulfilling job. And so they may turn their strongest suit into a losing hand by interfering with the cycle of necessary change. Also, because they do not have a designated season of their own, solstice/equinox people may live with heartrending questions always in the background, such as: "Where is time for me? When will my season come?"

The talk of a solstice/equinox person has a singing quality, as when you are talking to a baby. The walk has a relaxed, lyrical manner, a slow, rhythmic side-to-side sway, lightfooted as a deer. The sustaining mental state is compassion. The stress emotion is a codependent sympathy. In moving from one season to the next, the two seasons come into a resonance, a sympathy, as one transforms into the other. In times of transition, the ability to provide nurturing is no less than a survival tool, and no one does this better than the solstice/equinox person, whose archetype is the Earth Mother. In your own transitions, you must activate that archetype within yourself, supporting yourself through endings and new beginnings. The harvest of the season that is passing must be incorporated into the season that is coming. Solstice/equinox people instinctually help others in transition to transform past mistakes into lessons for the future. A transition person's generosity may be martyrish; with maturity, however, exaggerated sympathy ripens into a wise and balanced compassion.

The Rhythm of Autumn: Ending. The energy of an *autumn* or *metal* element person seems to be stretched between the heavens and Earth. Like a tall tree that has lost its leaves, the energies seem restrained yet serene, barren yet dignified. The rhythm glides like a ballet dancer—elongated, still, and graceful.

Autumn's rhythm embodies completion. Each day turns to night earlier than the last. The warmth fades. Yet autumn embodies the peace of completion, the meaning found in attainment, and faith that dying to the old makes way for the new. The leaves fall to earth, fertilizing the next cycle. This rhythm garners the meaning of the cycle that is coming to an end, evaluates what has been useful and what has not, and eliminates all that is not valuable so as to bring about a worthy completion.

Autumn people have the ability to mine truths out of their experiences and apply those truths. Living in the last cycle, there is an urge toward perfection, high achievements, and model results. Autumn people can see what needs to happen and are highly motivated to make it happen. Out of this vision of perfection grow standards of excellence that are true and pure, concerned with a higher good, and inspiring to others. That which is impure—whether ideas, behavior, or systems—is eliminated. As the last season of the cycle, autumn carries a sadness, and those whose rhythm vibrates with autumn carry a simpatico with the world's grief. From this affinity with sadness grows kindness, honesty, and integrity. They have a capacity to express themselves clearly, and they receive well the ideas and inspiration of others, for they have a gift for discerning the pure from the impure. They have an urgency to find meaning and serenity in what has been, for theirs is the final cycle. Forgive them their persistence. It is their rhythm.

Autumn people are vulnerable to becoming overly serious or sinking into depression. Shunning fun and lacking pleasure, they may find their energies becoming restrained and dry, like a tree without leaves. They may appear dreary and aloof. Living always in the energy of the final cycle, they may have difficulty with time, trying to cram more into each day than it can contain. Oriented toward the future, they see life through the lens of death, and they may become trapped in depression or in the pressure to reach perfection before the last grains of sand have emptied from the hourglass. Their ability to make pure judgments may be clouded by this despair or perfectionism, and their standards may be tarnished by hopelessness or inflated through unrealistic assessments. Either may paralyze them so that they become unable to let go into change, obsessively evaluating and reevaluating to the point of exhaustion, lacking the capacity to complete a cycle of their lives, again failing to reap the benefits of their strongest suit.

Elisabeth Kübler-Ross observed that the best way to prepare for your eventual dying is to meet with consciousness the "little deaths" life continually provides. When you are actually in the process of dying, you will be in autumn's rhythm. Each cycle of completion, each "little death," each autumn in your own life's flow is an opportunity to glean the lessons of the cycle that is ending, to create a meaningful completion, and to open the way for whatever is next to come. Each cycle trains you for all the autumns yet to be. So it was when you had to die to adolescence to be reborn into adulthood. So it is when your own children leave the nest. So it will be when your body comes to its final season.

The talk of an autumn person has a weeping sound. The walk is tall, straight, and subdued, gliding with head high and gaze forward. The sustaining mental state is reflectiveness. The stress emotion is grief. As the leaves fall and the wildflowers die, loss is in the air. The cycle draws to its close. When you come to the close of a cycle in your own life, there may be sadness for opportunities missed and for what must be left behind. If autumn is your primary rhythm, you are oriented toward completions, toward discerning what has been worthy and meaningful. There is a heaviness in these tasks, and you know the grief of what might have been but was not to be. With maturity, however, an autumn's grief transforms into an identification with the whole cycle, at peace with life, at peace with death.

From these descriptions, you may recognize within yourself one, two, or three of these clusters of traits. Fully understanding your rhythm and all its implications is a lifelong process. The system is that basic and that profound.

If you understand your own primary rhythm and its dynamics, you will know a great deal about your needs and your blind spots in all areas of your life, from your choice of a mate to your vulnerabilities for illness. Your primary rhythm manifests itself in the way you look, walk, sound, feel, act, and react. If you know the primary rhythms of a colleague, client, friend, or family member, you will be able to understand that person's behavior with greater insight and empathy.

How to Calm an Angry Spirit

For some people, depending on their energetic makeup, as stress builds in the body, anger builds up as well. This is not a voluntary response. It is driven by the hypothalamus. Chemicals designed to help you fight floo d your body. But since you can't regularly slug your fellow worker or other handy target, the chemicals are not burned off, and the anger continues to build. Muscles in the neck and shoulders can compress into a tight vise at the base of the skull. The pressure pushes for release. Groping for an outlet, anger may come out inappropriately at your children or your spouse. It may inflict emotional and physical trauma to your own body. You are confronted with a problem whose dimensions are not just psychological but also biochemical, so solutions must address this physical dimension as well. Before your inner rumbles begin to resemble Mount Vesuvius (time—about a minute):

1. *Stand, rub your hands together, and shake them off.*

2.. *With a very deep breath, circle your arms above your head, stretching them high.*

3. *Turn your hands toward you and clench them into the kind of fist you would make to hit someone.*

4. *Blowing the air out of your mouth, forcefully "pound" your arms downward, moving them quickly. After your arms have reached the bottom of this motion, open your hands.*

5. *Repeat this twice more, but now very slowly and deliberately. Feel the power in your biceps.*

6. *Repeat, "pounding" again with the swift movement, continuing until you have literally "cleared the air."*

7. *Finally, do a Zip Up (page 82–83), breathing in fresh air and energy.*

THE SEASONS OF YOUR LIFE, THE SEASONS OF YOUR SOUL

In addition to the obvious shifting seasons in our lives, we each vibrate to a core rhythm that reflects the current season of our soul. Saying the "season of the soul" assumes a belief in past lifetimes. I have heard two explanations of how we come into our core rhythm or season. One is that as we evolve from one lifetime to the next, we keep travelling through the five seasons, living within the energy of a particular season or rhythm for one or more lifetimes, garnering its lessons. We then continue on to the next rhythm, circling through these seasons for lifetimes upon lifetimes. The other explanation, which is deeply rooted in the Orient, is that your primary season corresponds to the season on the Earth at the point you make the decision to claim life. This is believed to occur any time from the moment of conception up to three months after birth. Perhaps both explanations are true. You may be drawn to claim life during the season you *need to assume* in your journey toward your soul's actualization on Earth.

In any case, you were born into a particular rhythm or combination of rhythms, and you will live from within this rhythm or blend of rhythms all your life. The rhythm of a newborn is already distinct. I've observed in children growing into young adulthood that the rhythm only becomes more deeply embedded. A baby's meridians, chakras, aura, and other energy systems do not yet completely vibrate with what I sense to be the core rhythm, but as the child grows, all of the energies calibrate themselves to it. I think of this as a soul-level inheritance. Unlike red hair or blue eyes, the core rhythm is not inherited from the parents but reflects rather the larger rhythm of the soul's journey.

Most of us seem to be imprinted more or less strongly by up to three of the five rhythms or seasons: the rhythm of our primary season, the rhythm of the season preceding it, and the rhythm of the season following it (see Figure 35). A spring person, for instance, may carry the rhythm of spring alone or may have some of the rhythm of winter and/or summer accompanying that of spring. While there are exceptions to this pattern and innumerable permutations, this overview gives you a good basic understanding of life's five seasons.

Figure 35.

Rhythms of the Seasons Wheel

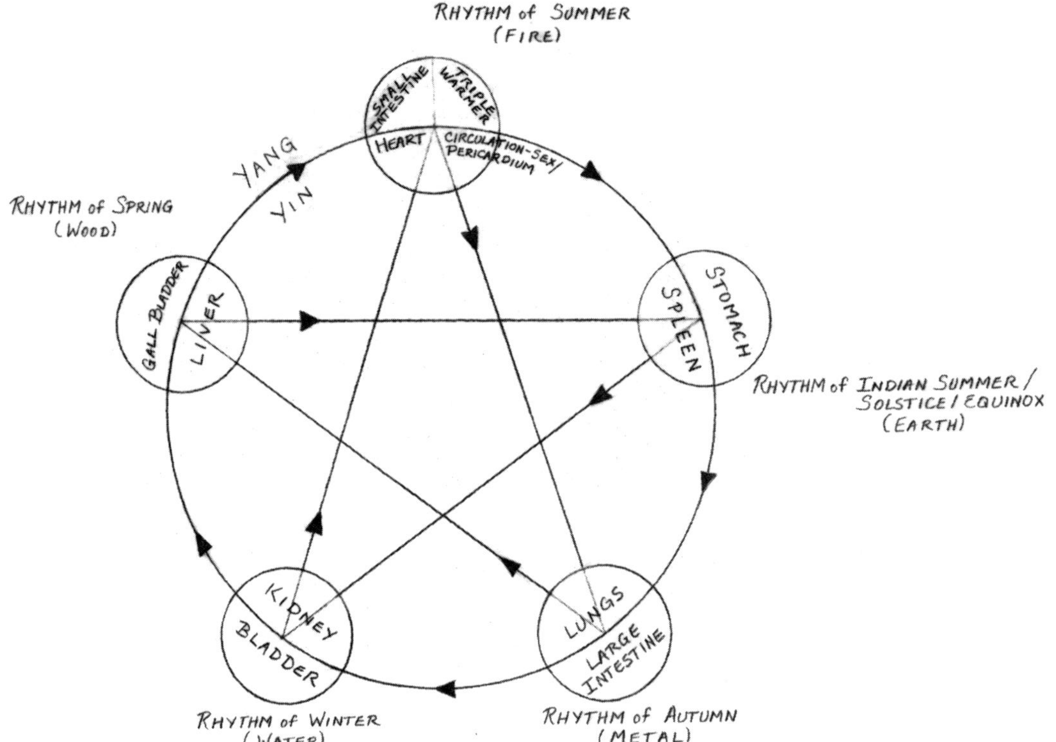

In summary, your body, your movements, and your disposition were formed within the *energy* of one of the seasons of nature, and they are imbued with a rhythm that blends that season's rhythm with the rhythms of the seasons preceding and following it. Beyond the rhythm that is embedded in your character, you move through all five rhythms many times during the cycles that comprise a lifetime. Both your core rhythm and the rhythm through which you are passing at any given moment profoundly influence the way you respond to the world.

IDENTIFYING RHYTHMS THAT ARE OUT OF BALANCE

Your own rhythm is made up of your primary rhythm and a unique combination of the presence or absence of qualities from the others. Many of your strengths as well as your weaknesses are inherent in your primary rhythm. Because it is the rhythm pushing your evolution, when you are unable to embrace it, your body will react. Your body in the ultimate biofeedback system. Illness, among other things, provides information about the fit between your body's needs and your lifestyle.

This is an altogether different concept from "New Age guilt," where the valid insight that we have responsibility for maintaining our health is boomeranged back as blame and self-blame when we are ill. It is simply untrue that you caused your disease. Blaming the patient for being sick is a way we distance ourselves from another person's pain and suffering. This judgment is naive, unhelpful, sometimes mean-spirited, and often self-righteous. Whether well-intentioned or self-protective, it is misguided and antihealing. The fact is that in the grand logic of body, mind, and soul, our physiological vulnerabilities may move into illness whenever we are physically, psychologically, or spiritually stretched beyond our means.

To grow is itself to be stretched, and our assignment on this planet seems to be to grow, to learn, to evolve. When the stretch is too sudden, extreme, or prolonged, pain or illness—whether physical, psychological, or spiritual—often follows. To burden yourself with merciless self-judgment or to be harshly judged by others "for your own good" is less than useless. Pain and illness are isolating and confusing enough. Judgment and self-judgment at the time you most need mercy blocks healing energy. The realization that we have stumbled off our path can help us garner the lessons of life and better adjust our way. We need to embrace and then be validated for our pain, yet not fall down the slippery slope toward self-pity or "woundology."[3] On the other side is another slippery slope, where the insight itself sets off a landslide of self-judgment that also blocks the path. We all deserve to have our pain touched with loving-kindness, and we are each capable of offering that for another's painful journey.[4]

The rhythms help you to understand your physical vulnerabilities. Each rhythm is prone to its own characteristic types of illness, and each offers lessons for avoiding or overcoming them. The lessons may be at the physical, psychological, or spiritual levels, and they may involve marshaling resources (winter), taking a stand (spring), savoring the moment (summer), integrating experience (solstice/equinox), or letting go of what has been (autumn).

While psychological and spiritual friction may cause you to take pause, physical problems stop you on your path. If you are a workaholic who is struggling with the lessons of summer's rhythm, for instance, you may know that it is important to develop the capacity to savor the moment, but your driven approach to life may have become firmly embedded in primitive brain centers. Such patterns are equated with survival at a level of your being that is deeper than your conscious intentions. This is why the New Year's Resolution theory of personal growth is so dismally inadequate. Physical illness can sometimes catalyze changes that all the willpower available to you could not. In its harsh way, illness sometimes serves as a spiritual teacher, bringing us to what we had been seeking but did not know how to find.

Energy Testing the Five Rhythms. Vulnerabilities in your primary rhythm are more likely to be chronic; problems with the other rhythms tend to come and go as your life cycle passes through their seasons. But any of the five rhythms may be out of balance, either overdeveloped and dominating the others or underdeveloped and not providing the resources you need from it. You can do an energy test that uses the Rhythms of the Seasons Wheel (Figure 35) to identify rhythms that are out of balance.

The energy test will give you a quick assessment. If all five rhythms test weak, however, it probably means that the energies moving through you are stressed, scrambled, or running backward. This needs to be corrected before you can get an accurate test. A good rule of thumb prior to embarking on any complex energy procedure is to begin by using the five-minute Daily Energy

Routine presented in Chapter 3. To determine which of your rhythms needs attention:

1. Imagine that the Rhythms of the Seasons Wheel has been placed on your stomach, with the middle point at your navel and summer's rhythm at the top.
2. Your partner places the middle finger of an open hand at your navel and pulls, with some pressure, down about four inches, toward your right hip bone. Keeping one hand at the hip bone, your partner uses the other hand to do a general indicator test (pages 65–67). Either arm may be used. Do not bend at the elbow. If you test strong, winter's rhythm is flowing properly within your energy field. If you test weak, you have an imbalance on winter's rhythm that deserves your attention.
3. For spring's rhythm, pull from the navel across the waistband to the right side of the body and test. Again, if you test strong, spring's rhythm is flowing well. If the test shows weak, you have an imbalance on spring's rhythm.
4. For summer's rhythm, pull from the navel straight up to the bottom of the sternum and test.
5. For the solstice/equinox rhythm, pull from the navel across the waistband to the left side of the body and test.
6. For autumn's rhythm, pull from the navel toward the left hip bone and test.

This simple test has an anatomical basis. Each rhythm governs several organs, and the pull of your fingers moves through the energy field of at least one organ that is on the rhythm being tested. Winter governs kidney and bladder. The ileocecal valve, which is down from the navel and to the right, is on the kidney meridian. Spring governs the liver and gall bladder, which are to the right of your navel. Summer governs the heart and pericardium, which you reach toward when you pull upward. Indian summer governs the spleen, stomach, and pancreas, which are to the left of your navel. Autumn governs the large intestine, whose endpoint is down from the navel and to the left. If the rhythm is in balance, moving your fingers through the field of the organ will not disrupt its energies; if the rhythm is not well balanced, the energy fields of its organs are easily unsettled.

How to Concentrate Amid Fear or Stress

Being afraid that you cannot learn creates stress that causes agitation or exhaustion.[5] *The mind does not stay focused. These techniques deal with the fear of failure without having to talk about it, and they also help with comprehension and concentration. They can also help kids with their homework (time—between 1 and 10 minutes):*

1. *Put on pleasant background music and sway to it. Move your hips in a natural figure 8 pattern. This will strengthen all energy crossovers in the body, all the way up to and through the brain.*

2. *On a blackboard or a large piece of butcher paper taped to the wall, draw large sideways figure 8s over and over, with the right hand, the left hand, and both hands (see page 184). The larger the drawing, the more completely the body incorporates the crossover. The smaller the figure 8s, the more intricately the brain and eyes are involved.*

3. *Hold the Wayne Cook posture (page 73), which optimizes the brain's ability to take in new information, counteracts dyslexia, and relieves other learning disabilities.*

4. *Massage the K-27 points (page 63–64) and do a vigorous Cross Crawl (page 69).*

ESTABLISHING BALANCE AMONG YOUR RHYTHMS

For any rhythm that needs balance, you may stimulate its lymphatic points, reprogram its stress response pattern, or do an exercise that brings you into harmony with it. Three techniques follow.

Stimulate the Rhythm's Lymphatic Points. The spinal flush (pages 79–82) stimulates your neurolymphatic reflex points. Unlike blood, lymph does not have a pump, but if you get your lymph flowing well using the spinal flush, toxins will be released, blood circulation will get a boost, your nervous system will be activated, your energies will move more freely, and your body will be more alert and responsive. For each rhythm that tested weak (time—1 to 2 minutes each):

1. Find the neurolymphatic points associated with that rhythm in Figure 9.
2. Massage these points with pressure. They are likely to be sensitive. As the massage clears toxins and helps the lymph, blood, and energy circulate, the pain should abate. If the blockage in that area has been fairly chronic, this may require several such treatments.
3. For points that are particularly painful, have a friend place a finger of one hand on the point and a finger of the other hand on the corresponding point at the back of your body, alternately pushing in on the one and then the other, creating a back-and-forth, seesaw motion.

Reprogram the Rhythm's Stress Pattern. Each rhythm has its own way of "seeing," and when balance among them is lost, it is like the proverb of the blind men and the elephant. You interpret the world through your primary rhythm, and you *misinterpret* it that way, just as the blind man who was brought to the elephant's side thought it was a wall, the man brought to its tail thought it was a snake, and the one brought to its leg thought it was a tree. Not cognizant of the other rhythms, you may find yourself out of step with everyone around you, stuck in an outlook you can't escape, crying without knowing why, withdrawing, being blocked, or feeling out of control. The stress emotion of your primary rhythm may become so pervasive that you become caught in its fear, anger, panic, codependent sympathy, or grief. You may not even recognize the feeling as an emotion—it may simply seem to be showing you *how things are.*

Working with a problematic stress emotion can free you from this myopia. You must understand that from your body's point of view, the emotion—whether fear, anger, panic, codependent sympathy, or grief—is a positive response. It is designed to help you survive. Your survival instinct plunges you into the emotion your body trusts the most. It is not concerned about balance; threats to your existence call for extreme reactions.

However, if a stress emotion controls you or has been repressed within you, this may in itself be creating havoc in the balance among your rhythms. In Chapter 3 you learned a technique for reprogramming your stress pattern by holding points that keep blood flowing to your forebrain while you experience a stressor. Here you will be holding additional neurovascular points *associated with a rhythm* that become overactivated when you are under stress. Practice using this exercise when you are highly stressed. You can also use it to defuse a traumatic memory. You can hold the points yourself, or you can have a friend hold them so you can fully relax into the experience (time—3 to 20 minutes):

1. Choose a scene in the present or from the past that carries an emotional charge for you.
2. Identify which of the following is closest to that emotion: fear, anger, panic, sympathy, or grief. If you are not sure, you can energy test it. Within the scene, imagine you are feeling fear. Energy test. Go through each of the five emotions. Your arm will be weakened when you test your stress emotion.
3. Find the two neurovascular holding points associated with this emotion (Figure 36). Stretch your fingers and thumbs so that with each hand you are gently touching one of the neurovascular points on your forehead (see the center of Figure 36) as well as one of the points for your stress emotion. Begin with a deep breath to make the connection.
4. As you or your partner continues to hold these points, bring the scene vividly into your awareness.
5. Your partner will eventually feel a strong pulse where the neurovascular points are being held. If you are working alone, you may or may not be aware of feeling the pulse, but feeling lifted up and out of the stress is usually proof enough. You can also energy test for the emotion once more. The pulse means that the blood supply has returned and you have stepped out of the stress response loop.

Repeating this sequence until the memory no longer evokes a stress response is a significant step in reprogramming the stress response loop. For a major psychological overhaul, take another memory and defuse it. And

Figure 36.

NEUROVASCULAR HOLDING POINTS

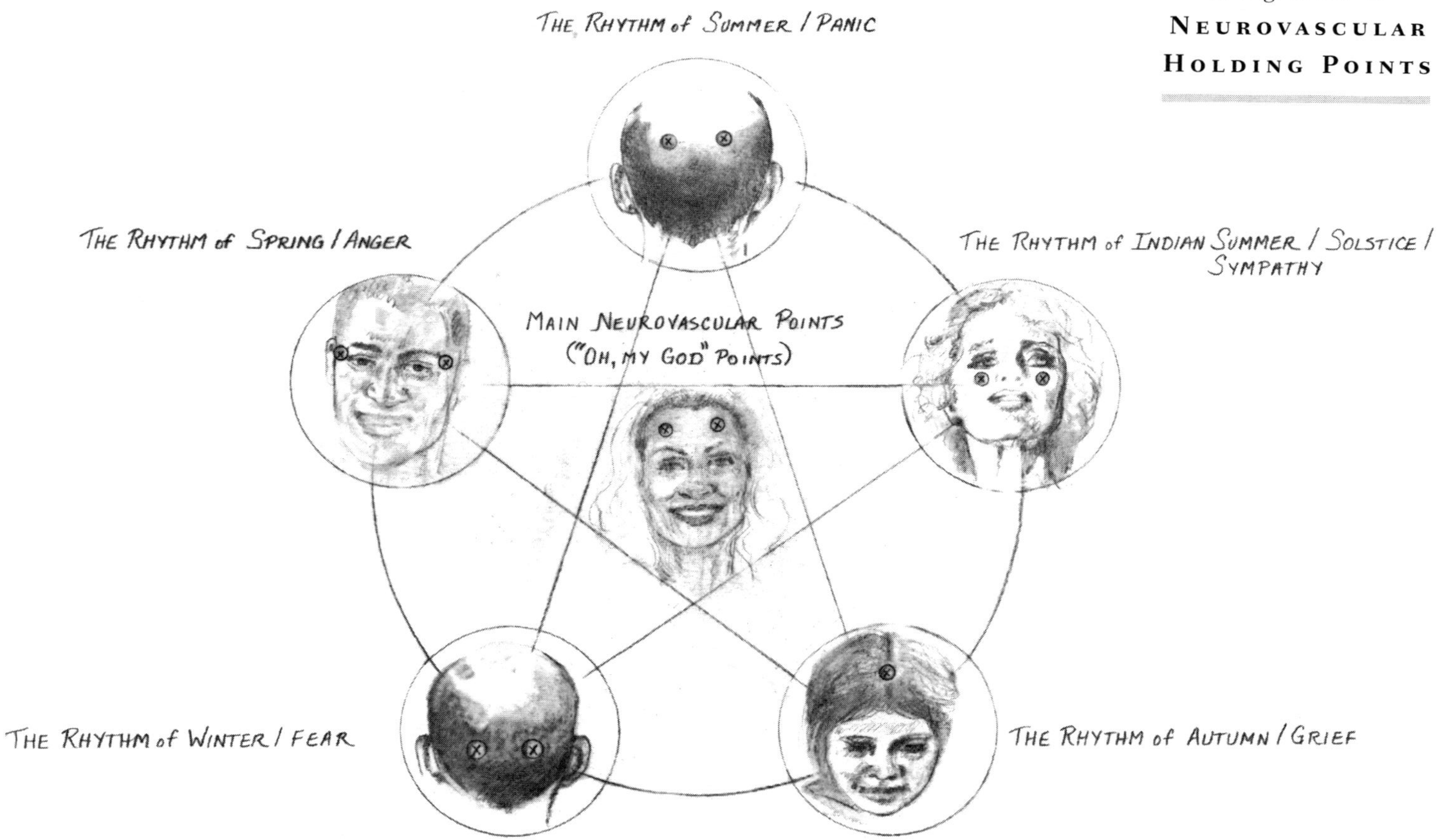

THE RHYTHM of SUMMER / PANIC

THE RHYTHM of SPRING / ANGER

THE RHYTHM of INDIAN SUMMER / SOLSTICE / SYMPATHY

MAIN NEUROVASCULAR POINTS ("OH, MY GOD" POINTS)

THE RHYTHM of WINTER / FEAR

THE RHYTHM of AUTUMN / GRIEF

another. The process will create a shift in the rhythm's stress response pattern.

Get into Rhythm with Your Rhythms. You can better attune yourself to each of your rhythms through the use of movements, images, and sounds designed to cultivate it and bring it into better balance with your other rhythms. I have borrowed from traditional practices from a variety of cultures in devising the following five exercises for enriching each of the rhythms within yourself. I suggest you choose one of these exercises and practice it daily for a time.

You may have, from the descriptions of the seasons, recognized in yourself a season that is problematic for you. This usually means that you vibrate to its rhythm too much or too little. Since the human journey seems to take us toward more fully embodying all five rhythms, an exercise that strengthens a rhythm that is weak in you or that balances a rhythm that dominates the others can be a very useful practice. Each of the following five exercises balances the rhythm of one of the seasons and pulsates its energy throughout your body.

A key in designing exercises for working with each rhythm is, again, the rhythm's stress emotion. As life puts us into its crucible, each rhythm's *sustaining mental state* tends to move into that rhythm's *stress emotion.* Winter's courage becomes fear, spring's assertiveness becomes anger, summer's passion becomes panic, Indian summer's compassion becomes a codependent sympathy, autumn's reflectiveness becomes grief. While humans may experience any of many stress emotions, these five tend to be primary. Others are often variations or combinations of them. Shame, for instance, is the terror (winter) of being cut off from a primary source of love combined with not having enough sympathy for self (solstice/equinox). Jealousy is a mixture of self-judgment (spring), fear (winter), and hysteria (summer).

Each exercise evokes and releases the rhythm's stress emotion. Even if you hardly ever feel this emotion, doing the exercise regularly may bring you in touch with hidden fear, anger, panic, sympathy, or grief. The exercise will also instill within you a primal way of releasing the emotion. If the emotion is a problematic one for you, the exercise can be valuable to use *at the time* you are feeling the emotion.

By stretching yourself into a stress emotion, you also make room for that emotion's healthy polarity. Becoming more viscerally familiar with fear makes space for courage; with anger makes space for assertiveness; with panic, joy; with codependent sympathy, compassion; with grief, reflectiveness.

Experiment each day with the exercise you select, and after a few days reread the description of its rhythm to see if you have experienced any changes in relation to that rhythm. Then either continue with the exercise for this rhythm or experiment with a different one. Let your intuition guide you. You may want to read the following instructions into a tape.

Winter/Fear: Blowing Out the Candle

When you have an unfounded fear (time—1 to 2 minutes):

1. Sitting on the floor, raise your legs to your chest, circle your arms around your knees, and hold. Or, sitting on a chair, bend over and hold your arms around your bent knees.
2. Take a deep breath. Rock back and forth, with your head lifted and staring into an imaginary blue flame.
3. On your exhalation, throughout the exercise, let the air pass through your puckered lips, making the soft sound of blowing out a candle, like trying to say "Woooooo" while only whispering.
4. Blow out the blue flame with each exhalation, feeling your fear going out with the flame.
5. Repeat until you do not feel as fearful.

Spring/Anger: Expelling the Venom

When you are feeling angry or judgmental or want to find the righteous anger within you (time—1 to 2 minutes):

1. Stand with your hands on your thighs, fingers spread, and take a deep breath.
2. With the exhalation, make a "Shhhhhhhhhh" sound, as if telling someone to be quiet.
3. With a deep inhalation, swing your arms out to your sides, and complete the circle by bringing them high above your head.
4. Turn your hands around so your palms are facing you and make fists.

5. With a loud "Shhhhhhhhhh" as you exhale, bring your fisted hands down swiftly, opening them as they drop near your thighs.
6. Bring to mind the source of your anger and make your movements quick and forceful. Repeat three times.
7. Feel your muscles and power unfolding as you slowly and deliberately pull down your arms.
8. Repeat until you feel the anger dissipate.

Summer/Panic: Taking Down the Flame

When you are panicked or hysterical or simply want to ground your energy (time—1 to 2 minutes):

1. Stand, placing your open hands on your thighs, fingers spread. Sense the energy travelling down your legs as you become more grounded.
2. Breathing deeply and slowly, make a sighing "Haaaaaaa" sound with each exhalation. Imagine the fire or chaos or clutter in your mind dissolving with each sigh.
3. Inhale deeply as you circle your arms widely over your head until the fingertips and thumbs of each hand meet.
4. Exhaling with the "Haaaaaaa" sound, bring your thumbs down to the crown chakra, above the center of your head, fingertips and thumbs still touching. Remain in this position as you inhale.
5. Exhaling with the "Haaaaaaa" sound, bring your thumbs to the point between your eyebrows—your third eye. Keep them there as you inhale.
6. Exhaling with the "Haaaaaaa" sound, bring your thumbs down to the point between your breasts, the "Sea of Tranquility," fingertips and thumbs still touching. Hands remain as you inhale.
7. Exhaling with the "Haaaaaaa" sound, bring your thumbs down to your navel, and roll your hands down so they form a pyramid beneath your navel. Keep them in this position with your next inhalation.
8. Exhaling with the "Haaaaaaa" sound, flatten your hands and stretch them down to the original position at your thighs. Keep them in this position as you inhale.

9. Exhaling with the "Haaaaaaa" sound, slowly move your open hands down your legs as you bend over, finally letting your hands hang down as you inhale.
10. Still hanging over, exhale with the "Haaaaaaa" sound. As you return to a full stand, inhaling, bring your hands up the inside of your legs and back to their original position. Exhale with a final "Haaaaaaa" sound.

Indian Summer/Sympathy: Cradling the Baby

When you find yourself compulsively helping others or are feeling too much compassion for others and not enough for yourself (time—1 to 2 minutes):

1. Stand, with your hands over your solar plexus, and tenderly cradle your midsection, as if holding a baby. Imagine you are bringing the mercy you show others back to yourself. Center yourself and take a deep breath in, exhaling with a controlled slow breath from the back of your throat, like a wind through a hollow, making a raspy sort of sound. Continue this breath until you feel centered.
2. Inhale, circling your arms above your head, stretching high.
3. Bend over on the exhalation, reaching your arms forward and down, finally placing the fingers of each hand underneath the inside arches of your feet. Pull your arches upward until you feel a stretch across your back.
4. Continue pulling up on your arches for one or two more slow, raspy exhalations.
5. Inhaling, return to a standing position, pulling your hands up the inside of your legs and then up the center of your body, finally stretching them high over your head once more.
6. Stretch your arms high, alternating each arm, continuing the slow, raspy breath.
7. Exhaling, bring your hands back to your solar plexus, again tenderly cradling your midsection with compassion.
8. Repeat until you feel a shift in your energy.

Autumn/Grief: Human Touching Divine

When feeling loss, grief, or loneliness or having difficulty letting go (time—1 to 2 minutes):

1. Stand erect. Round your arms in front of you, allowing your fingers almost to touch but not quite. Imagine you are holding the world and its problems, or your own world and its grief. Inhale deeply. With each exhalation throughout the exercise, make a "Sssssss" sound, like air being let out of a balloon.
2. Inhale, open your arms wide, releasing, surrendering, letting go of the world.
3. Exhaling, with another "Sssssss" sound, again round your arms in front of you and imagine you are holding the world. Your hands reach to touch one another but cannot.
4. Again, inhale, opening your arms wide, releasing, surrendering, letting go of it all.
5. Repeat, but this time your fingertips are able to reach one another, like in Michelangelo's fresco of God touching Adam.
6. Inhale, bringing your world close to your chest with your hands, one hand on top of the other, holding tight, and cherishing the world you are holding onto.
7. With a final exhalation, still making the "Sssssss" sound, let it all go, with your arms wide open. Letting go of the old, you open to the new.
8. Repeat until you feel a shift in your energy.

A RHYTHM RUNS THROUGH IT

Each of your body's energy systems vibrates to a specific rhythm. Each meridian, for instance, is associated with one of the five rhythms and reflects the qualities of that rhythm. The meridians that move with the rhythm of winter are bladder and kidney, with spring are gall bladder and liver, with equinox/solstice are spleen and stomach, with autumn are lungs and large intestine. Summer, in its abundance, is associated with four meridians: heart, small intestine, circulation-sex (which governs the pericardium), and triple warmer (which governs the thyroid and the adrenals). These relationships are precise and they are meaningful.

To give you a sense of the exquisite calculus that governs the interactions among your rhythms and your meridians, recall the chart (Figure 26, page 120) you used to increase the energies moving through your meridians by holding your acupuncture strengthening points. To understand the principles underlying the technique, look at the Rhythms of the Seasons Wheel (Figure 35, page 210). Your body's energy flows in the direction of the rhythms as they occur in nature (from winter to spring to summer to Indian summer to autumn). From these relationships, the system presented in Figure 26 was derived:

To strengthen a meridian, draw energy from a meridian that is of the season that precedes it. Because energy flows naturally in the direction shown on the Rhythms of the Seasons Wheel, you simply provide a jumper cable to expedite its natural movement. To strengthen a meridian that is of spring's rhythm, for instance, you hold the points that will draw energy from a meridian that is of winter's rhythm. Winter flows into spring. The specific points you would hold are described in Figure 26.

To sedate a meridian, release its energy into a meridian that is of the season that follows it. To sedate a meridian that is of spring's rhythm, for instance, you hold the points that will release energy into a meridian that is on summer's rhythm. Spring flows into summer.

Whether strengthening or sedating, jump a yang meridian with another yang meridian and a yin meridian with another yin meridian, like jumping a car battery (plus to plus and minus to minus). Each meridian is either yin or yang. Looking at Figure 35, you would strengthen liver (spring's yin meridian) by drawing from kidney (winter's yin meridian), and you would sedate it by letting it flow into heart (one of summer's yin meridians). From this elegant set of relationships, you can figure out how to strengthen any of the twelve basic meridians.

In this chapter you have explored the rhythms that run through your meridians, chakras, aura, Celtic weave, and basic grid. In the following chapter we will look at the final two energy systems, the triple warmer and the strange flows, each of which plays a vital role in the functioning of your immune system.

Chapter 8

Triple Warmer and the Strange Flows

The Energetic Arms of Your Immune System

Stealthy as a pirate slipping from a cove, the cancer cell severs the moorings that attach it to surrounding tissue. Slowly it extends one, two, three fingerlike probes and begins to creep. Then it detects the pulsating presence of a nearby capillary and darts between the cells that compose the blood-vessel wall. It dives into the red river that courses through lung and liver, breast and brain. An hour or so later, it surfaces on some tranquil shore [and] settles down. . . . It tricks nearby cells into forming food-bearing blood vessels, then compels them to churn out growth-spurring chemicals. To shield itself from patrolling immune cells, the cancer cell sprouts spiny armor like a sea urchin's. To expel the agents physicians send to kill it, the cancer cell deploys along its membrane a battery of tiny pumps. Is there a way to fight such a foe?

—Madeleine Nash
"Stopping Cancer in Its Tracks," *Time*

Is there a way to fight such a foe?" Your immune system is evolution's answer to that question. Scientists who study the immune system can only marvel at its capacity for memory, recognition, discrimination, anticipation, learning, and self-organization. But in spite of its remarkable sophistication, the immune system is operating on a battleground that has changed beyond recognition since it originally evolved. You can, however, greatly enhance the effectiveness of your immune system with a few well-chosen conscious actions.

Consciousness, you see, is evolution's answer to circumstances that change faster than our ability to adapt to those changes. In choosing a habitat, our predecessors were able to *consciously* factor in food availability, climate, other natural resources, and indigenous hazards. Preprogrammed instinctual responses just wouldn't do. Could nature have come up with anything better than to endow us with consciousness?

Since many of your body's preprogrammed responses are out of phase with today's world, you can add consciousness where your body lags behind. You can deliberately trace meridians, hold acupuncture points, and open energy blocks. And you can reprogram your immune system.

The immune system can, in fact, be conditioned like Pavlov's dogs. In one experiment, a group of people were given sherbet along with a shot of adrenaline, which increases the activity of the immune system. After several of these treatments, the injection was changed to an inert substance, yet the sherbet and the placebo injection continued to increase immune cell activity.[1] I know a woman who, as a girl, was eating red grapes at the moment she learned her mother had just died in a car crash. She has been violently allergic to red grapes ever since.

The immune system is responsive to emotional states, psychological interventions, and energy work. When you are angry, your antibody production increases.[2] Relaxation training can significantly improve the immune system's functioning.[3] So can directed imagery. Patients who prepared for surgery with guided imagery that helped them mentally rehearse active coping experienced less postoperative pain, requested less pain medication, and had more favorable blood chemistry measures than patients who did not use such imagery.[4]

Your immune system operates at the physical levels of the thymus, spleen, lymph, and bone marrow, but it is ruled by two energy systems. The Chinese practitioners named them the *triple warmer* and the *strange flows.* These energies are vastly different from one another in nature. The triple warmer aggressively mobilizes all the systems in your body to fight. The strange flows gently organize all the systems in your body to promote your health through a strategy of harmony and cooperation.

TRIPLE WARMER

My sense of triple warmer differs somewhat from traditional descriptions. Triple warmer is the meridian that networks the energies of the immune system to counter an invader, but it functions in ways that are beyond the range of any single meridian. It also operates as a strange flow. Strange flows do not follow specific pathways. They are more diffuse, and they intersect all of the meridians. Rather than staying on its own meridian line, triple warmer energy jumps its course and, like a strange flow, hooks up with the other meridians and organs.

The triple warmer meridian networks information to all the meridians and the organs they serve. It is governed by the hypothalamus gland, the body's thermostat and the instigator of the fight-or-flight response. In the states of fight, flight, or calm, triple warmer heats the body in three different ways. When the body is in relative balance, heat is distributed evenly throughout. When you become angry and are ready to fight, heat and energy rise. The chest, neck, face, and arms become red and flushed with blood in preparation for the battle. When your body prepares you to run away from danger, the heat and energy leave the upper part of your body and go to your legs so you can run faster. That is why your face turns white when you are terrified.

Though triple warmer seems to be a strange flow as well as a meridian, it is also conspicuously different from the other strange flows, which ensure cooperation, synergy, and peace. Triple warmer prepares the body for war!

The Militia Within. The cellular level of the immune system—with its lymphocytes, thymocytes, memory B cells, helper and killer T cells, antigens, and antibodies—is a military display that is dazzling in its design, complexity, and intelligence. It is the energy of triple warmer and the strange flows, however, that activates this extraordinary assembly. It is in the interplay of the triple warmer and the other strange flows that your immune system's strategy unfolds.

Triple warmer is like the king who conscripts an army from the various locales, organs, and systems. It has full authority, although it never conscripts energy from its queen, the heart. But if it so chooses, it can draft so much energy from any other part of the body that important systems become temporarily incapacitated—all presumably for the common good.

As commander-in-chief, the strategies available to the king are numerous and intricate. Passed down from one generation to the next, they evolved over many millions of years. Prototypes of the army's most basic maneuvers trace back at least to protozoa, which had, over two billion years ago, learned to recognize and destroy foreign invaders.

I used to resist trite military analogies when speaking of the immune system. I am a peaceful person, and I find the implication that a warlike intelligence in my body is necessary to keep me alive hard to embrace. The closer you look, however, the parallels become inescapable between the immune system and a nation's military (protecting against outside invaders) as well as its police force (patrolling the local inhabitants). I prefer other images, such as illness as a teacher or illness as a force to be embraced, and these are reflected throughout the book, but the military analogy can also be compelling.

The military analogy is even more apt today because triple warmer, like our exceedingly militarized civilization, has become as much a threat to the common good as the antagonists it was designed to oppose. Autoimmune diseases, for instance, where the immune system turns terrorist and attacks the body's cells and tissue, are a new breed of pestilence. Unless massive changes are made in the way we relate to the environment—whose pollutants overwhelm the immune system until it begins to treat almost everything as the enemy—autoimmune disease is on the docket as *the* illness of the future.

One of the most difficult challenges for modern armies is distinguishing between friend and foe. For most of the history of civilization, the main military problem was how to mobilize an army that was powerful enough to fight off flagrant enemies. But as the world has become more complex, international, and interdependent, destroying life and resources in any one part diminishes the whole.

Triple warmer's habit for millions of years has been to treat whatever it does not recognize as an enemy. While this strategy did sometimes result in kindred forces being decimated by "friendly fire," triple warmer was not confronted with much it did not already recognize. Today, however, we can transplant a kidney that the immune system can rightfully accept as friend or rightfully reject as invader. It did not evolve to make such distinctions. Moreover, we pump a greater variety of substances into the atmosphere in one day than, not so long ago, were generated by all of humanity in a century. Your body cannot possibly fight or even distinguish among all it encounters, and the job of triple warmer has become daunting. The immune system also faces unprecedented challenges in finding a balance between maintaining a strong enough protective force and perpetuating overkill and self-destruction.

This is not to say that triple warmer cannot rise to the occasion, but it needs allies. The forebrain, the seat of conscious thought, can join forces with the ancient and preprogrammed brain centers that control the immune system. Together they can counter the most convoluted hazards a body has ever faced. Some of triple warmer's most basic strategies are totally outmoded, as it tries to walk the line between overprotection and underprotection. Autoimmune and immune deficiency disorders are, respectively, fight-and-flight responses in extreme. In autoimmune disorders, an overactivated triple warmer fights its allies. In immune deficiency disorders, triple warmer has gone into retreat. Conscious action is required if we are to retrain our immune systems quickly enough. For the first time in history, we must consciously effect evolution. And we can!

Triple Warmer in Overkill. The threat of military insurrection increases when danger is perceived all around. We live in a world crowded with stressors which we can neither run away from nor overcome with brute force, where much of the food we eat and the medicines we take contain chemicals the human body did not evolve to assimilate, and where electromagnetic fields and industrial pollutants bombard us daily. Triple warmer energy, like an army abusing its authority to protect at all costs, often aims its weapons at friendly forces as well as invaders, at good citizens as well as traitors. Overwhelmed to the point of chaos, the immune system turns on the body it was designed to defend.

Triple warmer has no interest in your happiness or your spiritual development—only in keeping you alive. Unfortunately, its information about your survival has not had a major update in several million years. Yet triple warmer is still invested with the authority to override virtually any need in your body in order to mount an immune reaction. For your health, triple warmer ignites fevers and infections to fight disease. For your safety, triple warmer prompts the release of adrenaline to meet an emergency.

When I was twelve, the home of one of my classmates caught fire. After having safely fled from the house, she ran back into the kitchen, somehow encircled her family's new refrigerator within her four-foot, ten-inch frame, and *carried* it out of the house. Like a mother who lifts the front end of a car to save her child, triple warmer infuses the body with strength to accomplish seemingly impossible tasks.

Triple warmer is endowed with enormous power, but if it continually sets false alarms or sics the immune system's troops on the good guys, it can keep you on perpetual red alert. In chronic fatigue syndrome, the fight-or-flight response is perpetually engaged. Another response to threat is to freeze, a chemical reaction to the unexpected like that of a deer caught in headlights, commonly experienced when we were startled, yelled at, or attacked. Fight, flight, and freezing all exhaust the body.

Autoimmune Disorders. In illnesses that have an autoimmune component—such as rheumatoid arthritis, lupus, Crohn's disease, Addison's disease, and multiple sclerosis—triple warmer in its overvigilance directs the body to attack its own tissue. Beverly, a young woman who was crippled from multiple sclerosis (MS), had a session with me after she and her mother attended a class I taught in London. Beverly was on crutches. The trigger for multiple sclerosis can sometimes be an overwhelming shock to the system. Her illness traced back to a series of stresses during her teens, the crowning blow being the death of the grandmother who had raised her. It looked to me as if the nerves along the spine of this twenty-seven-year-old woman were still in shock, and this situation created a ripple effect in all her energy fields. It was as if her circuits were blown. Her triple warmer meridian was in a perpetual state of emergency, keeping her immune system, nervous system, and circulatory system oscillating among fight, flight, and freeze.

I wanted to interrupt this continuous emergency response. The entire first treatment was dedicated to communicating to her hypothalamus that she was safe. This would, in turn, bring triple warmer out of its perpetual state of overalertness. I taught her techniques she could use at home. While I knew it would take time for her body to build new healthy energy habits, her symptoms were noticeably reduced after just ten days of using these techniques twice daily. This provided strong incentive for her to learn more advanced techniques. After unscrambling her energies and helping them to cross over in a consistent manner, she began rebuilding the damaged nerve cells. Though conventional wisdom holds that such damage is irreparable, this has not been my experience.

Like many MS patients, Beverly was hypoglycemic. Consuming more protein proved to be an important dietary change, and through energy testing she was able to identify which forms of protein were best for her. Getting her cerebrospinal fluid pumping more vigorously was also important. After I returned to the United States, I monitored her over the phone about once each month, adding techniques as needed. When I saw her the following year, she was walking without crutches.

The year after that, Beverly was so markedly improved that the hospital that had once treated her was referring patients to her. She started holding MS self-help groups in her home to teach the techniques she had learned. She is now able to manage her illness so well that she is pursuing a career in energy kineseology, specializing in multiple sclerosis.

When you treat others who have your own vulnerabilities, certain hazards may arise. This is particularly true with autoimmune diseases, where the body often fails to make good distinctions about what is dangerous. Like many MS patients, Beverly had difficulties with boundaries, and this difficulty is mimicked by the disease. If Beverly is to continue to work with MS patients, she will be challenged from both sides. She will need to stay vigilant about her own boundaries while helping her patients with theirs.

How to Rid Yourself of an Allergy Headache

The One Nostril Breath works best on headaches caused by pollution or allergens (time—30 to 45 seconds):

1. *Close your mouth and close off one nostril with a finger.*
2. *Take four or five slow breaths in and out the open nostril.*
3. *Repeat through the other nostril.*

Allergies and Other Environmental Afflictions. While triple warmer imbalances are implicated in diseases ranging from MS to diabetes, simple allergies illustrate the basic dynamics of any immune system overreaction. Consider, for instance, an allergy to dust. Dust is a potential danger to the lungs. The respiratory system is equipped, through coughing or sneezing, to keep excessive amounts of dust from entering the nasal passages. In the case of an allergy to dust, triple warmer has registered the respiratory system's violent reaction. It sends an alert to the entire body that dust is a danger and should be avoided even at great cost. All systems go on alert. Dust on the skin or the hair then provokes the same response originally produced by dust in the nose.

An allergy to dust is based on flawed logic; it is a faulty generalization. Registering that dust is toxic to the respiratory system, triple warmer mounts a defensive response whenever it detects dust. Sometimes a substance that happens to have been present during a traumatic event becomes an allergen through conditioning, as with the girl who was eating red grapes when her mother died. Familiar as well as unfamiliar substances may be targeted in such immunologic overreactions.

John, at thirteen, was so incapacitated by hay fever every spring that he could not function in school. He was miserable, with swollen, red, itchy eyes, wheezing, and an inability to think, though he was generally an excellent student. He dreaded the approach of spring each year. I took him to a field near his home and energy tested every weed and plant that might be causing his allergy. On finding the offending plant, I brought both him and the plant back to my treatment room. I placed the plant on his stomach and in both of his hands. He was miserable. I held acupressure points that sedate triple warmer.

Immediately, some of the reactions calmed slightly, but triple warmer was still fighting like a warrior, battling to protect his body. My next step, while leaving the offending plant on his body, was to balance all his meridians. I also strengthened his triple warmer meridian by tracing it several times. This provides a shot of pure adrenaline. Triple warmer governs the adrenals. When it has been compulsively drawing energy from other meridians, funneling energy directly into triple warmer decreases its need to pull so much energy from the other meridians. The risk of his triple warmer treating me and these interventions as invaders was diminished because I had immediately established a direct alliance with it. As his meridians were rebalanced, triple warmer finally backed off. John lay there in my treatment room without any reaction to the plant whatsoever.

But allergies and environmental illnesses can rarely be corrected in a single treatment. Few problems are more challenging and elusive. Even though the person may seem fine for several days, triple warmer's habitual patterns tend to return, and new patterns may not hold. The procedures must be repeated daily, for about thirty consecutive days according to my experience, before the new habit becomes stable. To build a new pattern, John had the daily assignment of touching the plant while another person held triple warmer sedating and strengthening points, followed by tapping several acupuncture and lymphatic reflex points. He was without symptoms the following spring.

With all allergies and environmental disorders, you first have to figure out what triple warmer has falsely judged as foe, and then you have to outsmart it so it doesn't initiate a military alert. Exposing John to the plant while balancing all his energy systems established a new pattern that eliminated the message of danger. His immune system's reaction to the plant was, in the sense of Pavlovian conditioning, reconditioned.

Triple warmer has been one of the great success stories in species survival, and it does not have much incentive to alter its ways. Its habits are set, and to try to change them is a formidable undertaking. Triple warmer also tends to treat the healer or the healing intervention as a foreign invader. Beyond that, the toxic environment is always there, so the body is under real and present danger, even as you are trying to teach triple warmer to lay off the emergency buzzer.

How can you protect yourself? Individually and collectively, we of course need to become smarter about what we are doing to our habitat and about the foods we ingest. Energy testing provides instant information that can alert you to immediate perils. It is not necessary to eat food that is going to impair your energies. Energy test it. It is not necessary to buy a formaldehyde rug that is going to be continually draining your energies. Energy test it. Your immune system is challenged enough without your smothering it in a lifestyle that exposes you to unnecessary hazards. Our stereotypes about what is and what is not healthy are not, however, reliable guides. With my reputation in my hometown as a spokesperson for natural healing, it's very funny to me to see people's reactions when they "catch" me eating a hot dog at the local 7-Eleven. Sometimes hot dogs test weak on me, but with my hypoglycemia and a sodium deficiency, they are at other times nature's perfect food.

A deep uneasiness that comes with this admission, however, deserves comment. I would much prefer to be a vegetarian. No explanation or justification gives me comfort about the suffering inherent in the fact that all life sustains itself on a food chain composed of other sentient beings. But I am also inescapably aware that if I do not consume an adequate amount of animal protein, my particular body and mind function at a significantly reduced level of effectiveness and I am far more vulnerable to illness. All I know to do is to offer my humble and awkward prayers of gratitude for those creatures whose very flesh sustains my health and well-being and to maintain practices that help counterbalance the ecological costs of this choice.

Energy Techniques for a More Discriminating Immune System. You can convert your immune system from a mean, indiscriminate, fighting machine to an astute, discriminating, protective friend. To get a compromised immune system functioning well, you must first make sure your energies are crossing over from the left hemisphere to the right side of the body and from the right hemisphere to the left side of the body.

HOMOLATERAL CROSSOVER. When your energies are moving straight up and down each side of your body, like parallel lines, the pattern is referred to as homolateral. You are operating at less than 50 percent efficiency. *You cannot get well if your energies are homolateral.* It is that simple. It is hard to think clearly. You tend toward depression as all of your physical processes slow down. Your senses are less acute—you can't see, hear, smell, touch, or taste as well as at other times. You feel less alive. Your triple warmer may be hypervigilant, yet your immune system can't heal you from illness.

Whatever else you may be doing to improve your physical health, if your energies are homolateral the benefits won't last. In fact, the best exercises I know, even those I am suggesting wholeheartedly, may not work if you are in a homolateral state. Even walking will weaken you. Walking has a natural crossover effect, and if your energies are homolateral, walking goes against the flow of your energies. The very kinds of exercise that should benefit you wipe you out. Fortunately, you can get your energies crossing over again.

One way to determine if your energies are crossing over properly is to attempt a Cross Crawl (page 69). If you find that the Cross Crawl is difficult for you, that you cannot easily coordinate your opposite arms and legs, or that just starting to do the Cross Crawl confuses or exhausts you, you are probably in a homolateral state. If your energies are running in a homolateral pattern but you are walking or marching in the natural Cross-Crawl pattern, you are literally moving against your own flow. There is also an energy test to determine if your energies are homolateral:

1. Draw a large X on one piece of paper and two parallel lines on another.
2. Look at the X. Have someone energy test you.
3. Look at the parallel lines. Energy test again.

If your energies are crossing over properly, looking at the X will be in harmony with your internal state, and you will test strong. Looking at the parallel lines tests weak because it counters the flow of your energies. But if your energies are homolateral, the opposite will occur; the X will weaken you and the parallel lines will strengthen you.

Your energies are probably homolateral if you are chronically exhausted or ill and for unknown reasons can't get well. You can have someone energy test you if you are unsure. The strategy for getting your energies to cross over again begins by physically aligning your body with the parallel pathways in which its energies are already flowing. You begin to move in harmony with your reversed energies when you start with the homolateral crossover, and then your energies will gently start to cross over (time—about 4 minutes):

1. Begin with the Three Thumps (page 63), breathing deeply. They wake up energies that become stagnant when you are homolateral.
2. Do Separating Heaven and Earth (page 249) or the Wayne Cook posture (page 73).
3. March in place, lifting your right arm with your right leg and then your left arm with your left leg. Breathe deeply throughout the entire routine.
4. You can adapt these instructions for sitting or lying down. If you are too ill or too weak to move your limbs, another person can lift them for you. But find a way so you can do it even if you don't have some-

one to help you. If you are doing this exercise while lying down, you can place your legs on large pillows; if you are sitting, you can place them on a stool so you don't have to lift so high. This is *not* a "no pain, no gain" arrangement. Straining yourself sends your energies back into homolateral, so make it easy on yourself and rest whenever necessary.

5. After about twelve lifts of your arms and legs in this homolateral pattern, stop and change the pattern to a Cross Crawl—lifting the *opposite* arm and leg—again for about twelve lifts. If it is difficult to coordinate a cross crawl, you can touch your right hand to your left knee and your left hand to your right knee as you step.
6. Repeat the pattern twice more—twelve homolateral movements, then twelve Cross Crawls. Anchor it in by ending with an additional twelve Cross Crawls.
7. End with the Three Thumps. Now that your energies are moving in their natural direction, the benefits of tapping the K-27, spleen, and thymus points will be multiplied and your vitality enhanced.

Do this routine twice each day until the new pattern becomes established. It might take from ten to thirty days for a crossover pattern to stabilize, but you will also feel immediate benefits. And the long-term benefits can tip the balance from being ill to getting well or from being depressed to feeling hopeful.

SEDATING TRIPLE WARMER. Any time you are feeling overwhelmed or a little crazed, you can sedate the triple warmer meridian. A quick method is to flush it (time—under 20 seconds):

1. To flush triple warmer, trace it backward (see Figure 18, page 105). Breathing deeply, place the fingers of one hand on the opposite temple, trace around your ears and down your shoulders, and pull the energy off your fourth finger.
2. Repeat several times on each side.

A second method is to "smooth behind the ears," which traces a portion of the meridian backward and sedates triple warmer (time—1 to 2 minutes):

1. Rest your face in your hands, palms at your chin, fingers at the temples. Hold this position for two breaths.
2. Breathe in deeply and lift your fingers 2 or 3 inches, smoothing the skin from the temples to above the ears.
3. On the exhalation, circle your fingers around your ears, draw them down the sides of your neck, and hang your hands on the backs of your shoulders, pressing your fingers into your shoulders.
4. Stay in this position through at least two deep breaths. Then drag your fingers slowly over your shoulders with pressure. Once your fingers reach your clavicle, release them and allow them to drop where they may.

A third technique is to totally relax into the experience by having another person hold your triple warmer sedating points. Anxiety, anger, or terror melts away as you surrender to another's touch. Do not be overly concerned about being *exactly* on the point. By using the pads of three fingers, you won't miss. To sedate another's triple warmer (time—about 6 minutes):

1. Place your fingers on the "first" sedating points on either side of the body (see Figure 26, pages 120–123). One point is just above the elbow, in line with the fourth finger. The other point is located a hand's width beneath the knee, just outside the shinbone. Hold for up to 2 minutes.
2. Reverse the position, holding the "first" points on the other side. Again hold for up to 2 minutes.
3. For the "second" sedating points, place your middle finger in the indent at the outside of the little toe. At the same time, place the fingers of your other hand about an inch below the other person's fourth and fifth finger joint. Hold for about a minute.
4. Reverse the position, holding the "second" points on the other side. Again hold for about a minute.

REPROGRAMMING TRIPLE WARMER TO STOP ATTACKING FRIENDLY FORCES. If you are allergic to a food, a plant, or other substance that most people find to be user friendly, you can usually retrain your immune system. You will need a bit of that substance for this procedure. Use good sense—if you cannot at all tolerate placing the substance on your body, you will need a professional to help you reprogram your body's response to it. This method is presented in three variations, and they can be combined for a cumulative effect. The first does not require a partner. Begin with the Cross Crawl (page 69) or the homolateral crossover just described if your energies are running homolateral, and the Wayne Cook posture (page 73); then sedate triple warmer, as just described.

Variation 1 (time—less than 2 minutes):

1. Place the suspected allergen on your body.
2. Breathe deeply while vigorously doing the Three Thumps (page 63).
3. Tap the stomach points on the bone beneath your eyes (see Figure 37).
4. Tap the triple warmer points at the back side of each hand between the bones that separate the fourth and fifth fingers (see Figure 38c). Tap each point vigorously for about 20 to 30 seconds.

Variation 2 (time—about 5 minutes):

1. With someone else holding your triple warmer sedating points (page 122), place the suspected allergen on your body.
2. Turn over, place the allergen under your stomach or on your back, and have your partner treat you to a Spinal Flush (pages 79–80).
3. End by doing or repeating Variation 1.

Variation 3 (time—about 3 minutes):

1. Place the suspected allergen on one of your fourteen alarm points (Figure 25, page 112) and have someone energy test to find out which meridian is affected. Continue through all fourteen alarm points.
2. For each meridian that shows weak, tap its acupuncture strengthening point for 15 to 20 seconds (see Figure 38).
3. Tap the stomach and triple warmer points shown in figures 37 and 38c.
4. End with the Three Thumps (page 63).

Figure 37. ■

The Stomach Points

Figure 38. ■

Acupuncture Tapping Points

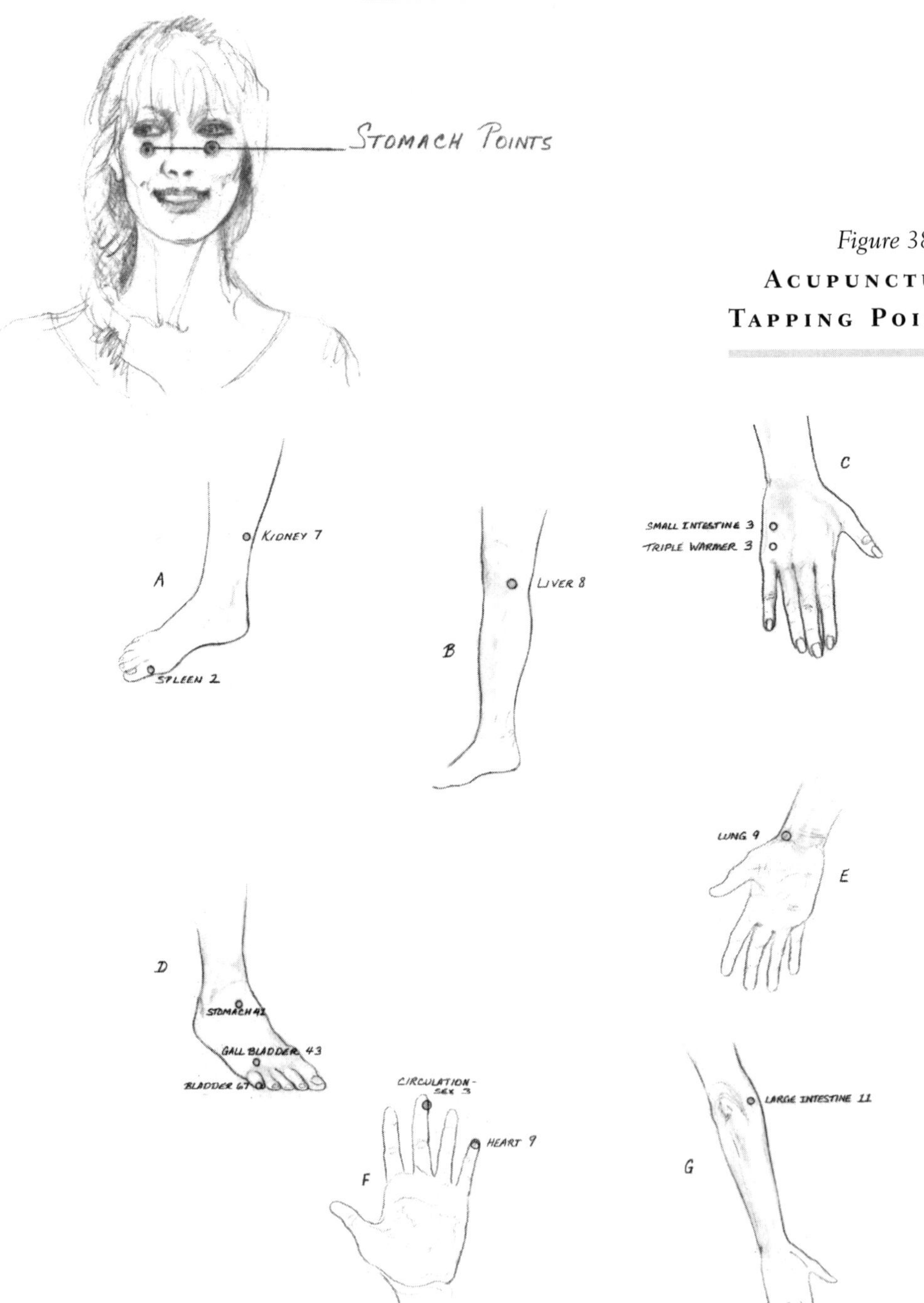

You may see immediate results or improvement within two or three days. I suggest, however, that you use the techniques daily and then retest after about ten days. It took me seventeen days using these methods to overcome an allergy to wheats and grains. If your health is frail or is deteriorating, your body may not reverse an allergy. Triple warmer may be too threatened, its priorities may be elsewhere, or it may know something you don't about the substance in relation to your body's unique chemical and energetic makeup. If you find you do not become desensitized, believe in your body's wisdom and know that this substance is, at least for the time being, not a friend.

REPROGRAMMING TRIPLE WARMER TO STOP ATTACKING YOUR BODY. Triple warmer governs hysteria. When you are hysterical, you may find yourself yelling at someone you love, or "losing it." Imagine triple warmer as an inner police chief who is working overtime, twenty-four hours every day, with no holidays, trying to protect you. He is giving his all to safeguard you, yet in the case of autoimmune disorders, he is receiving continual reports that the dangers facing you are becoming endemic. But the mayor and the city council (you and the other systems of your body) are just ignoring the problem. You are living your life the way you've always lived it. You've got the same job, the same spouse, the same kids, the same stresses, the same eating habits, the same polluted environment. So the police chief redoubles his efforts. But the crime scare continues, and the mayor says, "Sorry, just keep doing your job." Eventually, the police chief snaps. He has been giving his entire life caring for you. In charge of a deteriorating situation, he has been feeling utterly alone. Now he loses it. He turns his weapons on the city council. This gets your attention.

To begin reprogramming your immune system, the first thing you can adjust is your attitude. For instance, self-judgment triggers different biochemical pathways in your immune response than self-compassion. Activating compassion for yourself can be a critical step in reversing autoimmune disorders. Rather than being angry at your body because of an allergy or autoimmune disease, you can help a crazed triple warmer meridian loosen its grip by entering into conscious partnership with it. It is not enough to tell it to lay down its weapons. That only makes it crazier. But if you come in with energy techniques that are attuned to your body's needs, triple warmer senses that it is not the only one fighting in your behalf, and some of its panic immediately dissolves. Energy medicine in fact offers a better surveillance system than triple warmer's two-million-year-old equipment, and you can help it update its strategies and promote its sense of safety. Keeping your energies crossing over and balanced serves to ensure your immune system that a higher intelligence is on the job, assisting it with its critical mission.

Emotional Overload. A young woman burdened with a secret came for a session. Her secret was that she feared she would harm her children.

Extreme stress showed in her aura as a purple-black energy. It was surrounding her body and suffocating all her other energies. She was frightened and overwhelmed and unable to deal with the simplest things. She constantly screamed at her children, and she cried herself to sleep every night. Her husband had his own struggles trying to earn enough to support their three young children, so she did not feel she could complain to him. She put a lid on her self-expression, and I believed she was indeed a danger to her children.

I held her neurovascular points a long time, perhaps 30 to 40 minutes until the panic finally calmed and the stress began to stream out of her body. I taught her exercises to do at home to unscramble her force field and bolster her reserves. Along with the Daily Energy Routine, she began her mornings with the Hook Up (page 119) and Separating Heaven from Earth (page 249). When she felt extreme stress, she would breathe out the rage and anxiety and then sit down and smooth behind her ears. She held her neurovascular points whenever she took a bath. At the end of each day, her husband relieved her tremendously with the Spinal Flush and by holding her triple warmer sedating points. I have continued to see her on occasion for years now. From the time she first learned how to manage her triple warmer energies, she was no longer miserable and afraid, and I was confident that her children were no longer in danger.

Fight, Flee, Freeze, or Calm Triple Warmer. While violence toward a child or a spouse can never be condoned, the tendency toward violent outbursts is more complicated than a simple failure of discipline or moral strength, and preventive measures that do not recognize the broader dynamics perpetuate the problem.

When triple warmer stays in emergency alert and your body is unable to rid itself of the stress hormones by fighting or fleeing, the impulse is still to take the most immediate route possible, which often involves blowing up inappropriately. That this psychological response has such a strong physiological basis should afford some compassion for yourself and others.

A judge who had taken a class with me referred a twenty-five-year-old man who had, in a single episode, beat up his wife, hit his child, and attacked a neighbor who tried to intervene. Up to that point, no one had ever seen him even be angry. He was always the supportive one, the nice guy. The judge told me, "Before I sentence this man, I want to go out on a limb and ask you if this is one of those 'triple warmer' things." It was. The man was responsive to the routine I show you here, and having two more individual sessions and taking a class with me was his sentence. His recidivism rate has been zero, and he still schedules an occasional session when he thinks his stress level is getting out of hand.

If clients come to me distraught, terrified, overwhelmed, explosive, or suicidal, I generally sedate triple warmer and clear the neurolymphatic points early in the session. This provides the release they are needing without their having to scream, explode, run away, or beat somebody up. Once triple warmer has been sedated, all the other meridians become stronger, and the body responds as if a crisis has passed. While sedating triple warmer curbs its overvigilance, it does not compromise its effectiveness in protecting you.

Easing Emotional Overload. When you are feeling emotional overload (time—about 2 minutes):

1. Begin with either or both of the following: The Wayne Cook posture (page 73) or Separating Heaven from Earth (page 249).
2. Place the fingers of both hands so they meet at the back of your neck, touching the top of your shoulders. Push in and pull your fingers apart. Repeat and work up your neck, up over your head, culminating in the Crown Pull (page 77).
3. Place your thumbs under your cheekbones and your middle fingers at the bridge of your nose. Pushing up firmly to the middle of your forehead, stretch the skin apart.

When Triple Warmer Needs a Boost. Often, when an illness seems to be winning, the best interventions strengthen the immune system rather than attack the disease. When triple warmer disables you, leaving you in a stupor or fever, it is on the job. It is networking your thymus, spleen, and lymphatic systems into an acute immune response, while preventing you from lavishing your physical and emotional energy on anything but survival and healing. In autoimmune and environmental disorders, it is vigilantly engaged when it should be relaxing. At other times, triple warmer retreats from battles that need to be fought. Perhaps it collapses in exhaustion. Perhaps it has been fighting for so long with no victory in sight that it simply gives up. Perhaps it is patterning itself after your own disinclination to set boundaries around people or obligations that are invasive.

Immune Deficiency Disorders. Your immune system can go astray by *modeling* itself after your behavior. Some immune disorders are, in fact, exact analogies of a person's ways of relating to the world. People who are highly suspicious, for instance, tend toward an overly vigilant immune system. At the other extreme, people who cannot say no tend to have an ineffective inner guard. The immune system, by modeling itself after their behavior, can fail to protect them. Rather than a hypervigilant immune system, this is an *immune deficiency disorder.* To strengthen triple warmer:

1. Unscramble your energies, using the Daily Energy Routine (page 86).
2. Stretch your body in ways that feel good to you.
3. Hold the triple warmer neurovascular points (see Figure 42, page 274) and the acupuncture strengthening points for triple warmer (see Figure 26, pages 120–123).
4. Become familiar with the neurolymphatic points in Figure 9 (pages 84–85) and massage them.
5. Balance your chakras (pages 168–169).
6. Mobilize your strange flows as described in the following section.

By doing these techniques daily, you can build up your immune system and instill into it habits that will maximize its effectiveness.

Other Times to Strengthen Rather than Sedate Triple Warmer. While sedating an overactive triple warmer is helpful for managing most illnesses, there are times when strengthening triple warmer can save a life. If a person has gone into anaphylactic shock (a life-threatening allergic reaction), strengthening triple warmer can reverse it. In fact, in any situation where the medical treatment might involve a shot of adrenaline, such as an asthma attack or a bee sting, strengthening triple warmer can provide extra adrenaline. Another time to strengthen rather than sedate triple warmer is if a patient is "slipping away."

THE STRANGE FLOWS

If triple warmer mobilizes your "inner militia," the strange flows mobilize your "inner mom." They support, inspire, strengthen, and cajole all your organs and energy systems to function as a tightly knit family. Whereas triple warmer protects you using the principle of conflict, the strange flows protect you using the principle of harmony. Their idea of a good defense is a radiant and well-integrated community of organs, glands, vessels, and energies. Rather than relying on a military approach, the defensive strategy used by the strange flows resembles that of community policing, where police officers see their job as helping the local community become strong and naturally resistant to crime.

When a body is peaceful and content, the strange flows are in the background and not as noticeable. But let there be an inner disturbance, let there be a need to bring the body back into balance, and the strange flows leap into action, moving in all directions, jumping any chasm to coordinate energies among the rhythms, meridians, chakras, aura, Celtic weave, and basic grid. They are the advance men, ensuring that all systems will work for the common good, redistributing energies to where they are most needed, and preparing the way for a coordinated response to any happenstance the body might encounter.

The Strange Flows Are "Psychic Circuits." If we look through the lens of evolution, the strange flows have probably been around much longer than the meridians. I base this statement on the simple fact that in more primitive organism such as insects and the few amphibians I've examined, I see strange flows but not meridians. They presumably also existed prior to the more aggressive defensive strategies of triple warmer. The holistic strategy used by the strange flows is that by maintaining the body's integrity, the body is protected. Triple warmer's aggressive defense system is a more highly specialized development.

The energies that pulse through organisms that do not have meridians are diffuse rather than concentrated. As physiological systems became more differentiated, however, it was necessary for energy systems to become more specialized and focused as well. Where the strange flows can travel anywhere, the meridians have beaten a path to the organs they serve. Even though the strange flows have been around longer, the meridians have more form, so they were mapped earlier. When the Chinese sages later detected an energy that connected the body's energy systems but did not flow along the meridian lines, it must have seemed strange to them—hence the name. They also called them "curious conduits" because of their ability to turn on the senses. Traditionally they have been referred to as the "psychic circuits" because they are so responsive to what a person intuits or thinks. If you *think* about a particular place in your body, strange flow energy instantly moves there. My students have come up with other terms as well, including the "magic markers," "mystical maps," and "wondrous wires." Strange flows are the energetic bridge between a thought and the activity of the brain's neurotransmitters.

The central and governing meridians are probably the evolutionary link between the strange flows and the meridians. They jump their course like strange flows but are more efficient because they have established pathways like meridians. Whereas the primary action of the strange flows is to communicate and coordinate, the meridians added the function of *transporting* specific energies to targeted areas, which allowed for the development of more specialized organs. But this advance came with a price. The strange flows can instantly go wherever they are needed, but the meridians are tied to specific pathways.

The triple warmer and spleen meridians also retain a close resemblance to their heritage as strange flows, allowing for both efficiency and mobility. Central and governing presumably evolved in tandem with the central nervous system, and the triple warmer and spleen meridians in tandem with an increasingly sophisticated immune system. They represent the complementary but polarized self-protective strategies of aggression and cooperation.

They are polarities not only in their strategies but in their nature. Each of the fourteen meridians is paired with an opposite force; each yin meridian is balanced against a yang meridian (see Figure 28, page 126), and the triple warmer meridian (yang) is the polarity of the spleen meridian (yin).

The Most Ambitiously Peaceful Strange Flow. Unlike the spleen meridian, triple warmer departed from the peaceful demeanor of its strange flow brethren. It is as if one day long ago one of the strange flows figured out that it could protect the body by attacking outside invaders. This strategy succeeded in areas where the time-honored peaceful strategies had failed, and this strange flow was eventually promoted to the rank of the triple warmer meridian. Simultaneously, as if to keep a balance, the most ambitiously peaceful strange flow was promoted to the rank of the spleen meridian.

The spleen meridian governs the spleen and pancreas, and it is also involved with the thymus, lymph, lymph nodes, tonsils, and bone marrow. It oversees your blood supply, metabolism, homeostasis, the production of antibodies, and the distribution of nourishment throughout your body. There are ten strange flows: the belt flow, penetrating flow, left and right bridge flows, left and right regulator flows, and the four meridians that double as strange flows. With its dual citizenship as a strange flow and a meridian and its responsibilities for blood supply, metabolism, homeostasis, antibodies, and nourishment, spleen is the leader of the pack. Its model of protection is to foster a vital defense by maintaining a vital organism. And the same techniques that empower spleen also support each of the other strange flows.

If spleen is the paragon of peaceful defense, you might wonder what it is doing helping to manufacture all those warlike white blood cells. Like women working in a munitions plant or Red Cross nurses in a war zone, the genial spleen energy is a good citizen willing to do its part for the war effort. But its demeanor is so contrary to triple warmer's that it sometimes seems that evolution engineered a bad practical joke when it made spleen and triple warmer the two arms of the immune system. Yes, if they can blend their respective strategies—harmonizing and fighting—into a coordinated teamwork, it is a potent combination that makes for peace, health, and happiness in our lives. But the blunders that can and do happen are often disastrous.

How to Get Back in Control
If You Feel Hysterical

If you feel you are "losing it" (time—3 to 4 minutes):

1. *Do the Separating Heaven from Earth exercise (page 249).*

2. *Hold the pads of your fingers on your forehead and your thumbs on your temples for at least a minute (page 90).*

3. *Smooth behind your ears (page 236).*

4. *"Expel the Venom" (pages 219–220), Zip Up (page 82) with a deep breath, and let go.*

When danger is perceived, triple warmer pulls rank and conscripts energy from the spleen meridian. This immediately weakens the body, compromising the mechanisms that maintain metabolism, homeostasis, nourishment, and blood supply. Then triple warmer orders spleen to use its remaining energy to fight. But spleen, the "inner mom," can only try to love the enemy. Nothing else is within its repertoire. In this gambit, we have not evolved intelligently. When the spleen meridian is directed to do what it does not do well at the expense of performing the critical functions it does superbly, the body is more vulnerable to being invaded. Triple warmer senses this vulnerability, declares that a state of acute emergency exists, and pulls even more energy from spleen. The immune response can thus have the effect of severely hindering rather than optimizing the body's strength.

While triple warmer and spleen usually manage to establish an adequate if sometimes strained teamwork, in cases of immune dysfunction their tragicomic dance can be devastating. But it is much easier than you might think to come in as the coach who understands each player's strengths and weaknesses and reestablish elegant teamwork. The techniques you have learned in this chapter help triple warmer to relax, to improve its ways of making assessments, and to back off when its teammates can do the job better. You don't need to reprogram spleen, just speak to it with empathy and understanding. If you can support spleen energy, spleen energy will support you.

The spleen meridian and the other strange flows respond to your thoughts instantly and scrupulously. They are more affected by your ideas, images, and beliefs than any other energy system. Depending on the nature of your thoughts, however, this may be for better or for worse. The relationship between your conscious thoughts and the strange flows reminds me of what Colin Wilson calls "the Laurel and Hardy theory of consciousness."[5]

Based on the differences between the left and right cerebral hemispheres, he liken our two inner "selves" to the opinionated and willful character of Oliver Hardy and his emotional and immensely suggestible sidekick, Stan Laurel:

> *When you open your eyes on a wet Monday morning it is Ollie who assesses the situation and mutters, "Damn, it's Monday and it's raining. . . ." Stan overhears him and—being suggestible—is thrown into a state of alarm.* "Monday, *and it's* raining." *So he fails to send up any energy. And if you cut yourself while shaving and spill coffee down your shirt-front and trip over the mat in the hall, each mini-disaster causes Ollie to groan, "It's one of those days . . ." while Stan becomes practically hysterical with gloom.*
>
> *Consider, on the other hand, what happens when a child wakes up on Christmas morning. Ollie says, "Marvelous, it's Christmas," and Stan almost turns somersaults of delight. And of course sends up a spurt of energy, which produces a feeling of well-being. Everything reinforces the sense of delight: the Christmas presents, the lights on the tree, the smell of mince pies. . . . The result is that before the day is half over, the child can experience an almost mystical sense of sheer ecstatic happiness, a feeling that life is self-evidently marvelous.*

The symbiosis between your conscious thoughts and your strange flows can be as decisive for your health as this dialogue between your inner Ollie and Stan can be for your mood. The effectiveness of your immune system sometimes rests on your ability to play an upbeat Ollie opposite your spleen's Stan. The response of the spleen meridian and the other strange flows depends on whether your appraisal is "It's Monday, it's raining, and it's going to be a dreadful day" or "It's Monday, it's Christmas, and today will be glorious!" Your chemistry follows your thoughts,[6] and your strange flows are the energetic link.

Self-hypnosis, guided imagery, affirmations, or changes in attitude can impact your strange flows and bolster your immune system. Getting back to nature, seeing a sunset that staggers your senses, or sitting by a peaceful brook can all awaken your strange flows. So can good news or laughter. Norman Cousins's book, *Anatomy of an Illness,* elevated laughter to the status of medicine, even within many medical circles.[7] The book describes how Cousins decided that he had to find an alternative to the hospital regimen of lifeless food, powerful anti-inflammatory medications, and constant interruptions if he was to overcome a life-threatening connective tissue disease. His successful alternative regimen included megadoses of vitamin C and a steady diet of *Candid Camera* and Marx Brothers videos. Cousins calculated: "Ten minutes of genuine belly laughter had an anesthetic effect and would give me at least two hours of painfree sleep." Laughter signals to triple warmer that all is well; you are not in danger. Laughter is an elixir that heals.

When laughter, pleasure, or delight boosts your spirits, your strange flows start pumping. Suppose you are in despair, feeling hopeless, going down the tubes after your husband has once more angrily let you know how your shortcomings and blunders are responsible for his bad moods and are ruining the marriage. But then he calls and says, "My men's group just helped me see that it was all my fault, I've been undervaluing you for our whole relationship, I love you madly, let's get some couples' counseling with that therapist you like so much so I can understand you better." Suddenly your mood is elevated, the whole world looks good, and an amazing energy fills your body. This is how the strange flows work.

Keeping Your Strange Flows Flowing. Sometimes I depend on strange flows. I know if I don't have time to do all I want with a person, if I get the strange flows flowing, the doctor within can begin to make magic. Two powerful exercises for jump-starting the strange flows are Heaven Rushing In (pages 21–22) and Separating Heaven and Earth.

Separating Heaven and Earth. In addition to activating spleen and turning on your strange flows, this exercise is a powerful stretch that releases excess energies while bringing fresh oxygen to the cells. It opens the meridians, expels toxic energies, and stimulates fresh energy to flow through the joints. Most people tend to collect too much energy rather than having too little. We need to empty ourselves of this surplus or it becomes the etheric equivalent of sludge. Separating Heaven and Earth is an excellent exercise any time you are starting to feel poorly, perhaps threatened by a cold or the flu. If you do healing work, it moves energies out of your system that you may have picked up from a client. I thought I invented this exercise, but I have since found variations in other countries, including Egypt, China, and India. Perhaps it has been passed down through so many centuries and cultures because we have an instinct to use these motions for releasing accumulated energies and inviting in a fresh supply (time—under 2 minutes):

1. Stand with your hands on your thighs, fingers spread.
2. With a deep inhalation through your nose, circle your arms out, having your hands meet at chest level, fingers touching, in a cathedral or prayerful position. Exhale through your mouth.
3. Again, with a deep inhalation through your nose, separate your arms from one another, stretching one high above your head and flattening your hand back, as if pushing something above you. Stretch the other arm down, again flattening your hand back, as if pushing something toward the earth, as in Figure 39. Hold this position for as long as is comfortable.
4. Then release your breath through your mouth, returning your hands into the cathedral position. Repeat, switching the arm that raises and the arm that lowers. Do one or more additional lifts on each side.
5. Coming out of this pose, as you bring your arm down, allow your body to fold over at the waist. Hang there with your knees slightly bent as you take two deep breaths.
6. Slowly return to a standing position, with a backward roll of the shoulders.

Drying Off. I like to find ways to incorporate healthful practices into my daily routines. You can get your meridians and strange flows moving when you bathe or shower. You can do this either with a washcloth while in the water or with a towel while drying off. Even without your knowing how to trace your meridians, this towel rub keeps them flowing in the right direction.

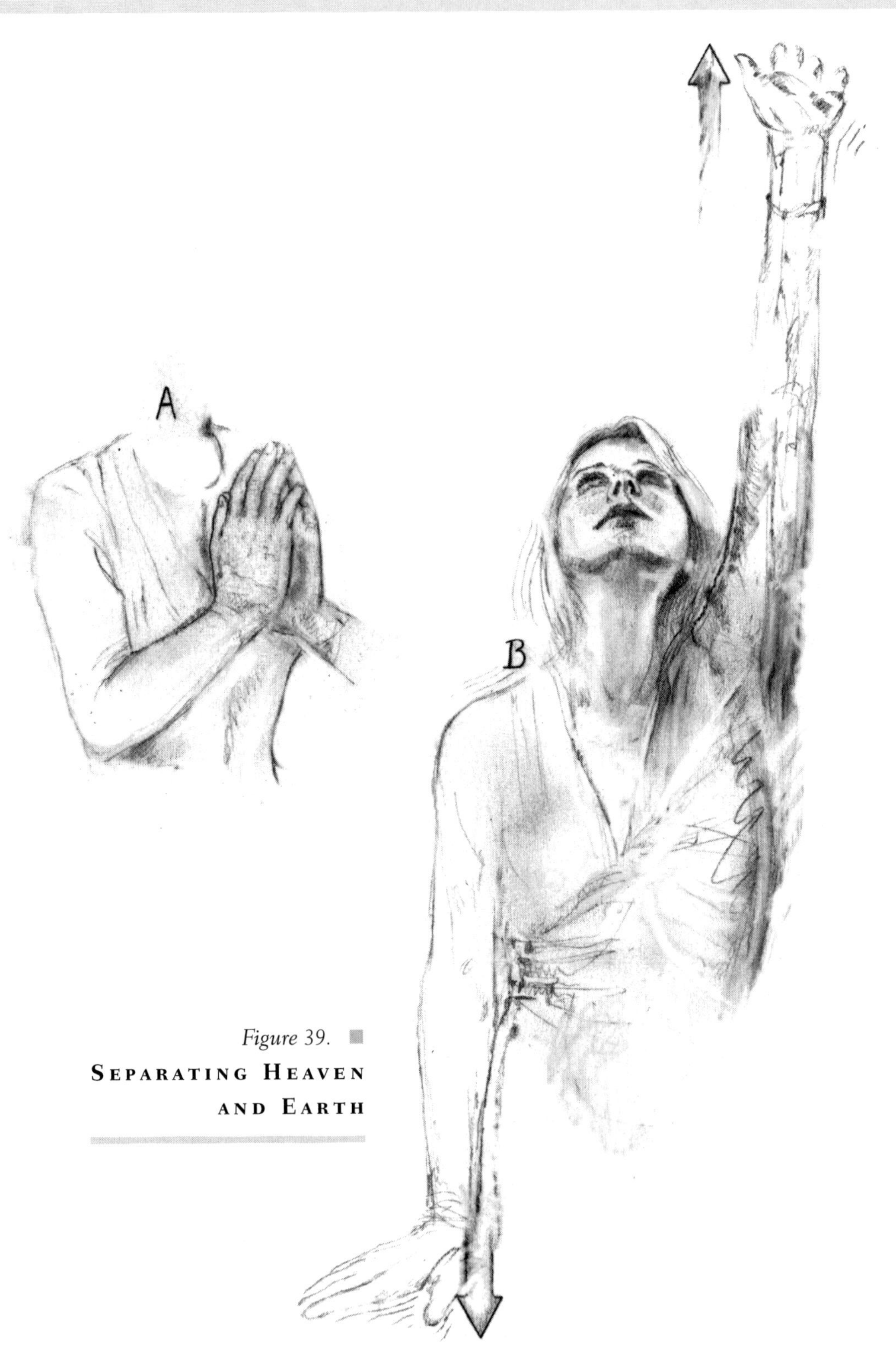

Figure 39.

Separating Heaven and Earth

1. Beginning with the bottom of one foot, deliberately and with some pressure, move the towel or washcloth up the inside of your leg, over the front of your body and, lifting your arm, up the inside of that arm and off your fingers. Repeat on the other side.
2. Starting at either hand, move up the back side of your arm, down your back, down the outsides of your leg, and off the top of your foot. Repeat on the other side.
3. Finish by drying or scrubbing your face, from the bottom pushing upward, over the top of the head, and down to the neck, and then pull your fingers to the side of the neck.

Jump-Starting Your Strange Flows. The following five exercises pump new life into your body by jump-starting your strange flows. They can be used individually or in combination to bring you back when you are "losing it." Begin with the Three Thumps and the Zip Up.

1. *Hook Up Your Yin and Your Yang* (time—between 1 and 2 minutes): Unlike the meridians, strange flows have no acupuncture points of their own. But by holding specific yin and yang acupuncture points, all the strange flows will be stimulated, which gives you a tremendous burst of healing energy. My favorite all-purpose yin-yang points are the belly button and third eye.
   a. Hook the middle finger of one hand into your belly button and the middle finger of your other hand between your eyebrows (third eye point).
   b. With a firm touch and pulling each finger up slightly, hold for a couple of minutes. This gets your strange flows moving, strengthens your auric field, and leaves you feeling whole again.

2. *Send Joy Through Your Body* (time—15 seconds): You know how your entire energy comes up when someone you are attracted to walks into the room. That is how quickly your strange flows can spring into action. Your internal images can also give a sudden boost to your strange flows.
   a. *Pretend* you are feeling happy. Pretend, and the impact on your strange flows can be instant. As Proverbs 17:22 counsels, "A

cheerful heart is good medicine, but a downcast spirit dries up the bones."

b. Or *imagine* someone who makes you feel alive and happy has just greeted you.

3. *Send Color Through Your Body* (time—15 seconds or longer): Bring to mind a color you love. Imagine the color flowing through and infusing every cell in your body. Again, the effect is usually immediate.
4. *Send Forgiveness Through Your Body* (time—a minute or longer): Bring to mind a situation where you hold resentment or anger toward yourself or toward another. The energies of resentment and anger are stored in your cells. Until there is forgiveness, you cannot be completely healthy. But in an instant, if you can imagine what it *would* be like to have a feeling of forgiveness, your strange flows will carry that feeling through your entire being. Do it often enough and you will find yourself becoming more forgiving.
5. *Send Gratitude Through Your Body* (time—as long as you want): Bring to mind something in your life about which you can feel thankful. Gratitude is among the most profound spiritual healers. Send this feeling of gratitude through your body. Say thank you to your heart, your lungs, your kidneys, all your organs. Thank your legs for walking you. Make it a practice to focus several times each day on feelings of gratitude. With gratitude in your heart, healing forces in your hands, and the Daily Energy Routine under your belt, you are on your way to better health.

The healing effects of optimism, enthusiasm, and a sense of humor have been documented in laboratory studies. It has been recognized for decades, for instance, that the immune system weakens and tumors grow faster in laboratory animals who have been put in a situation where they have a sense of helplessness. In a study conducted by psychologist Martin Seligman, in conjunction with the National Cancer Institute, cancer patients with "optimistic explanatory styles" had a significantly greater survival rate than others whose styles were less optimistic.[8] In another study, a group of sixty-six college students watched an inspiring film about Mother Teresa, and an equal number watched a documentary about power struggles during World War II. Chemistry measures showed significantly heightened immune responsiveness in the students who watched the inspiring film.[9] Studies of antibody production have also shown that the stronger a person's sense of humor, the more resistant that person's immune system will be to the effects of stress. While stress tends to impair immune functioning, subjects whose sense of humor was assessed on standardized psychological scales as being high were less affected by stress than those who scored low.[10] Humor jump-starts your strange flows.

THE TRIPLE WARMER–SPLEEN ARCHETYPE

I have for twenty years jokingly called myself a triple warmer–spleen archetype. When I figured out the huge role of this imbalance, it gave me a key to understanding the workings of my body, and it gave me profound relief. It was as if a light turned on where I had been in darkness about my health. I felt profound joy and gratitude.

No energy pattern can yank a person around as much as a triple warmer–spleen imbalance. Like a teeter-totter that is weighted on the triple warmer end, spleen is up in the air with no energy to do its job. This throws off the balance in every system of your body. It ensures metabolic, chemical, and hormonal disarray. It particularly affects the nervous and immune systems, and it leaves you with the feeling that your body's reactions make no sense and that you are powerless to help yourself. Some people are genetically prone to this imbalance. I am one of those people.

Spleen and pancreas, the major organs on the spleen meridian, begin the metabolizing process for the whole body. When they are perpetually being robbed of their energy, all the metabolic processes are disrupted. And it is not just in how you metabolize food and burn calories, affecting your weight, energy level, and disorders such as diabetes and hypoglycemia. It is also in how you metabolize feelings, ideas, worries, decisions. Triple warmer–spleen imbalances are also responsible for food cravings, the weight yo-yo, obsessive eating, anxiety, foggy thinking, mood swings, premenstrual syndrome, allergies, most autoimmune diseases, chemical sensitivities, and recurrent infections. Sometimes triple warmer will treat the chemicals in the healthiest of foods as invaders, inflating the white blood cells, causing bloating and swelling. My younger daughter, Dondi, can at times look like she gained twenty pounds after a simple meal. Her face swells, and she has to kick off her shoes as they no longer fit. She has had to learn her body well through energy testing the foods she eats.

My own version of the triple warmer–spleen imbalance creates many contradictions within my body. Fruits and vegetables cause me to gain weight, and they can trigger a hypoglycemic reaction in me. So does eating after 3 p.m. Sometimes if I am stuck on a plateau where nothing I do causes me to lose weight, a hot fudge sundae eaten at a strategic time can trigger my metabolism so the weight starts to drop. It is a metabolic wonder. My body does not follow the rules.

But nowhere is the triple warmer–spleen imbalance more evident than in PMS. The politics of PMS are as tricky as the condition. Even women can find it hard to comprehend the impact of severe PMS. They would like it not to be true, lest it give the patriarchy an excuse to keep women from having power. What if the commander-in-chief got us into a war because of PMS! PMS has been used as evidence that women are less capable than men to be in power positions. I have heard comments from women in leadership positions that women have severe PMS if they've been molested. Genetics, however, are strongly involved in PMS as well as in other triple warmer–spleen imbalances. While past abuse can complicate any physical symptom, manufacturing false "reasons" for a woman's severe PMS is neither healing nor helpful. Women who have severe PMS symptoms are told they are "hurting the cause" that women have been striving to achieve. It is isolating and painful to have such stigmas continue.

I, in fact, want to go on record as saying that there is something wonderful about PMS. It insists that you move into your own rhythm rather than staying within society's time frame. Living for a few days each month from within your natural rhythm is a powerful correction to the culture's alienation from nature. PMS drops you deep into your own being, and your own truth explodes forth. Whatever you have been successful at burying or denying to yourself bursts forth at this time of the month. It is a truth serum from which you cannot escape, and if you carve out space for it, as in the native "moon hut" traditions, PMS makes you wiser. If you do not, it can cause you to feel you are going stark raving mad. Rather than try to find a president whose hormones fit the structure of the job, why not structure the job to take advantage of someone whose hormones give them the full range of life's experiences? Policy decisions in most parts of the world could afford more compassion and family-oriented wisdom.

When PMS is exaggerated by an extreme triple warmer–spleen imbalance, however, it is near impossible to be in the world. Nothing has made a bigger difference to the quality of my life than bringing balance to the triple warmer–spleen teeter-totter. Just as the body needs sleep, this imbalance requires daily restoration. Sedating triple warmer and strengthening spleen are mainstays. Energy testing what I eat and what herbs I take, particularly during the PMS days, has been even more crucial. Many of the remedies that are advertised do not work for me. Energy testing, however, led me to progesterone, which is the best natural substance I have ever taken. For many women who have severe PMS, progesterone (not the synthetic progestin) works like magic because it potently strengthens spleen and sedates triple warmer. With my moon hut, my energy exercises, a resolve to abstain from major decisions because I will probably later have to get out of any commitments I make at this time, and my progesterone, PMS can take me into a world between the worlds, and I can emerge renewed, recharged, and reconnected with my spiritual roots.

FROM EXPLORATION TO APPLICATION

You have now completed this survey of the eight major energy systems on whose balance rests your good health. In the final part of this book you will learn how to draw on all we have explored so far and apply it to overcoming illness, relieving pain, and optimizing the electromagnetic energies in your environment and the subtle energies within your body.

Part III

Weaving It All Together

In being transformed, altered, and worked with, [our energies] become higher and more coherent.

—Jack Schwarz
Voluntary Controls

Chapter 9

Illness

Your Body's Reset Button

IF EVOLUTION BY NATURAL SELECTION CAN SHAPE SOPHISTICATED MECHANISMS SUCH AS THE EYE, HEART, AND BRAIN, WHY HASN'T IT SHAPED WAYS TO PREVENT NEARSIGHTEDNESS, HEART ATTACKS, AND ALZHEIMER'S DISEASE? . . . WHY HASN'T IT SELECTED FOR GENES THAT WOULD PERFECT OUR ABILITY TO RESIST DAMAGE AND ENHANCE REPAIRS SO AS TO ELIMINATE AGING? THE COMMON ANSWER—THAT NATURAL SELECTION JUST ISN'T POWERFUL ENOUGH—IS USUALLY WRONG. INSTEAD . . . THE BODY IS A BUNDLE OF CAREFUL COMPROMISES.

—RANDOLPH NESSE AND GEORGE WILLIAMS
Why We Get Sick

If evolution has created a human body that is "a bundle of careful compromises," energy medicine helps those compromises succeed in today's world. Energy medicine shows you how to assist your body to thrive within an evolving drama. At this point in the story, each organ, gland, and energy system in your body is figuring out how to use a two-million-year-old game plan on an overcrowded, technologically sophisticated playing field. It is hardly possible to monitor every activity governed by your autonomic nervous system, nor is it possible to avoid illness, but you can be consciously involved as your body encounters challenge after challenge. Illness, as Dr. Bernie Siegel has wryly observed, is nature's reset button. It shuts down your circuits and provides an opportunity to resume the program on a clear screen. If your body's attempts at compromise have jammed your system, which is one way to look at illness, you can get your system back on track by clearing, strengthening, and balancing your energies.

Energy medicine draws on the principle that by establishing greater teamwork among body, mind, and spirit, you can transcend the automated strategies designed by evolution. While instinctual strategies tend to be effective when the circumstances in which they evolved don't change, the circumstances that affect your health are changing at a bewildering pace.

About a month before writing this chapter, in Tasmania, I witnessed a remarkable and poignant scene that occurs nowhere else in the world. Thousands of small birds, Tasmania's entire population of mutton birds, take flight at the crack of dawn, headed toward Antarctica. At some point on this journey, their internal navigation system apparently signals to them that the destination is too far away. They turn around en masse and fly back to the beach from which they embarked, landing after dark, drunk with exhaustion, bumping into one another, and falling or crash-landing on the ground. They finally burrow their way back to their homes in the sand, spend the night, and commence the same impossible mission the next morning. Day after day. The only other place in the world this species of mutton birds is found is Antarctica. Apparently the Tasmanian mutton birds' hapless expedition is an attempt to return to a home of days long gone. They have not figured out that shifting land masses have increased the distance separating Tasmania and Antarctica since their homing instinct evolved.

I wished I could somehow explain the problem to these unfortunate little creatures so they might change their ill-fated ways, but of course this was not to be. Without your conscious participation, strategies designed by evolution millions of years ago to keep you healthy may ultimately be rendered as ineffective as the mutton bird's gambit to return to Antarctica.

When you are sick, your body is ripe for finding another strategy than the one it has been using to maintain its equilibrium. Energy medicine is the art of helping it find such a strategy. Only by bringing another level of awareness to its daily fiasco would the mutton bird be able to regroup. The mutton bird is probably doomed to repeat its exhausting and futile marathon into perpetuity. You are not.

Social support and education can, by shifting your consciousness, help your body to more effectively meet disease.[1] Each time you successfully rebalance your energy body with tests and remedies such as the ones in this book, you are nudging it toward a better game plan.

In addressing disorders such as diabetes, cancer, and heart disease, all the standard disclaimers apply. Self-help techniques do not substitute for professional health care when it is needed. Long before such disorders become acute, however, you can detect trouble brewing by simply attempting to keep your meridians and chakras balanced. Distress anywhere in your body will be reflected in the corresponding chakras and meridians. What you need to do to rebalance them—which may also involve work with other systems such as the aura, the Celtic weave, the rhythms, and the basic grid—will address the deeper problem. You can circumvent a great deal of suffering with relatively simple preventive measures. And when illnesses do hit—as they will, because of genetics, environment, misfortune, or time, no matter how diligent you have been—there is much you can do to establish an energy field that promotes healing.

A STRATEGY FOR GETTING WELL

The energy tests and corrections presented to this point are the basic tools you need to create an internal environment that is conducive to health and healing. They can be applied routinely, and they will make a difference. For illnesses and chronic conditions, however, you may be called on to transcend the routines and find more sophisticated strategies.

This chapter offers a framework for figuring out the next step if health problems still persist. It will help you counter the thousand and one ways your health and vitality may be diminished as your body negotiates compromise after compromise within an environment and lifestyle that stretch its capacities. Its logic is simple: *If every day you make sure your energies are not scrambled and are flowing freely, your body will have strong medicine for countering virtually any illness.*

PRELIMINARIES: UNSCRAMBLE ENERGIES, SET FIELD, OPEN VALVES, FREE DIAPHRAGM, CLEAR LYMPHS

Let's face it, when you are sick, you are not inclined to do *anything*. You don't have energy or confidence. So I will make this section as simple as I can. If you push through and do these preliminaries, you will experience a difference, and you will gain incentive. You can use simple movements and your own hands to create an atmosphere that will help you heal. While the specific techniques have not for the most part been scientifically studied, I have seen each perform as claimed hundreds of times and with a wide variety of individuals. And numerous laboratory studies support the general principles involved.

For instance, identical incisions were inflicted to the shoulders of twenty-three male college students. After receiving Therapeutic Touch treatment, where the healer's hands are placed over the injury without touching the body, the wounds had at one point shrunk an average of 93.5 percent. In a control group that had been similarly inflicted but did not receive the healing treatments, the wounds shrank only 67.3 percent during the same time period.[2] After reviewing a series of experiments investigating Therapeutic Touch, the medical investigator William Collinge concluded: "Techniques such as smoothing, clearing, rebalancing, or otherwise working with the energy field can speed wound healing, reduce acute and chronic pain (including in burn victims), and help other conditions."[3]

Energy approaches to illness will work better if your energies have been unscrambled, your intention is strong and clear, your valves are opening and shutting properly, your diaphragm is free, and your lymphatics are unblocked. The following preliminaries, which can be accomplished by adding five to fifteen minutes to your Daily Energy Routine, address each of these considerations. Done daily, this alone will potently support your healing.

How to Safeguard Your Breasts

The exponential increase in breast cancer has left many women feeling powerless against the whims of the disease. Bringing an energetic perspective to breast self-care can not only overcome problems before they advance into symptoms, it can prevent them.[4] *It is essential, for instance, that lymph flow freely and be able to drain from the breasts. Yet bras that are tight or that have a wire or plastic underbra tend to clog the lymphatics under the breasts. A few very simple massage techniques done each time you remove your bra can help keep the energies flowing in your breasts.*

1. *A set of lymphatic reflex points follows the half-moon shape under each breast, along the line of an underbra (the liver points, beneath the right breast, are shown in Figure 9b; the stomach points, beneath the left breast, are shown in Figure 9d). Massage each of these points vigorously (time—30 seconds).*

2. *Another pair of lymphatic reflex points are at the level of the nipples (the circulation-sex points in Figure 9c). Place your middle finger or thumb below each breast and lift the breast slightly. These points may be quite sore, and massaging them releases toxins and gets energy moving throughout the breast (time—15 seconds). The points at the level of the nipples are paired with points on your back that are at the tip of your shoulder blades. Having a friend rub those points while or after you rub the points on the front is a powerful combination for breast care.*

3. *Massage the points along the spleen meridian at the sides of your breast (Figure 40). This allows any congested energy or toxins that are released to have a place to go. Massage these points with one hand while your other hand massages your large intestine lymphatic points, which run along the outside of your legs (Figure 40). Use a rotating motion while moving downward on both areas, one side of the body at a time (time—45 seconds).*

4. *Periodically flushing (page 108) the stomach (Figure 23, page 109) and circulation-sex (Figure 17, page 106) meridians is also good for your breasts. Stomach meridian travels through the middle of each breast. The circulation-sex meridian is its polarity and begins at the outside of the nipple. Flush each meridian by tracing it once backward and forward three times (time—less than a minute).*

5. *"Chakra clear" each breast, using the same counterclockwise and then clockwise rotations you used when spinning the energies of your chakras (page 167). You can often feel energy building up in your hand during the counterclockwise circling, which means you are removing stale energies. "Shake out" your hands and finish with the clockwise circling, which restabilizes the energies (time—about a minute).*

If you come upon an unfamiliar lump that can't be massaged away over a few days, waste no time in having it checked out medically.

Figure 40.

Spleen and Large Intestine Drainage Points

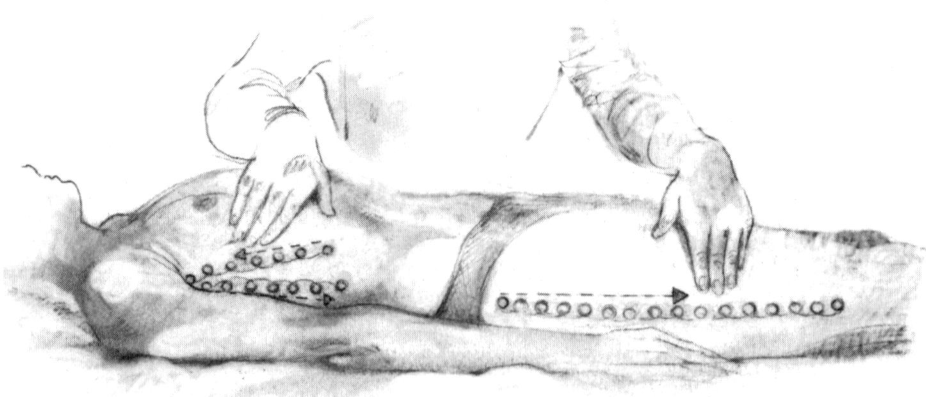

Unscrambling Energies. The Daily Energy Routine (Chapter 3) is a good way to unscramble your energies. In addition, consider using the homolateral crossover (pages 233–235), the Celtic weave (page 180), Separating Heaven and Earth (page 249), and the Hook Up (page 119). The Hook Up strengthens all your energies and provides protection from other people's illnesses. If you are quite ill, holding the hook-up points for between three and five minutes, or having someone else hold them, can refresh your entire system.

Setting an Intention for Healing. As you mobilize yourself to confront an illness, set a clear intention for becoming well. Ask yourself, "What is the most basic change that will help me?" Possible answers to the question, stated in the present tense, include: "My tumor is dissolving." "My body is strong." "My immune system is awesome."

Once you have settled on a simple statement describing what you consider an important change, "zip" it in with a deep breath (page 82). After doing this each day for a week, ask yourself the same question about the most basic change that will benefit you. People often find that their understanding of an illness shifts with time, and asking yourself about the most basic change required can help track and facilitate the development of your understanding. If you get a new answer, adjust the Zip Up statement to fit it. Review periodically.

Opening the Ileocecal Valve. A valve that seems to play havoc with the body monitors the stress response. Called the ileocecal valve, it is situated between the small and large intestines, on the right side of your body, and it is governed by the kidney meridian. It rhythmically opens and shuts to process waste material, not only from the foods you eat but also from the hormones and other chemicals that continually pour into your bloodstream.

If the ileocecal valve's rhythm of opening and shutting is thrown off, a series of health calamities can follow. Problems with the valve can mimic over two dozen illnesses. If toxins aren't eliminated, they can back up into your system, and difficulties may appear almost anywhere, yet the source is the malfunctioning valve. Digestive ailments, lower backaches, toxic buildup in the kidneys, bronchitis, and even eczema may serendipitously clear after the ileocecal valve's functioning is restored. When other remedies aren't working, it is worth the effort to check and reset the ileocecal valve. Fortunately, it is fairly easy to reset, and you can energy test to see if it needs resetting. You do not, however, need to energy test before you reset the valve as it is always helpful in the same way a good massage is always beneficial. And if it is needed, the benefits can be dramatic. To reset your ileocecal valve (time—less than half a minute):

1. Place your right hand on your right hip bone with your little finger at its inside edge (see Figure 41). Your hand is over the ileocecal valve.
2. Place your left hand at the corresponding spot at the inside edge of your left hip bone. This is the Houston valve, and resetting both valves creates a symmetry between them.
3. Exert pressure as you massage and slowly drag the fingers of each hand up six to seven inches with a deep inhalation.
4. Shake the energy off your fingers with the outbreath and return to the original position. Repeat about four times.
5. End by dragging your fingers downward one time with pressure.
6. Close with the Three Thumps (page 63).

The energy test for determining if your ileocecal valve needs to be reset is to place your right hand, flat and firm, over this valve and test it using the general indicator test (pages 65–67). Placing your hand over the valve is called therapy localizing. You can actually test any part of your body by putting either your hand or three fingers notched together over an area that concerns you. Then use the spleen-pancreas or the general indicator energy test. If the test shows weak, there is a problem with the energy in that area.

If after using the above method for resetting the ileocecal valve it still tests weak, a more advanced technique can be done lying down. Push the fingers of both hands very deeply into the ileocecal valve and lift your right knee very slowly and deliberately toward the trunk of your body, up to about a ninety-degree angle. Then very slowly straighten your leg, still pushing in deeply on the valve. This may feel uncomfortable, but it has the effect of abruptly changing the energy in a sluggish valve.

Freeing the Diaphragm. Several years ago, I noticed that after working with hiatal hernias, a burst of energy would go out in every direction from the midsection. Positive changes in the energies throughout the body would oc-

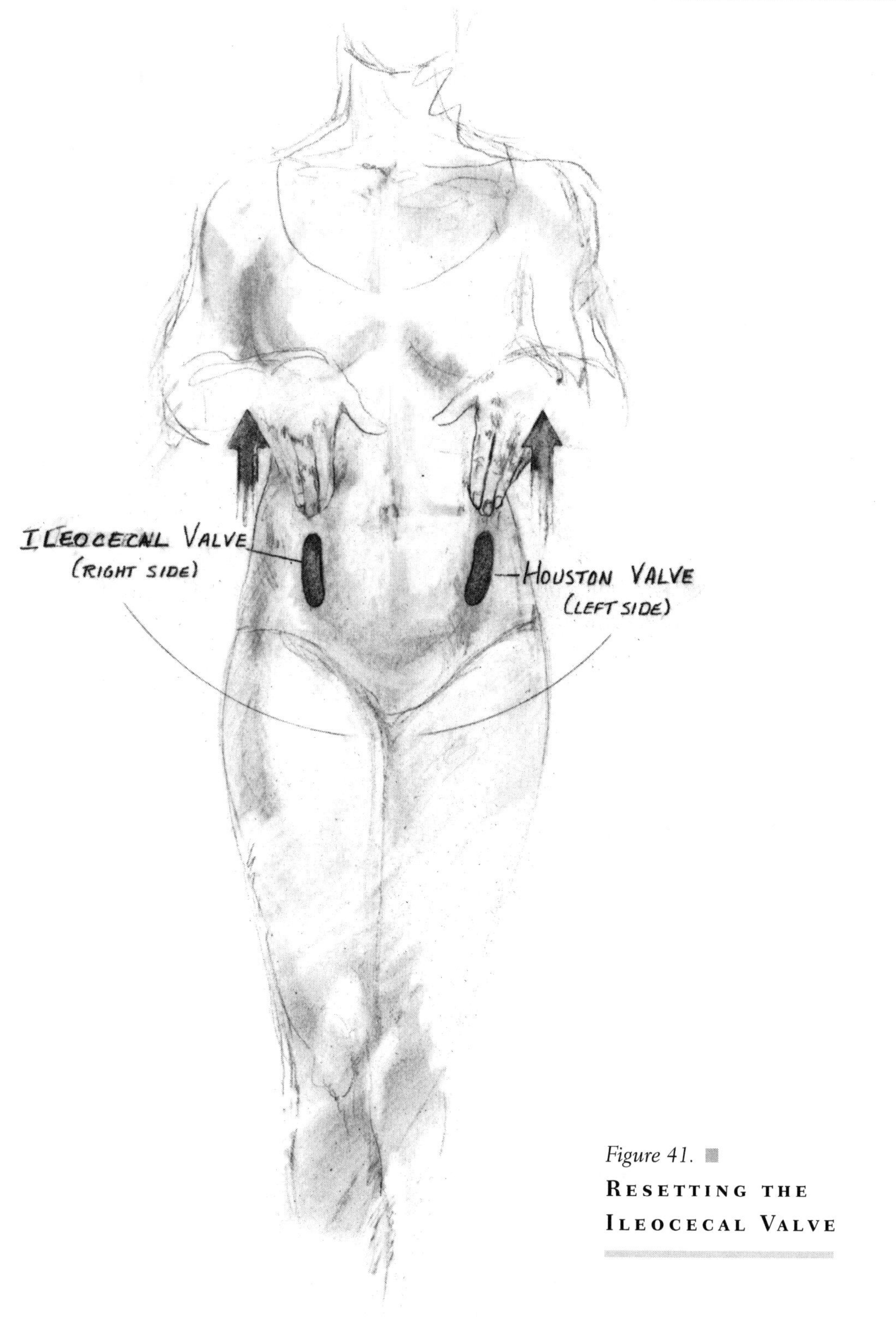

Figure 41.
Resetting the Ileocecal Valve

cur. The changes I was witnessing seemed related to an improvement in how oxygen was being distributed. The diaphragm is a strong, thin muscle partition between the chest and the abdomen. Like a bellows, it literally fans oxygen throughout the body. The technique to help hiatal hernias opens space for oxygen and energy, pushing the auric field out further and sometimes gives the client a feeling similar to a runner's high.

In fact, the functioning of every cell, gland, and organ is enhanced. The diaphragm, however, becomes very conservative in distributing oxygen during a stress overload. This makes you more vulnerable to many illnesses, from heart disease to cancer. Cancer, for instance, cannot thrive in a well-oxygenated area. So I adapted the hernia technique into a more general method for freeing the diaphragm. This very simple exercise, which anyone can do in less than a minute, instantly moves oxygen throughout the body. People sometimes look surprised when they feel the power of the oxygen pushing into a part of their body they usually do not feel. The technique also improves circulation, pulsates the energy field, and restores the "belt flow," the strange flow that keeps energies connected between the upper and lower parts of the body.

This is a good exercise to do every day, but it is particularly valuable if you are not well. While energy testing is not required, you can therapy localize to determine whether your diaphragm is functioning properly: Bring the thumb and first two fingers of one hand together and push them into your diaphragm just under the bottom of your rib cage. Have your partner energy test your other arm. To free the diaphragm (time—about a minute):

1. Firmly place your left hand under the center of your rib cage and place your right hand on top of it. With your hands flat, pull your elbows close to your body so you are hugging your midsection.
2. Inhale deeply and push your body toward your hands while your hands push back against your body. Hold your breath and push hard. Although there is no set amount of time, the longer you hold your breath and push, the better.
3. Release your breath naturally, along with your hands. Relax. Repeat about three times.
4. On an in-breath, reach your right hand around the left side of your waist with your fingers spread; on the out-breath, pull your fingers to your navel with pressure. Repeat several times.

5. Repeat several more times with your left hand crossing your right side.

Neurolymphatics. If your lymph isn't flowing well, your energy isn't flowing well. While the lymph, unlike the circulatory system, has no internal pump to keep it moving, breathing pumps it, exercise pumps it, and stretching pumps it.

Tapping or massaging the neurolymphatic reflex points (Figure 9, pages 84–85) also pumps your lymph and frees lymph that is blocked. If you are sick, lymph flow is impeded in part because you are getting less exercise, and as your lymph fights the illness, your glands and lymph nodes fill with toxins. Therefore, with any illness, massaging or tapping your neurolymphatic reflex points is not only a powerful intervention but is also good preparation for any other treatment. If you build up too many toxins, you are more vulnerable to an illness, but if you keep clearing the toxins through neurolymphatic massage, the lymph can pump it out. I've many times been on the verge of a cold that never materialized. I attribute this to having had my lymphatic points massaged.

You can massage your own neurolymphatic points (Figure 9, pages 84–85), although it is always nicer to lie there and receive, plus a partner can reach all the points on your back. If you are sick and on your own, without the strength to massage all your frontal neurolymphatic reflex points, there are two sets of points that I recommend because they will get the lymph moving throughout your body—the central and large intestine points (time—at least a minute):

1. The central meridian neurolymphatic points are on the outer rim of your chest. To find these points, start massaging K-27 under your clavicle corners (see Figure 3, page 51). Continue to massage as you move out to the sides underneath the clavicle. You will arrive at the indent where the clavicle meets the shoulder bone.
2. Massage deeply. Follow the semicircles where your arms are attached to your body (see Figure 9f). Take all the time you need. These points are on the central meridian, and because that meridian is also a strange flow, massaging them will affect the flow of the entire lymph system.

3. The large intestine neurolymphatic points are on the sides of your legs above the knees (see Figure 9e). Massaging these points moves the lymph and helps the large intestine move toxins. These points form a straight line from the knob at the top side of your leg (the femur), down to the side of your knee. While lying flat, you can massage down both sides of your legs, slowly, deliberately, and with strong pressure. Bend your legs so you can reach the lower points near your knees. If you don't have much strength in your hands, find an object such as a dowel that you can use to push in on the points.

Massaging your lymph points usually works to improve an illness, although on rare occasions you may feel a bit worse before you feel better as the massage releases trapped toxins into your bloodstream. If this occurs, just back off for a while so your kidneys have a chance to filter out the toxins, and then return to the massage. As you will see in the following section, you may also target specific lymphatic points for special attention.

How to Energy Bathe an Unborn Child

Every time I've done a chakra clearing on a pregnant woman, I can feel the child responding. The energies of the unborn child are distinct from those of the mother, and the child's response makes me happy.

A woman was overdue for delivery, and her child was in a breach position. The mother was paralyzed with fear and did not want to have a Cesarean section. She asked me if I could find anything in her energy that was keeping her expected daughter, already named Melissa, from turning to the right position. Melissa's energy appeared healthy, but she was in no rush to get out. The only problem I could determine was that the mother's abdominal muscles were tightly constricted from stress.

I relaxed this area with a chakra clearing over the womb. After about two minutes of counterclockwise motion, it looked and felt to me as though Melissa's energy was getting in synch with mine and picking up the rhythm of my hand. After about fifteen minutes, she began to turn, as if following the energy of my hand, until she was upside down, ready for birth. Another person in the room was jumping up and down and screaming, "I can see her turning." The mother went directly from my office to the hospital, where she gave birth to a beautiful baby girl. While you cannot always count on this technique to prevent a breach delivery, I have since had nearly identical experiences two more times.

THE MERIDIAN FLOW WHEEL

Your meridians are a Rosetta stone for your health. They monitor the energy balances minute to minute and offer a key for deciphering the indigenous energy language of disease. The disease process is altered by corrections in the meridian system. In Chapter 4 you learned how to restore balance and strength in your meridians by tracing them, flushing congested meridians, twisting and stretching the alarm points associated with each meridian, massaging the meridian's neurolymphatic points, and holding or tapping specific acupuncture points that are located toward the ends of each meridian. We will take up here where we left off in Chapter 4. The following techniques give you the additional tools to restore balance and strength to any meridian.

The Meridian Flow Wheel (Figure 28, page 126), simple as it looks, provides an abundant amount of useful information. It names twelve of the meridians. Again, there is actually only one long meridian that surfaces twelve times, plus the central and governing meridians, which, as strange flows, connect equally with all the meridian segments. The wheel shows how each of the twelve segments flows into the next. The energy normally moves in a clockwise direction. The wheel also lists the two-hour time period when each meridian's energies are strongest.

Viewing the meridians as tributaries of an underground river that surfaces twelve times is a good analogy. If the tributary that is immediately upstream or immediately downstream becomes polluted, the ecology of the neighboring segment is often affected, and even a meridian that is several segments away can be impacted. When the river runs high through a meridian segment, it can run low through the meridian segment that is across the way, which is its polarity on the wheel.

If you are ill, find which meridians are out of balance using the alarm point energy test (pages 114–115). Use the techniques you learned in Chapter 4 to balance each of these meridians. If a meridian cannot be strengthened with these techniques, the problem may not lie in the meridian itself but in its relationship to other meridians. The three most likely candidates are:

1. *The Meridian It Feeds:* the meridian that immediately follows it on the wheel (clockwise). Think of a river. If the flow is blocked downstream, energy may be backing up into the meridian, so it tests weak. But this meridian may not be the source of the problem, just the place the problem shows up.
2. *The Meridian That Feeds It:* the meridian that immediately precedes it on the wheel. If the flow is blocked upstream, the meridian isn't receiving the energy it needs, so it tests weak, but again, the problem may originate with another meridian.
3. *The Meridian That Opposes It:* the meridian that is directly opposite it on the wheel. The universe is organized around polarities—day and night, hot and cold, yin and yang—and each meridian is matched by an opposing force. The push-pull effect of these polarized energies keeps us in balance. You saw in Chapter 8 how the teeter-totter relationship of triple warmer and spleen balances the warrior and peacemaker strategies of the immune system. Half your meridians are yin, half are yang. You can see in the Meridian Flow Wheel that opposite each yang meridian is the yin meridian that is its polarity. When one has too much energy, the other is often being drained.

Specific Neurolymphatic Points. If a meridian is consistently out of balance, I first look to see that the lymph is moving throughout the body. If your lymph is blocked, your energies will be blocked as well. First massage each lymphatic point on the problematic meridian for about fifteen seconds (see Figure 9, pages 84–85). If this massage does not strengthen the meridian on an energy test, then massage the lymphatic points on the following meridians.

1. *The Meridian That Feeds It:* the meridian before it (clockwise) on the wheel (time—12 to 15 seconds). If this massage does not strengthen the meridian, massage the lymphatic points on
2. *The Meridian It Feeds:* the meridian that follows it on the wheel (time—12 to 15 seconds). If this massage does not strengthen the meridian, massage the lymphatic points on
3. *The Meridian That Opposes It:* the meridian that is opposite it on the wheel (time—12 to 15 seconds).

These tests and remedies are quick, and the corrections usually hold. Once you have restored a meridian's flow, you can seal it in place by doing the Celtic weave (page 180), making figure 8s over your torso, your head, and your legs.

If the correction does not hold, the following sequence of simple techniques will show you how, through the process of elimination, to find a technique that will strengthen the meridian.

The "Inchworm" Meridian Trace. In Chapter 4 you learned how to trace each meridian. If a particular meridian is chronically weak, get to know its pathway (see Figures 10–23). Here you will use the same basic method you used in Chapter 4 to trace the meridian, except you will be exerting pressure instead of smoothing over the meridian pathway.

1. Begin by finding the diagram that shows how to trace the meridian you have selected (Figures 10–23).
2. Rather than simply passing your hands over the meridian pathway as you did in Chapter 4, "walk" your fingers over it in the way an inchworm walks, exerting pressure with each "step."
3. Particularly work any point that has tenderness and pain. Breathe deeply as you move along the meridian.

Specific Neurovascular Points. Blood circulation is a primary concern in illness. Illness and stress both interfere with it. It takes very little energy to hold your own neurovascular points, and doing so will boost your circulation. To find the neurovascular points, you can follow them as if they form a trail over your head, as they are numbered in Figure 42. Numbers 1 through 4 are on the centerline of the head; numbers 5 through 13 are each on the right

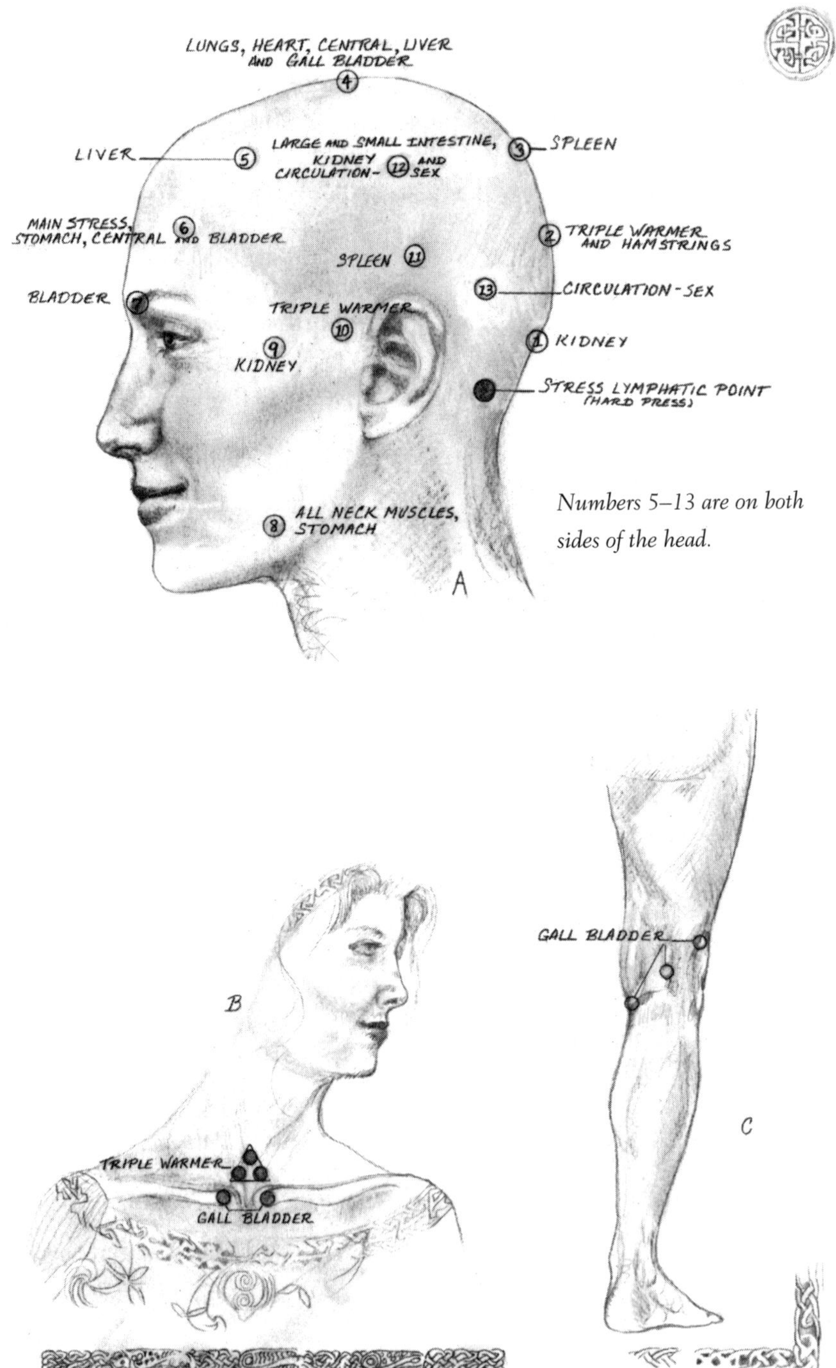

■ *Figure 42.* ■

The Meridians and the Neurovascular Points

and left sides of the head. Once you are familiar with this "trail," find the specific point(s) that correspond with the meridian that has been out of balance (time—2 to 3 minutes per meridian):

1. Locate the neurovascular points associated with the meridian in question (see Figure 42).
2. Lightly hold the pads of your fingertips on these points for between 2 and 3 minutes. Using several fingers ensures that you will cover all the right points.
3. Combining the specific neurovascular points with the main stress points (number 6 in Figure 41) is even more effective.

The darkened circle at the bottom of the head in figure 42 is a neurolymphatic point that you can massage to open the energy and blood flow of the neurovascular points. You might massage this point before holding the neurovascular points.

How to Restore Peace After a Bad Dream

Nightmares are a shock to the psyche, but, like earthquakes, they also play a role in its ecology and its equilibrium. If you are wakened by a nightmare, the following techniques will realign your body's energies, making it easier for you to examine the meaning of the dream or simply to get back to sleep.

1. *Stretch the skin across your forehead and follow with the Crown Pull (page 77).*

2. *Do the Hook Up (page 119).*

3. *Hold the main neurovascular stress points on your forehead (number 6 in Figure 41).*

4. *With your hands wide open, slowly and deliberately pull the energies down your legs, beginning at the sides of your hips. Draw the energies off your feet and finally rub your feet and toes.*

5. *If you are willing to get out of bed, do Separating Heaven and Earth (page 249).*

If your child has just been awakened by a bad dream or is afraid of having nightmares, after holding and reassuring him or her, you can administer the first four of these exercises and then do Separating Heaven and Earth together. While holding the neurovascular points, describe a sweet memory or sing a song. Rather than trying to negate the bad dream, you will be building a storehouse of treasures into your child's psyche. All this provides an unconscious reminder that you are there and that you are connected into his or her distress about the bad dream. By the time your child goes back to sleep, he or she will feel safer, more bonded, and less alone.

BALANCING YOUR CHAKRAS AND KEEPING YOUR STRANGE FLOWS FLOWING

With this chapter you have been learning how to determine which meridians are involved with an illness and how to clear and strengthen them. The chakras may also be involved in illness. The meridians feed the chakras and the chakras feed the meridians. If one or more meridians will not take or keep a correction, or if an illness persists though your meridians are strong, balance your chakras (Chapter 5). In dealing with a longstanding illness, you might also reread the discussion of the immune system (Chapter 8) and apply what fits. Particularly use the simple techniques under the heading "Keeping Your Strange Flows Flowing" (page 248). If you make just a single conscious intervention to improve your energies each day, you are turning on healing forces and building a momentum that allows your entire system to rebalance itself. The following chapter focuses on pain, your body's most insistent way of urging your conscious mind to become involved with your health.

Chapter 10

Pain

Nature's Idea of Tough Love

ILLNESS IS THE MOST HEEDED OF DOCTORS:
TO GOODNESS AND WISDOM WE ONLY MAKE PROMISES;
WE *OBEY* PAIN.

—MARCEL PROUST
Remembrance of Things Past

Like the notion that the best love you can give addicts who are destroying themselves is to be tough about not enabling their destructive behaviors, pain is your body's version of tough love. Pain forces you to stop what you are doing and shift your attention. It requires you to recognize when your body has been damaged or when its ability to care for itself is being overwhelmed.

Pain is evolution's oldest device for getting your conscious mind involved in your body's health and safety. Nowhere is the interdependence of mind and body more conspicuous. Pain is more than just a distress signal. It is a distress signal with teeth. It insists that you fix the problem and that, if you don't know how, you figure it out. This chapter will help you figure it out. However motivated you may have been to get your energies humming, to bring them into balance, even to overcome illness, if you are in pain, your desire to succeed with these techniques will be passionate. As Proust starkly noted, we obey pain.

Many of us, however, are also ashamed to be in pain. If we are consumed by pain, we have a nagging suspicion that we are psychologically feeble, morally inferior, or are otherwise "damaged goods." Other difficult emotions also complicate physical pain. Pain leaves us feeling isolated, helpless, and out of control. Compelled to get rid of the feeling, we immediately reach for our painkillers. But you can do much else that is constructive when you are in pain.

How to Feel Joyful More of the Time

You can reprogram your nervous system to be more joyful, uplifting, and life-affirming! The next time you feel fabulous, riveted, connected in, on a high, or otherwise happy, reinforce this energy by "tapping in" the joy at your third eye, the point between your eyebrows just above the bridge of your nose. Tap with a steady beat for 10 to 12 seconds.

The methods you learned in Chapter 8 for working with your strange flows, such as sending joy, forgiveness, and gratitude throughout your body (pages 251–252), can also change patterns within your nervous system, leaving you more upbeat in mood and outlook.

THE NATURE OF PAIN

One of the great things about energy work is that pain often diminishes as energy that has been congested is freed, and it is much easier to move the energy in a way that relieves pain than most people ever realize. While techniques that use massage, tapping, and holding specific acupuncture points can be as effective as puncturing those points, acupuncture has been researched more thorougly than the other methods. In 1997 the U.S. National Institute of Health published a consensus statement on acupuncture summarizing evidence that opioid peptides are often released during acupuncture sessions. This explains, at least in part, the evidence for the effectiveness of acupuncture in relieving pain in fibromyalgia, postoperative conditions, menstrual cramps, tennis elbow, and low back discomfort. The report also suggested that acupuncture may be helpful in dealing with carpal tunnel syndrome, stroke rehabilitation, headaches, addictions, asthma, and nausea. According to the report, "The data in support of acupuncture are as strong as those for many accepted Western medical therapies."[1] More dramatically, on the Bill Moyers television series *Healing and the Mind,* through the use of acupuncture anesthesia, a patient stayed awake and aware during brain surgery and was able to engage in dialogue during the operation.[2]

Before learning how to relieve pain, it is helpful to know a bit about its psychology and physiology[3]. Pain involves a complex interaction of physiological and psychological factors, and people in different cultures respond to it differently. David Bresler, former director of UCLA's Pain Control Unit, reflects: "Pain is a sensation, a perception, an emotion, a cognition, a motivation, and an energy."[4]

Suppose that walking along the street one dark night you slip on a not-so-proverbial banana peel and twist your knee. A remarkable series of events is set in motion. Nerve cells dedicated to identifying anything that is potentially harmful are activated. Such sensors are distributed over your skin and throughout the tissue that covers your bones, organs, blood vessels, fascia, and muscles. These nerve cells send impulses that collectively form an inner "picture" of the injury.

As your knee twists, the nerve fibers that conduct pain impulses produce two kinds of experiences: intense short-lasting pain right at your knee and a slow, persistent, diffuse pain. The piercing pain provides information about the location and severity of the injury and stimulates a reflex so that muscles contract, causing you to shift your balance to protect the leg. Such protective circuits are wired into your nervous system at birth. These short-lasting sensations are also forwarded to higher centers in the brain. Meanwhile, the longer-lasting sensations alert you to the need to care for and protect the injury.

The *suffering* component of pain—with the accompanying emotions of fear, anxiety, anger, or annoyance—interacts with your temperament and outlook. If you are still experiencing pain a year after your fall, it may signal that the injury never healed properly and still requires your attention, or it may mean that the pain has become a chronic condition.

Chronic pain is one of nature's less easily forgivable bloopers. The nervous system is misfiring, providing you with incorrect information and a false alarm. Chronic pain can steal your life force and debilitate you. Pain experts speak of the "terrible triad" of chronic pain: suffering, insomnia, and depression. In addition to the anguish for the victim and the calamity for the family, chronic pain may lead to drug dependence and other desperate measures.

The annual costs for chronic pain in the United States has been estimated at fifty billion dollars.

While one part of the mystery of pain is how to stop it, another part of the mystery is why so many seemingly unrelated techniques can help with chronic pain, and why so many successful interventions last for only a short while. Among the treatments that have successfully been used to treat chronic pain are acupuncture, antidepressants, antiepileptic medication, aspirin, beta-endorphin injections, electrical stimulation of the skin or brain, exercise, narcotic painkillers, surgery that disengages nerve fibers, and psychological methods such as hypnosis, relaxation, meditation, biofeedback, behavior modification, and placebo cures. Many of the successful treatments increase endorphins, the body's natural morphine-like proteins, which suppress pain. Others are believed to close the "gateway" that allows pain signals to reach the brain. Surgery destroys the nerves that carry pain signals. Still, not much is known. Why, for instance, does chronic pain often return six months to a year after surgery that has obliterated the physical mechanisms of pain?

Phantom limb pain, discussed in Chapter 2, suggests an energetic counterpart to the physiology of pain. The "limb" of the energy body is apparently sending pain signals to the brain. This relationship of the energy body and the physical body allows an energy test to reveal problems before they can be detected by standard medical tests. Healers often go right to the area of pain without any cues from the patient. In one study, the energy fields of fifty-two chronic pain patients were independently assessed by practitioners of noncontact Therapeutic Touch. Agreement among the practitioners and the patients regarding the location of the pain was highly significant.[5]

A corollary to "Matter follows energy" is "Sensation reflects energy." You only feel as good as your energies. By taking an energy approach to the treatment of pain, you not only avoid the side effects that almost always tarnish invasive pain treatments, you also address more basic problems. The same energy techniques that address the pain also bring about healing. Holding sedating points to free trapped or stale energies, for instance, both relieves pain and makes room for fresh healing energies to enter the area.

How To Overcome Digestive Problems

If your stomach is upset after a meal, massage up and down the inside length of your thighs, pushing in hard with your fingers. If more is needed, try the following, one by one (time—about a half-minute each):

1. *Stretch your abdomen with your hands or by bending backward.*

2. *Clear your ileocecal valve (page 266).*

3. *Firmly massage the large intestine points on the outside of your upper thighs (Figure 9e, page 84).*

4. *Firmly massage the small intestine points that run along the edge of your rib cage (Figure 9c, page 85). Begin at the center of your rib cage and move down along its edges.*

ENERGY TECHNIQUES FOR RELIEVING PAIN

In the first energy healing class I ever taught, my co-teacher, Hazel Ullrich, wanted to demonstrate a technique called pain chasing we had just learned in our Touch for Health instructor training program. She asked for a volunteer who was dealing with longstanding discomfort. A woman in the class was in obvious pain whenever she stood up or sat down. She was reluctant, however, to volunteer, finally admitting that she didn't want to make the teachers fail. She had been coping with this pain for years, and she had been disappointed by too many promising treatments. Hazel responded, "This will help you. It doesn't matter what you believe. This isn't one of those things where you need to have a positive attitude."

With this, the woman agreed to be the guinea pig. Energy testing showed that her pain was on her bladder meridian, the longest meridian in the body. Hazel "pain chased" the pain along the entire meridian. It took a long time, and when the pain finally got to the end of the meridian, it was pulsing hard and it wouldn't leave. I had an intuitive flash that Hazel might try holding the sedating points of the bladder meridian. She did, and the pain left completely. The woman became almost hysterical, crying and unable to catch her breath. She could hardly believe that she was pain free. It was the first time in years. The change was obvious to everyone. She hadn't been able to stand up straight because the pain hurt her back so badly. Now she was not only standing straight, she had far greater range of motion. I've since used or taught pain chasing (pages 289 and 292) hundreds of times with encouraging results.

Additional methods presented in this chapter include "breathing out the pain," tapping, stretching, and pinching the area of the pain, sedating meridians related to the pain, working with diffuse pain, and zone tapping. Each is presented succinctly and in the order of increasing complexity, so you can get right to the business of seeking relief. In the next chapter, you will also learn how to reduce pain through the use of tiny magnets. That and a good massage table are about as high-tech as I get.

Breathing Out the Pain. Breath moves energy. By focusing your breath, you can gain some control over your pain. Millions of women have been helped through the pain of childbirth using the Lamaze breathing technique, and it can be adapted to any situation. The procedure is simple. Breathe in through your nostrils, mouth closed, as if smelling a rose. Breathe out through your mouth, as if blowing out a candle. Make each breath slightly slower than the breath prior to it. As you exhale, you are releasing some of the energy involved with the pain. It is helpful to have a simple reminder because when pain hits, your mind is not at peak performance. The mantra taught to guide people through an acute attack of pain is simple: smell the rose, blow out the candle.

Tapping, Stretching, Pinching, and Siphoning the Pain. We all have an instinct to touch and massage ourselves when we hurt. This is an instinct to be listened to! You can use the following techniques anywhere, and sometimes they are all you need to relieve pain.

Tapping. For sore muscles, such as shoulder pain, tap the area. Use either your fingers or something with many little "tappers," such as a hairbrush with plastic prongs. Tap for as long as it feels right, perhaps about a minute or until you notice that something has jarred loose and the energy can move more freely.

Stretching. As long as it is not an open wound, stretching the area surrounding the pain can also bring relief. Pain involves too much energy at its site. Stretching the areas about an inch or two beyond the sides of the pain relieves some of the congestion. Then press in deeply around the perimeter of the pain and stretch the muscle in every direction.

Pinching. There is a way of lightly pinching the skin that signals to the brain that the pain is no longer necessary, like hitting the reset button after a breaker has been thrown. Again, this cannot be done over an open wound. Come to the center of the painful area with your thumb and forefinger and very, very lightly pinch the skin one time. Pinching has a resetting effect, triggering what is called the "spindle cell" mechanism. This frees energy that is clogged in the area of the pain. The slight pinch sends a signal into your nervous system that releases trapped energy.

Siphoning. Place your left hand on the area of the pain of someone you want to help. Be aware that your left hand, which pulls energy, is siphoning off the other's pain. Then hold your right hand down, out, and away from your body. Be aware that the pain is draining off your right hand. If you feel the energy getting caught in your own body instead of siphoning out of your right hand, stop immediately, and vigorously shake both hands. If you ever feel someone else's energy is caught in you, place your arms, to above the elbows, in cold running water. Usually you will be able to siphon the other's pain without getting it caught within you, and I have noticed in my classes that this technique seems instinctual. When class participants have been working on one another, I have seen students who have never heard of pain siphoning, when they sensed they were moving into an area that is painful to their partner, automatically put their right arm out, down, and away from their own body.

Sedating the Pain. The technique I use most frequently for pain relief is to hold the acupuncture sedating points associated with the pain. Sedating a meridian has the effect of releasing excess energy while calming it. This does not mean, however, that you are weakening the meridian or the muscle.

The Muscle Meridian Chart (Figure 43) shows all fourteen meridians and the muscles related to each. To hold a meridian's acupuncture sedating points (time—about 4 minutes):

1. Identify a muscle that is in the area of the pain, using Figure 43.
2. Find the name of the meridian that corresponds with this muscle.
3. Look at Figure 26 (pages 120–123) to find the figure that corresponds with the meridian. Note the sedating points labeled "first" and the sedating points labeled "second."
4. Lightly but firmly hold the sedating points labeled "first" for 2 to 3 minutes.
5. Hold the sedating points labeled "second" for 1½ to 2 minutes.

Suppose your pain is on the left side of your body and you find in Figure 43 that the muscle going through the area of the pain is governed by the stomach meridian. In Figure 26 you would see that for the first acupuncture sedating points on stomach meridian you need to hold the index finger and the second toe on the left side of your body. To reach them, sit down, get comfortable, and take your left hand down to your second toe and place your thumb and middle finger around the toe, leaving your index finger sticking out. Hold the points on your index finger with your right hand. After 2 or 3 minutes, go to the second points, on the top of your left foot. If the pain is on the right side of your body, simply reverse these directions.

Taking the Teeth out of a Toothache. The same strategy for sedating the meridian that governs the muscle going through a painful area can be used for working with toothaches. Look at the dental chart, Figure 44. Each tooth is identified by the meridian that governs it. Sedating these meridians will not only relieve pain, it initiates a healing process. I have shown this technique to numerous clients who were seeing me for something else but who were also experiencing pain caused by a cavity that had not yet been treated. In several of these cases, when they did get to the dentist to have the cavity filled, the cavity could not be found. Teeth can be healed! I have had three cavities that were identified by my dentist heal after I sedated the meridian associated with the afflicted tooth daily for about four weeks. At a minimum, using this technique to alleviate a toothache can keep you from being debilitated until you can get to your dentist.

A Hopi Healing Technique for Diffuse Pain. For people whose pain is dispersed throughout large areas of the body, particularly if it results from an autoimmune disease such as fibromyalgia or lupus, a Hopi practice provides relief. It is also helpful for other kinds of diffuse pain, such as when all your muscles ache or your back is sore. The technique simply tells the brain to turn off the pain (time—2 to 3 minutes):

1. Have the person lie face down. Do a spinal flush (pages 79–80).
2. Curl your fingers and place the fingers of one hand on the side of the spine that is closer to you. You can start at the bottom of the spine if you want to move the energy up; start at the top of the spine if the person tends towards headaches.
3. Place your other hand, fingers curled, on top of the first hand, reaching across the spine so that it is between the two sets of curled fingers.
4. With your fingers exerting a bit of pressure, move your hands along the length of the spine. As you do, mentally "laser beam" energy through your fingers and into the person's back.
5. An alternative is to put both thumbs on the side of the spine closest to you and all your fingers on the opposite side of the spine, travelling its length in either direction. Then move to the other side of the person and repeat the process.
6. A variation from travelling straight up or down the spine is to make spirals along the spine. Let one set of curled fingers spiral the skin in a clockwise direction on the far side of the spine and the fingers of the other hand spiral the skin in the opposite direction on the near side of the spine.
7. Before finishing, pinch and lift the skin over the spine itself, beginning at the waist and traveling up the spine. Stop at the point where your fingers cannot get enough skin to pinch and lift.

I was doing a demonstration on chakra clearing in front of a class in Australia. I sensed that the circuitry was disconnected in the energies of the woman who had volunteered. She complained of pain throughout her body, and the energy in her nervous system had a particular feel that I had encountered before. I asked if she had fibromyalgia. She confirmed that she did. I had her turn over, and I did the preceding procedure. By the end of the

■ *Figure 43.* ■

MUSCLE MERIDIAN CHART

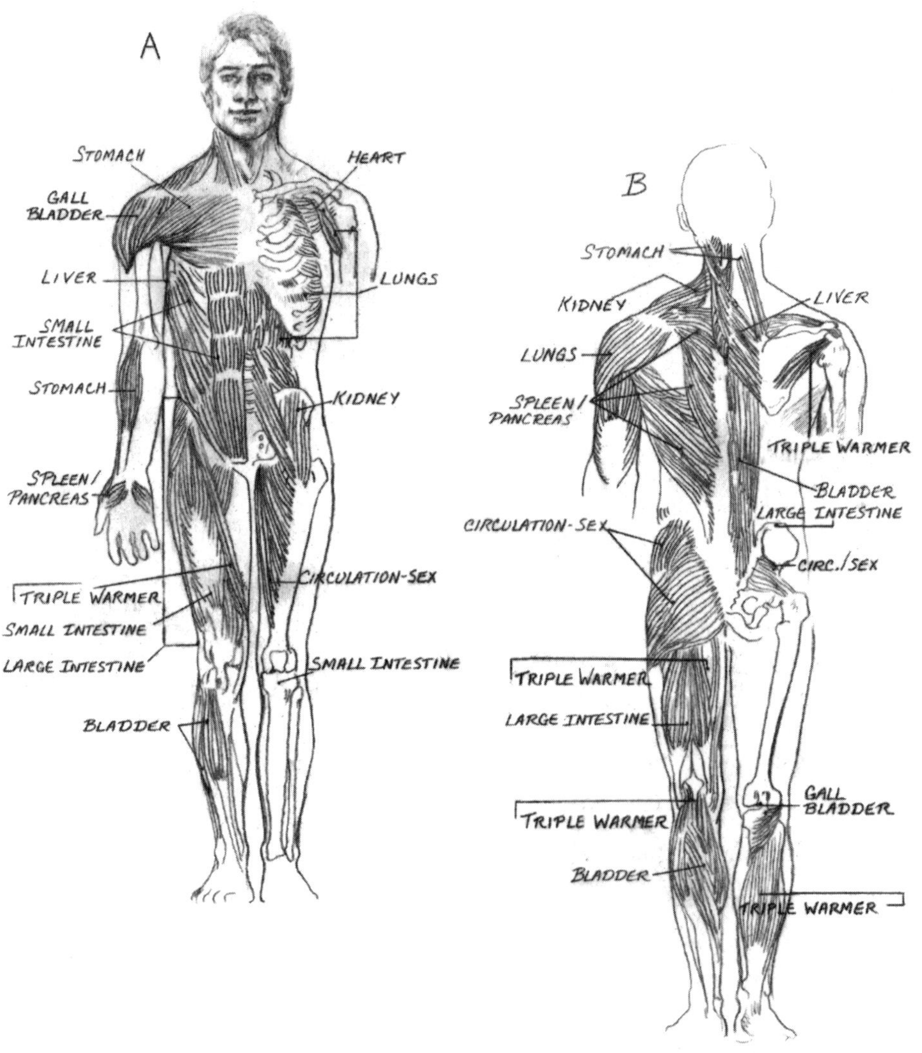

Figure 44.

DENTAL CHART

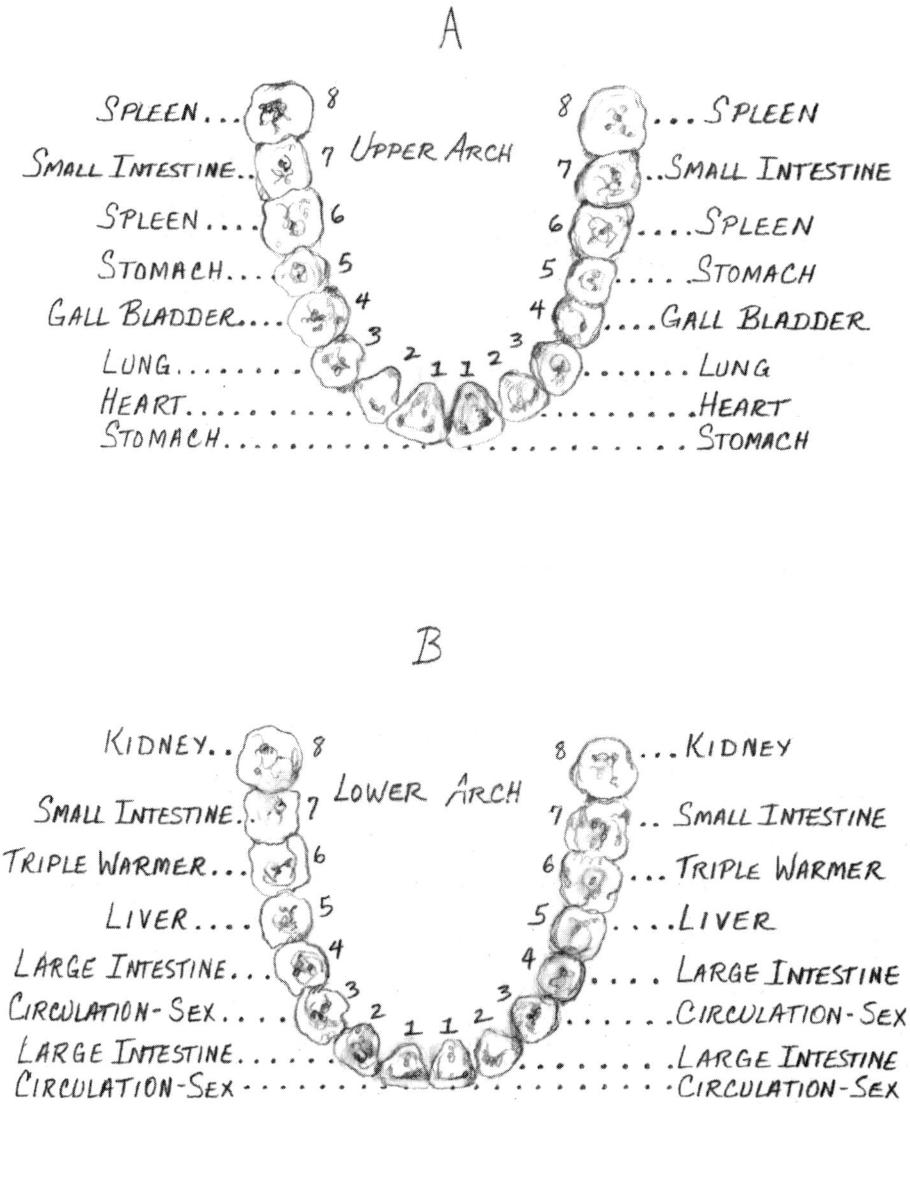

A

UPPER ARCH

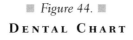

SPLEEN... 8 | 8 ...SPLEEN
SMALL INTESTINE.. 7 | 7 ..SMALL INTESTINE
SPLEEN.... 6 | 6 ....SPLEEN
STOMACH... 5 | 5 ....STOMACH
GALL BLADDER.... 4 | 4 ....GALL BLADDER
LUNG........ 3 | 3 ....LUNG
HEART......... 2 | 2 ....HEART
STOMACH........ 1 | 1 ....STOMACH

B

LOWER ARCH

KIDNEY.. 8 | 8 ...KIDNEY
SMALL INTESTINE. 7 | 7 .. SMALL INTESTINE
TRIPLE WARMER... 6 | 6 ... TRIPLE WARMER
LIVER.... 5 | 5 ....LIVER
LARGE INTESTINE... 4 | 4 .... LARGE INTESTINE
CIRCULATION-SEX.... 3 | 3 ......CIRCULATION-SEX
LARGE INTESTINE...... 2 | 2 ....LARGE INTESTINE
CIRCULATION-SEX........ 1 | 1 ....CIRCULATION-SEX

treatment, her pain had subsided. She was not surprised, as she was familiar with the technique. It was I who became surprised as I thought I had invented it, but she told me that with some minor variation it was a Hopi healing technique that had been handed down through the oral tradition and taught to her by a Hopi elder.

Zone Tapping. Zone tapping is based on foot reflexology, which maps the body into the zones that are shown in Figure 45. With zone tapping you locate points to tap based on the zone within which your pain sits. Zone tapping is most useful for localized rather than diffuse pain. I would estimate that between 70 and 80 percent of the people I know who have tried zone tapping have reported that it relieved some or all of the pain.

The zones are on both the front and back of your body. Your softer parts, which are in the front and inside the limbs, are considered to be yin. On your arms and legs, yin is where hair doesn't grow as easily. By this map (Figure 45), the front or yin side of your body includes: the bottom of your feet, the palms of your hands, the inside of your arms and legs, the front of your neck, and your face. The more rough-skinned areas of your body, mostly on the back, are considered yang. The yang side of your body also includes the top of your feet, the backs of your hands, the outside of your arms and legs, and the back of your head and neck. For zone tapping (time—about 2 minutes):

1. Locate where your pain is in one of the zones (spaces within the lines) in Figure 45.
2. Find the number that is in the same zone as your pain. If the pain is below your waist, you will be tapping at your ankle ("A" on the diagram). If the pain is above your waist, you will be tapping at your wrist ("W" on the diagram).
3. If the pain is on the front of your body (yin), tap on the inside of your ankle or wrist (yin). If the pain is on the back of your body (yang), tap on the outside of your ankle or wrist (yang).
4. Find the point on the diagrams of the wrists or ankles that corresponds with your pain. "A3" means your pain was in zone 3 below the waist.
5. Tap the point about ten times. Stop for about 10 seconds. Then tap for about 90 seconds more. That is enough. The uninterrupted tap-

ping calms the slow, persistent pain impulses, while interrupting the rhythm engages and calms the faster, shooting impulses.

6. The pain will continue to diminish for about 10 minutes after you have stopped tapping. If most of the pain has disappeared, but not all of it, tap in the same zone on the opposite side of the body.

You can energy test to be certain you are on the right point, although you will not harm yourself if you tap the wrong point. Suppose you are not certain whether your pain is in zones 1, 2, or 3 of the inside ankle. Energy test while touching each point. You will lose energy on the zone that needs to be tapped and will stay strong on the other two zones.

Pain Chasing. If you have chronic pain and the pain falls along a meridian line, pain chasing helps. You can think of it as squeezing the pain out of the meridian like squeezing toothpaste out of a tube. You can do it on yourself, although it is always nice to have a partner do it for you (time—5 to 30 minutes):

1. Identify the meridian the pain is on (Figures 10–23).
2. Place a finger from one hand at the center of the pain and a finger from the other hand on the point at the beginning of the meridian. Press both fingers fairly hard.
3. Pain always has a twin—at least one matching point of pain on the meridian—though you may not be aware of it until it is stimulated. Keeping your one finger on the pain, inch along the meridian a finger's width at a time with the other hand until you reach a place that hurts.
4. Once found, simultaneously hold that point and the pain point firmly. The pain on one of the points usually will have vanished in less than 3 minutes. This occurs because you are opening a channel for blocked energy to flow.
5. Move the finger from the point that no longer hurts. If it is the second point, continue to move your finger toward the original point, a finger's width at a time. If it is the original point that stops hurting, move it a finger's width *away* from the other finger, toward the end

■ *Figure 45.* ■

Pain Zones

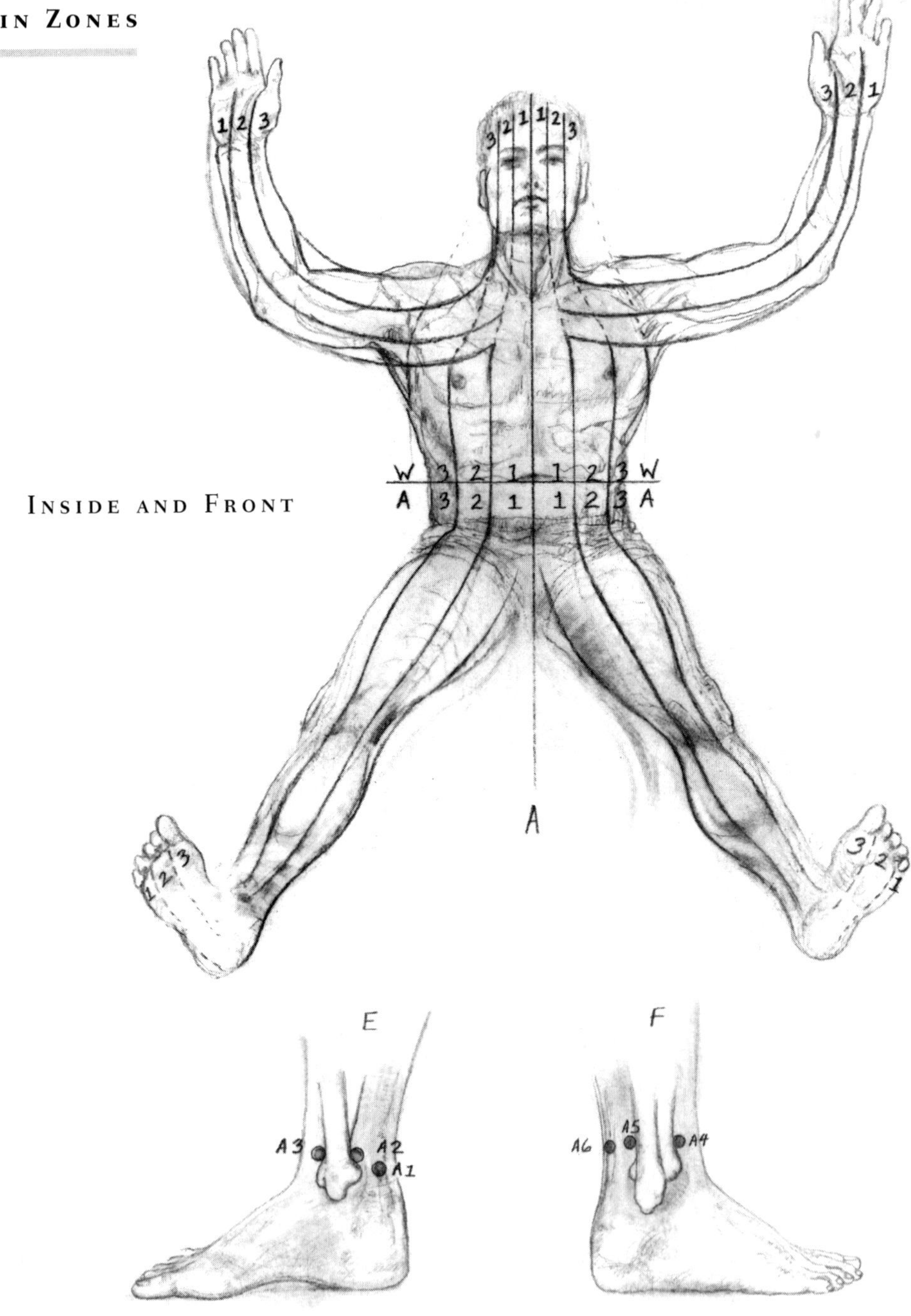

Inside and Front

Inside of Foot

Outside of Foot

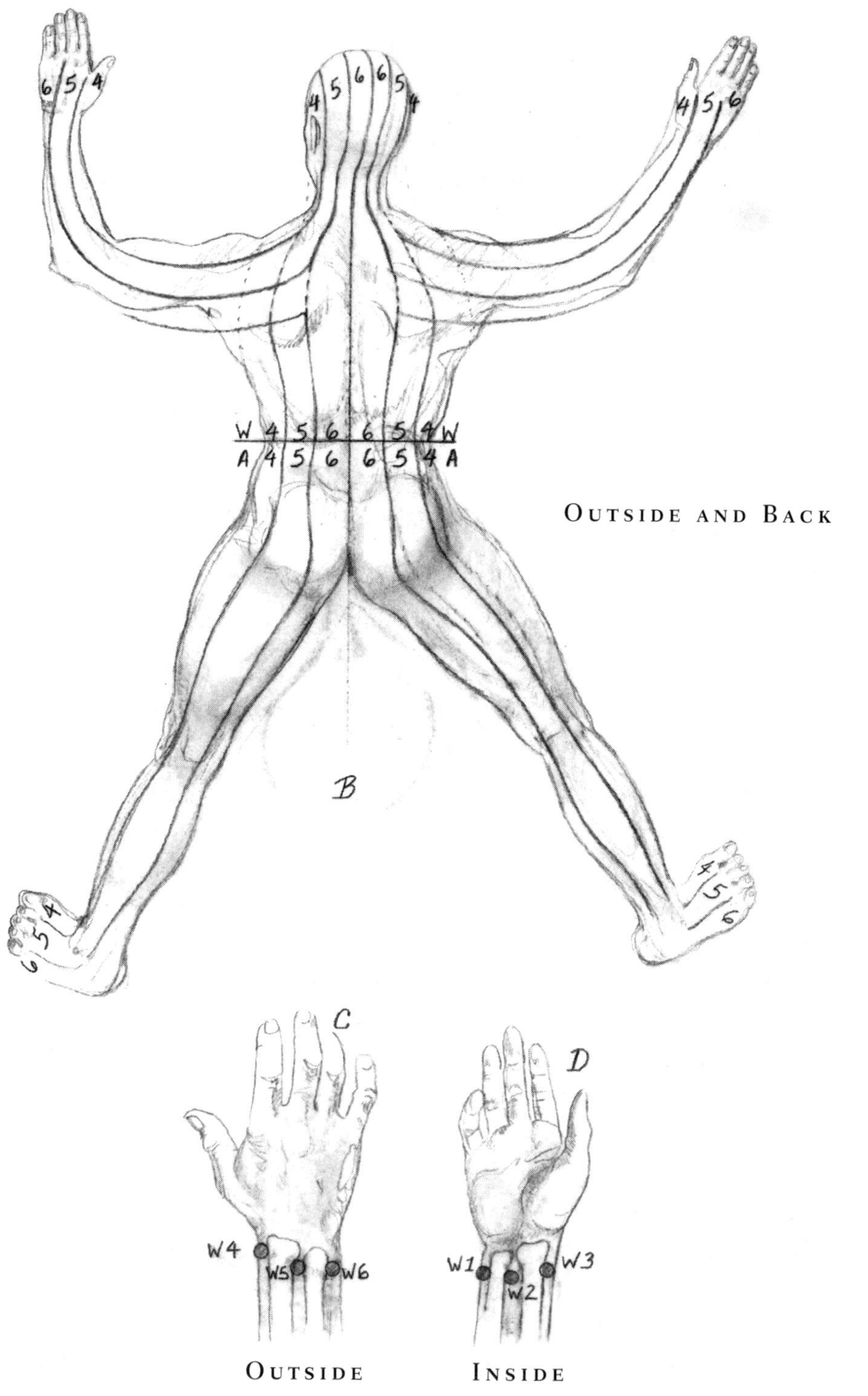
6 5 4
4 5 6 6 5 4
4 5 6
W 4 5 6 6 5 4 W
A 4 5 6 6 5 4 A
Outside and Back
B
4 5 6
4 5 6
C
D
W4
W5
W6
W1
W2
W3
Outside
Inside

of the meridian. Again press in fairly hard. Continue until you find a second point that is painful. Then repeat step 4.

6. Eventually, one finger will get to the end of the meridian, or the fingers will be side by side. If the fingers are side by side, move the finger from the point that doesn't hurt a finger's width toward the opposite end of the meridian from which you started.
7. Finally, one finger will get to the end of the meridian. When it does, just "chase it" with the other finger, a finger's width at a time.
8. Continue until you have chased the pain off one end of the meridian. If the pain gets "stuck," hold the acupuncture sedating points at the end of the meridian (Figure 26, pages 120–123).

These techniques—which employ breathing, stretching, pinching, holding, and tapping—used solo or in combination, can provide substantial relief for a wide range of pain conditions.

THE INTRICATE DANCE OF PHYSICAL PAIN AND EMOTIONAL PAIN

Physical pain can cause emotional pain. Emotional pain can cause physical pain. Their interplay can be complicated and confusing.

My daughter Tanya had been suffering for several days with intense pain around her ovaries. Nothing she did gave her relief, and she was sometimes doubled over and writhing. As luck would have it, I had a stopover at just this time in the city where she was living, and she met me at the airport. She suspected that something major was wrong with her body, perhaps a serious illness. It seemed a mystery that the obvious remedies that should have taken her pain away had not worked, so in a busy airport and needing quick answers, I simply sat her down and began to energy test her. You can find a way to energy test anywhere. I went through her alarm points. Every one of them tested weak, which told me that all her meridians were in disarray and I had to look deeper for the source of her problem. Testing her rhythms, winter was weak and the others were strong. This narrowed the problem down to the two meridians that are governed by winter: bladder and kidney. Because kidney governs the ovaries, I decided to hold the acupuncture sedating points for the kidney meridian.

The moment I put my fingers on the points, I felt fear stream through my own body. Fear is the stress emotion of winter, and I told her what I was experiencing. She protested that she was not feeling afraid of anything. Her face had been pale, with black circles around her eyes. Within minutes the color began to return to her face and the circles around her eyes became lighter. As I held a bit longer, she began to feel waves of fear draining from her body. In a flash, she recognized that beneath the outer layer of her stress, she had grown very frightened that she would forever be trapped in the mountain of stressful issues with which she was grappling. Her fear had blocked the energy around her ovaries. As I continued to hold the points, the fear that had intruded into her awareness had released. Within the next five minutes, the energies began to flow along the meridian, and the pain left completely. She was out of fear, out of pain, feeling confident again, and I was able to make my flight.

Whether the source of the pain is physical or emotional, you can sometimes within as little as ten minutes free yourself or another of considerable suffering (time—about 10 to 15 minutes):

1. Begin by determining which of the five rhythms is out of balance, using the general indicator test as you pull in each of five directions across your midsection (pages 65–67).
2. Look at Figure 35 (page 210) to find the rhythm's stress emotion and the meridians this rhythm governs.
3. Determine which of this rhythm's meridians is in trouble. If you identify with the stress emotion, ask yourself if it is more inwardly directed (fear about self, anger toward self, and so on), or more outwardly directed (fear about the world, anger toward others, and so on). It is the yin meridian if the emotion is inwardly directed and the yang meridian if the emotion is more outwardly directed. You can also energy test the meridians that are on the rhythm using the alarm points (pages 114–115).
4. Once you have identified the meridian, hold its acupressure strengthening or sedating points (figure 26).
   a. Hold the sedating points if the pain involves congestion, sorrow, pent-up emotions, or difficulties letting go or relaxing.
   b. Hold the strengthening points if empowerment or strength needs to be evoked. If the meridian in question is heart or spleen, use only the strengthening points.

5. In general, holding the neurovascular points (page 216) is always helpful when emotions are involved.

Your emotions are a product of your biochemistry, as well, of course, as your cognitive style, perceptions, and the events in your life. Working with the stress emotion of a meridian that is involved with your physical or emotional pain results in a more favorable biochemistry for overcoming that pain.

A CHILD'S QUESTION

While this is one of the shortest chapters in the book, it will for some people be the most appreciated. I used to wonder, "Why would the same God who created joy and ecstasy also make us suffer pain?" Couldn't God or nature have figured out a less ghastly signal system to get our attention? And, of course, there are many. Before the pain of illness are signs in your complexion, your digestion, your mood, your energy level, and other minor symptoms that are all much friendlier than pain. Given an informed choice, most people would far prefer to know how to read their subtle energies as a built-in gauge of potential problems than to have to rely on pain. It is strange that our culture doesn't, from kindergarten onward, teach us how to read that gauge and how to make the necessary adjustments. Then pain could be reserved for emergencies. Like a mother's love, pain in an emergency is able, with its unambiguous and compelling force, to save your life. Pain as the court of last resort sometimes cannot be avoided. But by maintaining an awareness of your body as an energy system and keeping that system balanced and vital, there is much you can do so you are called into that court less frequently.

Chapter 11

Swimming in Electromagnetic Currents

Making the Best of an Ecosystem in Distress

OUR MODERN WORLD BEGAN A LITTLE OVER A HUNDRED YEARS AGO WHEN THOMAS EDISON FIRST DEMONSTRATED HIS ELECTRIC LAMP. . . . TODAY, WE SWIM IN A SEA OF ENERGY THAT IS ALMOST TOTALLY MAN-MADE.

—ROBERT BECKER
Cross Currents

At the core of the Earth, spinning molten iron creates a magnetic field that surrounds the planet. This field interacts with high-energy atomic particles radiating out from the sun. Together they form the geomagnetic field that shields the Earth from the full force of the sun's energies. Without this shield, life could not exist, and even minor fluctuations in it affect both behavior and biological systems. Extreme fluctuations, such as when the North and South poles trade places (this has occurred numerous times in the Earth's history, but not since humanity evolved), have corresponded with the extinction of many species. Within the Earth's geomagnetic shield, we have not only learned to generate and manipulate electromagnetic forces, we have saturated the atmosphere with humanmade frequencies that now fill the entire electromagnetic spectrum. We evolved within an electromagnetic ecosystem that is vastly different from the one we live in now.[1]

While the dangers of "electropollution" are being fiercely debated, it is clear that electromagnetic energy in the environment influences your meridians, chakras, and other energy systems. And it is clear that its impact may be for better or for worse. This chapter will show you how to minimize the hazards while using magnetic forces for your benefit.

CREATING AN ELECTROMAGNETIC ENVIRONMENT THAT SUPPORTS YOUR HEALTH

Your body needs the Earth's electromagnetic field. A disorder that developed in Japanese industrial workers who spent long hours in metal buildings that shielded them from the Earth's natural magnetic field, called magnetic field deficiency syndrome, included such symptoms as insomnia, decreased energy, and generalized aches and pains. The application of artificial magnetic fields alleviated these symptoms.[2] A similar syndrome was seen in Russian cosmonauts who, after over a year in space, had lost 80 percent of their bone density. After strong artificial magnetic fields were introduced into spacecraft, this ailment was no longer a problem in space travel.[3]

We usually don't think of biological tissue as being affected by magnetic fields. Iron fillings, steel spoons, and compass needles seem to be made of an altogether different substance than human flesh. It turns out, however, that in addition to the iron in your blood, all life forms contain tiny crystals of magnetite, a natural magnetic mineral comprised of black iron oxide. This magnetite continually registers changes in your relationship to the Earth's electromagnetic field and helps orient you to that field.[4] Magnetite is the original lodestone used by some preliterate people in healing potions and ceremonies. The magnetite crystals in your body are so small that they can only be seen through an electron microscope.

The concentration of the magnetite crystals found in the human brain is considerably less than the concentration of magnetite found in birds, bees, and fish. Scientists have speculated that magnetite in animals helps them with navigation. The magnetite in bacteria, in fact, turns them into "swimming compass needles that orient with respect to the earth's magnetic field."[5] The relatively small concentration in humans may explain why we cannot, by sense alone, navigate as well as the birds and the bees, but it is strong enough to offer part of the explanation for the health effects of natural and artificial magnetic fields.

Vested interests in government and industry have been more than a little reluctant to acknowledge the potential adverse impact of artificially generated electromagnetic fields on health, but in my practice I frequently see the fallout on human lives. While these fields do not lead to appreciable harm in most people, sensitive individuals can be devastated. At one time, miners would send a canary into a mineshaft to test for poisonous gases. If the canary stopped singing, they knew it had died and the shaft was contaminated. People who are particularly sensitive to electromagnetic fields have been the "canaries," but the problem is accelerating.

Hundreds of studies over the past two decades looking into the health impact of electromagnetic fields have produced mixed findings. As with the mixed research results on the dangers of smoking, vested interests shape outcomes. A highly publicized and influential report released in 1996 by a panel of the prestigious National Academy of Sciences' National Research Council concluded that electromagnetic fields pose no real health threat.[6] Less publicized, however, is the panel's finding that children living near power lines do indeed have elevated rates of leukemia. When asked what is responsible, if not electromagnetic fields, "the committee members said they didn't know."[7] The National Research Council's findings have been criticized for having excluded some of the most important existing studies that do reveal the health-related dangers of electromagnetic fields. An earlier report by the U.S. Environmental Protection Agency (EPA) did indeed find a link between electromagnetic fields and cancer. The Air Force and White House apparently tried to suppress this report because it "might be unnecessarily alarming to the public," but some EPA staff members were themselves so alarmed that they leaked a draft copy of the findings to the press. The suppressed report concluded:

> *Studies showing leukemia, lymphoma and cancer of the nervous system in children exposed to magnetic fields from residential 60 Hz electrical power distribution systems, supported by similar findings in adults in several occupational studies also involving electrical power frequency exposures, show a consistent pattern of response that suggests, but does not prove, a causal link.*[8]

In short, many illnesses probably are related to harmful electromagnetic fields. Video displays, power lines, transformers, electric blankets, fluorescent lights, and microwave ovens are all possible culprits. Countering this problem has become an increasingly important challenge within my own practice. While it is still a largely uncharted area, I can suggest a few strategies if you suspect that electromagnetic fields—whether artificially generated or created by the earth's mineral deposits or "energy vortexes"—are adversely affecting your health.

FIGHTING MAGNETISM WITH MAGNETS

Electrical currents run through every cell and tissue of the body. Wherever electrical currents exist, a magnetic field is formed in the surrounding area. Electric fields are due to the *presence* of charged particles, such as electrons, while magnetic fields are due to the *movement* of these charged particles, such as an electron current.[9] Until recently, the magnetic field of the human body could not be measured, but new sensitive instruments are now capable of measuring it. John Zimmerman of the Bio-Electro-Magnetics Institute notes the following advantages of working with the body's magnetic fields as contrasted with working with its electrical fields: the information acquired is more precise; magnetic fields pass through bone and all other tissues virtually unaltered while electrical currents are blocked by bone; and no contact is needed to obtain information from a person's magnetic field. Unlike X rays, dyes, radioactive substances, and exposure to electrical currents, magnetic imaging procedures are completely noninvasive.[10]

The evidence that magnetic fields are involved in healing is growing. The electromagnetic radiation given off by the hands of healers increases markedly when they are projecting healing energy toward a patient.[11] Electromagnetic energy directed through damaged cells under precise conditions causes the cells to regenerate, a principle that has been applied for accelerating the rate at which broken bones mend.[12] Even the testimonials in the popular media that magnets have provided athletes and other celebrities with pain relief and improved performance are being backed now with scientific evidence that magnets do indeed possess healing properties.

For instance, a study at Baylor College of Medicine has demonstrated that magnetic therapy can reduce pain. Fifty patients diagnosed with post-polio syndrome and suffering with muscular or arthritis-like pain were divided into two groups. Active magnets were applied for forty-five minutes to the area of pain in one group. An inert placebo device was applied to the area of pain in the other group. Neither the patient nor the doctor knew if the device had an active magnetic field. The pain score for patients treated with the active magnet decreased an average of 4.4 on the ten-point McGill Pain Questionnaire. The score for the placebo group decreased an average of 1.1.[13] Studies summarized by the International Council of Magnetic Therapists suggest that magnets have produced improvements in a wide range of medical conditions, including tendonitis, blood circulation, diabetic neuropathy, bone cysts, hypertension, optic nerve atrophy, facial paralysis, and fracture healing.[14]

While I had dabbled in the therapeutic use of magnets for years, the first experience that caused me to understand that magnets can be more than just a marginal healing strategy was when a woman who was having heart problems consulted me. She had high blood pressure, fibrillation, and her heart was skipping beats. An electrocardiogram showed that her heart was severely stressed, and her doctor was concerned that she was a prime candidate for a heart attack.

As I moved my finger along her heart meridian, using very little pressure, she felt so much pain that she nearly jumped off the table. The energy along the meridian, especially in the section from her elbow to the top of her arm (where the meridian begins), was so thick that it didn't move very well. I could clear it to a degree in the session, but I knew that substantial improvement would require ongoing work. My schedule didn't allow me to take on another regular client, and I offered her the best plan I could. It included a number of energy exercises for freeing the energies around her heart, and with hope more than confidence, I taped four weak magnets to four different acupuncture points on her heart meridian. I suggested she wear the magnets half a day on, half a day off, for about a week, but to remove them instantly if anything about their use worried her.

She called a few days later to say that she, her doctor, and her husband all were surprised at how well she was doing. The doctor wanted to know more about the treatment. Not only had the pain around her heart ceased, so had the fibrillations and irregular heartbeat, and her blood pressure was going down. While I cannot prescribe magnets as a cure for heart ailments, and I would not ever put a magnet directly over the heart, I certainly believe this ought to be studied.

I have also used magnets with good success for pain reduction, unlocking "frozen" energies, and helping bones to knit. A potentially life-saving application, however, has been in shielding people from electromagnetic fields that are causing illness.

A Few Basics. The therapeutic use of magnets is not so simple as just taping a magnet to the body or placing a magnetic patch over a problematic area. The strength of the magnet, the pole that is placed against the skin, and the area on which the magnet is placed all make a difference. If these factors are calculated incorrectly, the use of magnets can be harmful rather than beneficial. For instance, in one study, people with their heads in the north direction of a pulsed magnetic field became restless, suffered mental confusion, and the electrical activity of their brains became inhibited. With their heads in the east direction, however, the subjects became calm and peaceful.[15]

As more companies are touting the wonders of magnet therapy, commercialism rather than education seems to be driving the field. Unfortunately many of the commercial items you can buy for magnet therapy, such as elbow sleeves, wrist bands, ankle bands, back braces, and sleeping pads, either sew the magnets in randomly, with no concept that the polarity matters, or they make it so the south side will be against the skin. The south side of a magnet may help athletes and others with their circulation, for instance, but it can also increase pain, spread infection, or promote tumor growth. A basic understanding of magnets and how to use them is important.

Another tricky aspect of working with magnets is that the formula that will help you one day may not help after the magnets have shifted your energy field. If you are conscious of this dilemma, you can counter it. For instance, if a magnet that has been helping you feel better begins to make you feel worse, you can flip it over. This will balance the overcorrection, usually within a minute. Then remove the magnet. If you mistakenly place the south rather than the north side of the magnet against your pain, your pain and discomfort will immediately increase, so turn the magnet over.

How to Test the Polarity of a Magnet. Magnets prelabeled "north" or "south" may be misleading because there are two contradicting systems for labeling magnets. One, used most commonly in navigation and industry, labels as the north side the side that would point north if the magnet were used as a compass. But since *opposites* attract, it is actually the *south* side of the magnet that points to the Earth's magnetic north. So in navigation the south side, because it points to the north, is thought of as the magnet's north side. Physicists, on the other hand, call the north-seeking side the south side, which is more technically correct. This is becoming the standard among magnet therapists, and it is the system I use here.

The easiest way to find out which is the north or south side of a magnet is to use a compass. The side of the needle that points to the Earth's magnetic north will also point to the north pole of your magnet. To mark the north side of the magnet for future reference, oil-based paint or fingernail polish works well.

Without a compass, you can energy test a magnet's effect on your body to determine which is the north and the south side. Place the magnet above your left ear and have someone energy test you. You will lose your energy when the north side of the magnet is against your skin. The magnet is not really weakening you, but since the north side contracts energy, it is pulling the energy away from your extremities. The south side will appear to strengthen you as it disperses the energy to your extremities. You can later use a compass to prove to yourself the validity of this energy test.

How to Mend a Broken Arm or Leg

Determine which are the south and north sides of three round magnets as just described. Tape the north pole of one of the magnets against your skin directly over the break. Tape the north pole of a second magnet against your skin below the break. Tape the south pole of the third magnet about an inch above the break, toward your head. This creates a closed circuit of energy that reduces pain and helps knit the bones. Allow the body to return to its normal magnetic state by removing the magnets for a few hours each day.

Magnets can be used as a supplement to the other techniques for pain control presented in the previous chapter, and the same principles apply when working with electromagnetic energies in the environment.

There are some **contraindications** when working with magnets. Do not place a magnet over a pacemaker or over the womb during pregnancy! Nor should the south side of a magnet, which stimulates growth and increases circulation, be placed over an infection, a swelling, or a tumor, or on a person with diabetes!

MAGNETS IN PAIN MANAGEMENT

A man approached me after the first day of a three-day class. He was in the class to learn all he could to care for his wife, who was debilitated with terrible phantom limb pain at the site of an amputated leg. He didn't feel he could come back for the rest of the class unless he went home with some way of providing relief for the pain. I tried to explain a rather complex though effective pain relief technique, but he was getting confused and going into panic. Finally, I gave him a tiny magnet, marked the north side with lipstick, and told him to tape the north side to the stump of his wife's leg. The next day he not only returned, he had a testimonial for the class about how much relief she experienced.

The first consideration before taping a magnet to the body is the strength of the magnet. I suggest you begin with a small magnet that has a weak force field. More is *not* necessarily better when it comes to exposing yourself to corrective magnetic fields. You can usually find weak magnets that are about one-quarter inch in diameter in craft stores, nature shops, or electronic stores. Be aware that nature shops sometimes have tiny magnets that are still quite powerful. When in doubt, ask. Once you are familiar with the effects of these weaker magnets, you may wish to experiment with magnets that are somewhat more powerful.

Another consideration is whether to place the north side of the magnet or the south side against your skin. Every cell in your body is a miniature battery; its positive pole is like the south side of a magnet, and its negative pole is like the north side. When you wake up after a good night's sleep, there is a balance between these two poles. As you become active, you generate more positive energy, and the cells in your body become more like the south side of a magnet.

Since the north polarity has a negative charge, taping the north side of a magnet to your skin can sedate the buildup of positive energy in your cells.

This calms you. The north side is also generally the stronger healing force. Its tendency is to keep bringing us back into balance. It provides a restful, restorative energy and is associated with taking away pain, swelling, and infections, lowering blood pressure, and inhibiting tumor growth. It is also used in the treatment of sprains, broken bones, arthritis, and toothaches.

The south side of a magnet makes things grow. For better or worse, it activates, stimulates, and increases—whether blood circulation or cancer. For this reason it is trickier to work with the south side of a magnet. You don't know what you may be activating. The south side will stimulate and amplify energies, disperse fluids, and increase blood flow. Because of its effect on the production of insulin, *the south side of a magnet should not be placed on someone who is diabetic.* I generally limit my use of the south side to working with burns, broken bones, sprains, and blood clots, or to create a closed circuit in conjunction with the north side (as explained in the box on mending broken bones). I will also hold the south side of a magnet to an area for about thirty seconds to restore balance after the north side has been taped to the skin for a period of time. If your experience differs—perhaps the south side continues to be beneficial over many hours—trust your experience. It is possible for a person's polarities to be reversed or to reverse from time to time, so stay alert.

A third consideration is where to place the magnet. Put the north side of the magnet on the skin directly over an area of pain, unless there is an open wound or other contraindication. In that case, place it immediately below the area of the pain (i.e., away from the head and toward the toes). Do not place the south side of the magnet directly on a painful area. You can, however, place the south side above the pain (toward your head). This can help with blood flow, digestion, and overall energy. To create a closed circuit that helps heal a broken bone, a sprain, or a blood clot, the north pole of a magnet is placed on the injury, the south pole is placed above the injured area, and the north pole of a third magnet is placed below the injured area. The exception to the principle that you should never place the south side of a magnet directly *on* a painful area is in the case of a burn. First place the north side below the burn (away from the head) for pain relief and fighting infection. Once there is a scab, laying the south side on the burn can help skin heal faster, but if it begins to increase the pain, turn the magnet over for a few seconds. Then remove it or try again.

You can also use an interplay between the north and south sides of a magnet. For instance, suppose you want to hold the south side over your stomach because you have indigestion. But you also feel some pain in the area. You could hold the north side over your stomach until the pain goes away and then turn it over so the south side can work with the indigestion. For a headache, you could put the north side on the top of your head to sedate the energetic congestion, and once you feel some relief, use the south side for about thirty seconds to disburse the energy.

Use Band-aid–type tape to attach the magnet to your skin. It is not possible to provide exact guidelines for how long to keep the magnet taped to your body, so stay attuned to your intuition. You may remove the magnet as soon as the pain abates, perhaps in as little as a minute. You may wear it all night. You may wear it for a few minutes several times each day to change the field of a chronic pain condition. When the pain stops, remove the magnet. If the pain returns, replace the magnet. As a rule of thumb, never keep the north side of the magnet taped on for more than twelve hours or the south side for more than half an hour.

How to Relieve Arthritic Pain

Lay the north side of a small, weak magnet against your pain, and you will feel relief. Taping the north side of a magnet against a neurolymphatic or acupuncture point that is tender or painful for six to eight hours will open the related meridians and bring healing to the systems they govern.

USING MAGNETS TO COUNTER ELECTROMAGNETIC FIELDS

How do you know if a health problem is being caused or aggravated by an electromagnetic field? The effects of these fields are often imperceptible until obvious symptoms occur, such as a sudden irritable mood, pain, nervous system disorders, tumors, or leukemia. But less severe consequences harass many people who never connect the problem with an electromagnetic source. I have seen an office worker's productivity fall dramatically after fluorescent lighting was installed, and I know a student whose grades plummeted after she was transferred to a school that used such lighting. I've also seen vast improvements in classroom behavioral problems after a teacher was persuaded to allow a student to sit next to the window for light while turning off the strip of fluorescent lighting directly overhead.

Other electromagnetic fields can be equally disorienting. A client who had some training in yoga interpreted the sensation of energies shooting up his spine as a kundalini experience, believing that a spiritual breakthrough was occurring. Like a true kundalini experience in someone who is not adequately prepared, having external electromagnetic energy assaulting your body can make you feel crazy.[16] My client concluded he wasn't ready for this opening, and that he needed help in stopping it. What was really going on, it turned out, is that he had been coming into contact with new electromagnetic fields in his environment. Showing him how to identify and protect himself from the fields that were hurting him resolved the problem.

A tip-off if you suspect you are being harmed by external electrical fields is that if you keep falling asleep or losing your concentration or are for no reason feeling lethargic, the electromagnetics may set your energies in a homolateral direction (p. 233). If your energies aren't crossing over or if they keep returning to a homolateral pattern, it is almost always for one of five reasons: (1) you are ill, (2) you are in extreme stress, (3) you are having an allergic reaction, (4) your hormones are out of balance, or (5) your environment is scrambling your energies. Simple questioning usually sorts out which of these is occurring. You can also energy test your environment to determine if and where discordant energies are assaulting you. Notice if there are places that leave you drained or unable to think straight. Energy test when you are away from them and then when you are near them. Be alert to the possible effects of power lines, transformers, satellite dishes, video displays, wall sockets, electric blankets, microwave ovens, and fluorescent lights. Some buildings or rooms in buildings are built over an underground mineral or water deposit or in an energy vortex that may also adversely affect a person's energies.

If you can identify areas of your environment that are causing electromagnetic problems, avoid them if you can. A young woman who had an extremely irregular menstrual cycle and debilitating cramps happened to move into the woods, where she lived in a cabin that had no utilities. Exposed only to natural light and no electricity, her cycle spontaneously and unexpectedly readjusted, and her cramps ceased. Although most of us are bound to a lifestyle that is inundated with technology, it is possible to use magnets and grounding devices to counter many of the adverse effects of electromagnetic fields.

How to Relieve a Cramped Muscle

While there are many possible causes for muscle cramps, from magnesium deficiency to strain, you can address the energetic dimension of the problem by experimenting with a magnet taped over the area of the cramp. Hold the north side (see p. 300) of a small, weak magnet against the area of the cramp for up to two minutes. If the cramp does not immediately begin to release, try the south side. With a cramp, two opposing processes are involved. You have pain, which the north side of the magnet addresses, and you have a problem with circulation, which the south side of the magnet addresses. You may find it necessary to use one pole for a time and then the other, allowing your sensations or an energy test to provide biofeedback.

Start Where You Sleep. To begin assessing your environment for electromagnetic problems, first consider your bed, the single place you spend the most time. Find out if an energy test shows that you are weakened when you climb into bed. The problem may be with the bed itself. Perhaps you need another mattress. But even if you test strong, there may still be a problem with the electromagnetic field in the room. If you test strong right after getting into bed but test weak after being there an hour, the problem is probably with the electromagnetic field. If this is the case, you can protect yourself from the field by grounding your bed. Have someone who knows how to work with electricity wrap bare copper wire around the metal innersprings in your mattress and connect the other end to a groundwire in the house's electrical system. You can also place a magnetic pad on top of your mattress. Again, energy testing can help you determine what will work best for you. Many manufacturers of these pads recommend sleeping with the north side up. But some people do better with a mix of north- and south-oriented magnets or "concentric" magnets, which have poles on each side. The north and south fields cancel each other out so the polarity of your own field of energy is not affected, yet the pad protects you by creating a kind of shield or electromagnetic boundary. Again, the arrangement that is most helpful not only may differ from person to person, it can shift from day to day. Some people also have an adverse reaction to any magnetic pad. Use energy testing liberally when you are making potent interventions into your energy field.

Twirling a Magnet over Your Body. People vary in how much energy they absorb from the environment. You can energy test to see if you draw in so much energy that it is causing a problem for your body. You will need a small round magnet tied to string or to dental floss. The magnets I use for this are about an inch in diameter, have a hole in the center, and are available at Radio Shack. To test:

1. Have a friend energy test your spleen meridian (page 50). Note its relative strength or weakness, but do not correct it at this time.
2. Have your friend hold the string about one and a half to two inches above the magnet, place it between the thumb and forefinger of either hand, and twirl the magnet by rubbing the thumb and forefinger together.
3. With the magnet swirling in a clockwise direction, have your friend move the magnet from about arm's length in front of you toward your chest. Then energy test again.
4. Next, with the magnet swirling in a counterclockwise direction, have your friend move it from near your chest out to at least arm's length in front of you. Energy test once again.

As with balancing the chakras (pages 168–169), the counterclockwise motion tends to draw energy out, and the clockwise motion tends to bring it in. If the meridian grew stronger when the magnet was spinning in a clockwise direction and moving toward you (bringing energy into you), this means your body wanted to absorb more energy and you are probably not suffering from having taken in excess energies from your environment. If the meridian grew stronger when the magnet was spinning in a counterclockwise direction and moving away from you (drawing energy out), this means your body needed to have energies removed, and you may well be suffering from excess electromagnetic energy.

To remove excess energies, use the same magnet, twirling it in a counterclockwise direction as you or your friend move it away from your body. Holding it away from your body, let it spin in the clockwise direction. While there is no rule about where to pull out excess energy, if the energy test showed that you have absorbed excess amounts of energy, I suggest you remove excess energy in this way from all seven chakras (Figure 30). Spin the energy out of each chakra, pulling the magnet, as it spins in a counterclockwise direction, away from the chakra. Make one or two such passes at each chakra. Then use your intuition or energy test to determine what other parts of your body need energy drawn from them and use the magnet to pull it out. The feet and hands are likely areas. This is a simple yet powerful way of removing toxic energies from your body, and most people can feel its effects.

Taping Magnets to Your Body to Protect You from Electromagnetic Fields. You can tape magnets to your body not only for pain control and healing but also to build a field that creates a magnetic boundary that keeps other energies out. Use small, weak magnets, no larger than one-quarter inch in diameter. While you will need to experiment with where to place the magnets, which polarity, and what strength to use, begin with four magnets placed in the following manner:

1. Place a small magnet on each hand at the soft spot beneath the thumb, north side against your skin.
2. Place additional magnets at the bottom of each foot, on the Wellspring of Life points (Figure 24, page 111), again, with the north side against your skin.
3. The magnets will shift your energies in the night. If you notice yourself feeling uncomfortable during the night, simply remove them.

While most people do not need magnets to protect them as they sleep, if you are a person who is dramatically affected by electromagnetic fields each night, this basic strategy can over the long run be a life-saver. If you wake with unexplained pains or mental confusion, the possible influence of electromagnetic fields should be considered, and the methods recommended here will not harm you. You will need to experiment to find the arrangement that works best for you in your particular environment as well as each particular evening, at least until you understand the patterns that are affecting you. In the experimenting, you will gain a great deal of knowledge about your own energies and the energies around you. Often you will need to use an interplay between the north and south sides of the magnet, as just described. The south pole against your skin stimulates your muscles, tendons, and blood flow, but overstimulation can keep you awake, increase swelling, or boost the growth of microorganisms. The north pole, on the other hand, will restrict unnatural growth and infections while reducing pain and tenderness.

We tend to absorb the greatest amount of energy through our hands and then through our feet, and these are the areas that need the greatest amount of protection. If the magnets are providing partial but not complete relief, consider placing additional magnets at the back of your wrists, below your elbows, above your elbows, and/or behind your knees. Experiment and follow your instincts. If every night invasive energies seem to be entering through your spine, causing pain, tape the north side of a tiny magnet against your skin at the bottom of your tailbone. Turn it over at any sign of discomfort during the evening because your own polarities may be switching. Remove it in the morning. I have known clients who were regularly waking up with headaches to report that their headaches ceased after they attached tiny magnets to their cheeks or the point where the back of the neck meets the head. Another variation is to place the north pole of the magnet against your skin on one side of your hand, foot, wrist, or ankle, and the south pole on the other side. Magnet bracelets and wrist, elbow, and knee sleeves are available commercially, and they can help build a more balanced protective field around you.

The Direction You Work and Sleep. Does the direction in which your body is faced affect your energies? An architect consulted me because he was having difficulty concentrating. He loved his work and never before had problems with productivity, but every time he went to his drawing board he was hit with a malaise and loss of energy that would undermine him. He also reported that whenever he was in his office he felt a cold coming on, but then when he left work, he would feel better.

This was the clue that led me to suspect that the building where he was working was toxic, and I scheduled a meeting with him there. I carefully watched his energies as we entered the building. His aura stayed strong, his meridians continued to flow well, and his energies did not go into a homolateral pattern. This was not what I expected, and I had to conclude that the building was not toxic for him. Then, however, I noticed that when he turned to the direction of his drawing board, his auric field collapsed and dimmed, his meridians slowed down, and his energies stopped crossing over from one side of his body to the other. He looked like he was in a haze of scrambled energies. To demonstrate to him what I was witnessing, I energy tested him while he was standing in several different directions. He was strong in every direction except one, which happened to be the same direction that the chair next to his drawing table faced. When he stood in this direction, he lost his strength on the energy test, which reflected what I was observing. This made no sense to him or to his co-workers, who were witnessing this strange demonstration. I, however, had a theory for what might be occurring.

The Canadian scientist Frances Nixon had investigated the relationship of the body's energetic axis to the energetic axis of the Earth.[17] She concluded that because of the vast difference between the inside of the womb and the outside world, the body sets a magnetized field, a "vivaxis," which serves as a bridging force between the womb and the external environment. The shock of entering the new electromagnetic environment causes the body to form this protective energy field, creating a bond that electromagnetically aligns the infant's physical direction at the moment of birth with the planet's energies. This protects the infant and, according to the theory, orients the child during the first three months after birth. Moreover, the direction in which you were born, and in which your system received the first shock of the environment's electromagnetic energies, supposedly determines a physical direction in which you may be vulnerable throughout your life.

Whatever the origin of the vivaxis, I know that some people are weakened when facing in a specific direction. When the architect was facing his drawing board, several of his meridians that were strong in every other direction became weak. While having him stand in that direction, I strengthened each of his meridians, using the technique you are about to learn for demagnetizing your vivaxis field. I also suggested that, to stabilize this change in his vivaxis, he pour a small box of baking soda into a bath each day. While he shifted the direction of his desk immediately and found that this solved the problem, demagnetizing his vivaxis field was a more basic correction. I suggested that, as an experiment, he move his desk back to its original position after that first treatment. While he was reluctant, he did so with no adverse effects.

The vivaxis technique reminds me of the way you have to periodically clean a cassette deck because it has become too magnetized. I suspect that the technique is needed because we are exposed to so many artificial magnet fields. If your vivaxis field is detracting from your health or interfering with your healing—and I will show you an energy test to find out if it is—there is a fairly simple routine for demagnetizing the field that can be extremely beneficial.

A young woman who had three toddlers came to me because of her perpetual exhaustion. While a full-time nanny was the treatment of choice, that was not within her means, so I taught her a number of techniques to maintain her strength and reduce her stress level. In a second session, she reported that the exercises would work for a time, but she noticed that whenever she was pulling meals together, the exhaustion would come over her again.

I wondered if her vulnerability was in her vivaxis field. I did a home visit. She had a long kitchen. Her stove, sink, refrigerator, and the place in which she cooked were all next to one another along the same wall, so she was always standing in the same direction when preparing meals. Sure enough, this was the direction of the vulnerability. After a treatment to demagnetize the field followed by a week of baking soda baths to stabilize her vivaxis, she no longer had this problem in her kitchen.

Demagnetizing Your Vivaxis Field. In working with the vivaxis field, you can energy test to see if a particular direction drains or scrambles your energy. If one does, the simplest correction is to be sure you are not spending excessive amounts of time facing in this direction. You can make changes in the placement of your desk, the arrangement of your workplace, or the direction your head is facing when you sleep. The advice to sleep with your head to the north, which many people get from energy "experts," is a prescription for disaster if the direction of your vivaxis vulnerability is to the north. In addition to becoming aware of the effects of the direction in which you stand, if the energy test shows a vulnerability in your vivaxis field, you can take some simple steps to demagnetize it. To test for a magnetic problem in your vivaxis field (time—1 to 2 minutes):

1. Massage K-27 (page 64) and do a vigorous Cross Crawl (page 69) to be sure your meridians are not flowing backward.

2. With your feet together, have a friend do a general indicator energy test (pages 65–67). Feet still together, pivot about 20 degrees (one-eighteenth of a full circle). Test again. Turn again. Test. Turn. Test. You will be testing eighteen times as you make a full circle. If you lose your strength in any particular direction, that is the direction of vulnerability in your vivaxis field. Here are the possibilities:
   a. If you test weak in every direction, there is an energy problem that needs to be corrected before you can test your vivaxis field. Empower your force fields with the Daily Energy Routine from Chapter 3 and strengthen your auric field with the Celtic weave (pages 177–180). Retest.
   b. If you test weak in one direction and strong in the other seventeen, you have identified the direction of vulnerability in your vivaxis field. I have never seen a person be weak in more than one direction unless the weakness was in every direction, as just described. You may, however, also be slightly weaker when facing in the exact opposite direction of your vivaxis vulnerability.
   c. If you test strong in all eighteen directions, your vivaxis field is not overly magnetized.

Just as you can use magnets to protect yourself from invasive electromagnetic fields, you can use them to fortify the protective field that surrounds your body. In addition to strengthening your aura by scanning, fluffing, massaging, and weaving (pages 177–184), you can further build your resilience by taking a round magnet and twirling it from a string above the ends of your meridians. If you have found a direction of vulnerability in your vivaxis field, you can correct it by following these steps (time—about 2 minutes):

1. Stand or lie in the direction that tested weak.
2. Spin the magnet over the beginning point and end point of each of your fourteen meridians (see Figure 46), about an inch above the meridian. These points are located at the ends of each of your fingers and toes, at the outside of both nipples, a hand's width above and to the outside of each breast, at the sides of your temples, at the bottom of your cheekbones, next to your nostrils, beneath your eyes, and between your eyebrows.

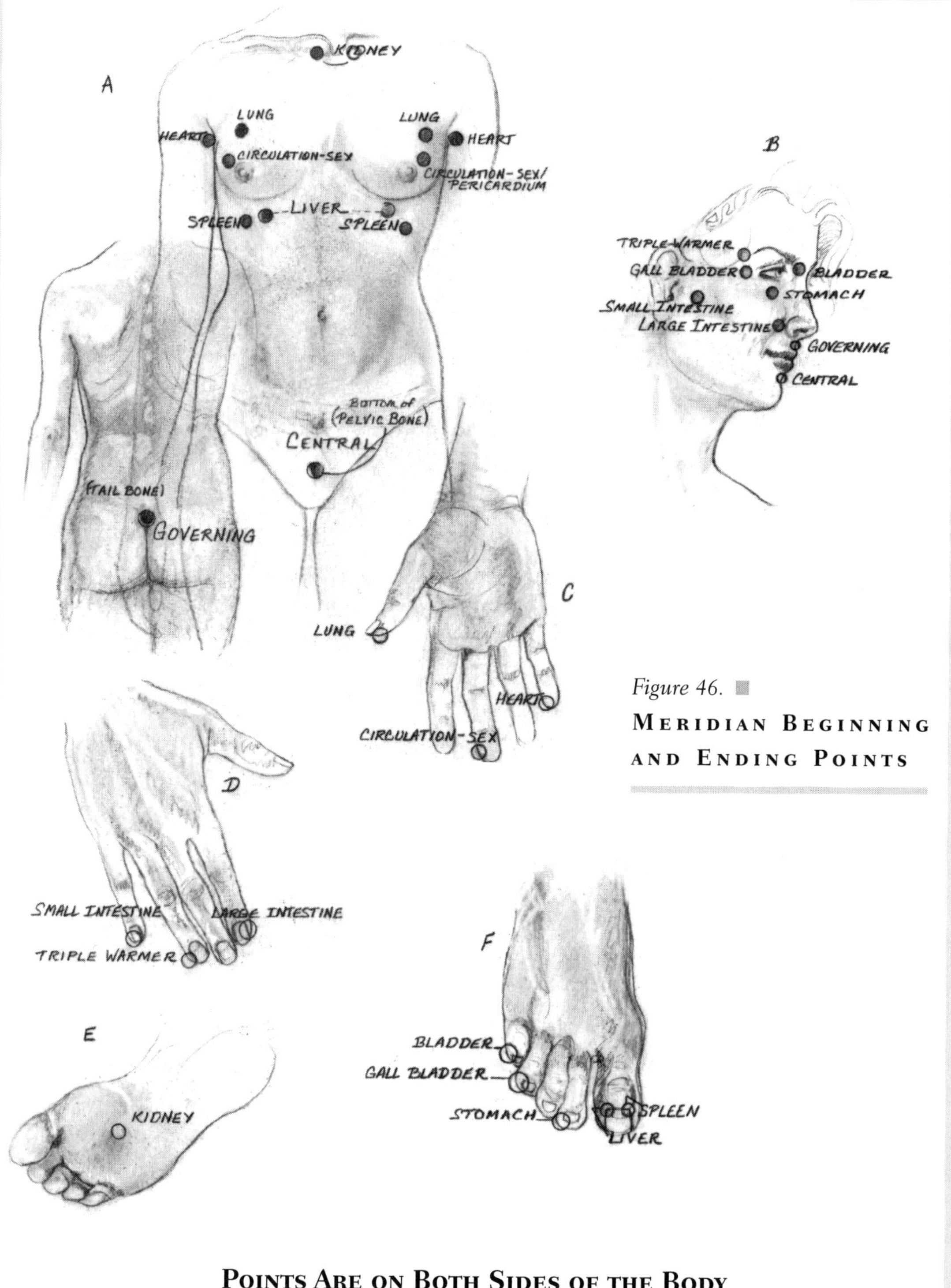

Figure 46. **Meridian Beginning and Ending Points**

Points Are on Both Sides of the Body Except for Governing and Central

3. Grasp the string between your thumb and your forefinger about two inches away from the magnet. Move your thumb and forefinger back and forth so the magnet spins in one direction until it stops. Then spin it in the other direction. Spin in each direction about five times.
4. This procedure will demagnetize your vivaxis field. Verify by repeating the test.
5. You can then stabilize the changes by pouring an eight-ounce box of baking soda into a bath each day for about a week.

Treating the Water You Drink. In carefully controlled laboratory experiments, Bernard Grad had healers do a laying-on-of-hands treatment to packets of water that were to be used to germinate barley seeds. Half the packets were given this treatment, half were not. The seeds were placed in the water by lab technicians who did not know which packets had been treated and which had not. Seeds grown in the water that had been treated by the healers were more likely to sprout, reached a greater height, and had a greater chlorophyll content. In addition, Grad found that treating the water with a common magnet would also accelerate a plant's growth rate.[18]

Treating the water you drink with the north pole of a magnet supposedly changes the water's molecular structure and is associated with a variety of benefits. Among them are that it prevents cholesterol from depositing on the inner surface of blood vessels, improves digestion, and improves the blood supply and nutrition to the various organs, including the heart. I believe that this treatment can have health benefits for many people, including providing an added measure of protection against electromagnetic fields in the environment.

One way to magnetize drinking water is to place a glass jug of water on top of a large magnet with the north side facing up. Keep it there for between ten minutes and twenty-four hours. I have also found that a quick pick-me-up in the morning is to lean or tape the north pole of a magnet on one side of a glass and the south pole on the other side for about five minutes before drinking it.

AN ATTITUDE OF EXPERIMENTATION

It is often not until we have a personal experience about something that we *really* learn about it. Although I have been working with and teaching about magnets for over twenty years, I felt little need to use them on myself, except to tape a magnet on each side of my waist when travelling across many time zones (see page 129) or spinning a magnet over the ends of my meridians if I felt I'd been "zapped" by too many energies. A few years ago, however, my own nervous system got into serious trouble, and the use of magnets was the only treatment that helped.

I have been very comfortable allowing the energies of 10,000 individual clients to filter through my body. Because energies do easily travel through me, I've always thought I could take in others' energies with no consequence, and for many years it was true because I knew how to move these energies out of me. Over time, however, this openness left an imprint on my energy field. It was as if I had inadvertently trained my nervous system to always be open to receiving the energies around me. In my work, this served me because it created a powerful empathic connection with my clients that allowed me to know a great deal about what was going on inside of them, but it began to take a toll. My body no longer knew how *not* to be an open channel. When I would go to sleep at night, instead of being restored, I was bombarded by electromagnetic energies that do not bother most people. I would be awakened by sharp pain shooting along the meridian lines, sometimes into my joints and sometimes into my eyes. The strain on my system was so intense that my vertebrae would go measurably out of alignment after the stress of a supposedly restful night in bed. We bought three top-of-the-line mattresses from different companies over a two-year period. They did not help. I was slow to realize that I had become unable to counter an electromagnetic assault when in an altered state such as when I work with clients or when I sleep.

Unable to find a remedy in my own toolbox, I became scared. I consulted several physicians and other health care professionals, both conventional and alternative, but to no avail. An MRI revealed aberrant neurological patterns that resembled my old multiple sclerosis, but an effective remedy was not found. I was on my own. While I used everything I knew to heal myself, "physician heal thyself" is sometimes a tall order. I was being assaulted rather than restored at night, and I was taking on my clients' symptoms. All the excellent preventive measures I knew weren't working. My own electromagnetic field had been disrupted, and I wasn't able to protect myself.

The key to my recovery was the use of magnets as shown in this chapter. But I could not rely on my previous understanding about them. I could find no book that teaches the specifics of building a stronger, protective magnetic boundary in one's aura. Magnetic pads from even the best companies made things worse. I had to experiment night after night, learning at the level of my own body. I am still learning about the use of magnets and about the natural and artificial electromagnetic energy coming from the earth and the atmosphere. The guidelines offered here come from my experience and they are hard-earned, but they are hardly the final word. I am deeply certain that magnets properly used can be an extraordinarily valuable tool in energy medicine, and I encourage both an open mind and an attitude of experimentation.

Once I had correctly diagnosed the problem, working with small magnets, the Celtic weave, and my vivaxis helped tremendously. Particularly if you work on other people's bodies professionally or frequently, I hope that you remain conscious of the necessity of deeply restoring your energies during sleep at night. If you do not feel fresh and vibrantly alive when you awaken in the morning, question why. Look to the possibility that electromagnetic fields may be disturbing you. Check to be sure your auric field, Celtic weave, and vivaxis are in good order; keep yourself zipped up; and do your three thumps regularly.

Chapter 12

Setting Your Habit Field for Optimum Health and Performance

ALL SYSTEMS ARE REGULATED NOT ONLY BY KNOWN ENERGY AND MATERIAL FACTORS BUT ALSO BY INVISIBLE ORGANIZING FIELDS.

—BARBARA ANN BRENNAN
Hands of Light

The energy systems you have been exploring are eight highways to better health. Underlying each of them is an organizing field that regulates your energies, your physiology, and much of your behavior. This chapter shows you how to focus your efforts so you can cut to the energetic quick and adjust this field directly. I refer to the organizing field as the habit field because it maintains the habits of body, mind, and energy.

The habit field is a force in nature that is as real as gravity or electromagnetic fields. A habit field surrounds and permeates a living system—a body, an organ, a cell. It is a "region of influence" that carries information, organizes energy, and gives form to living organisms. A habit field is a highly stable energy. It is the die or cast that gives form to the other energy systems and ultimately to all biological development. Like the ocean pounding on the rocks that form its shore, the energies of the meridians, chakras, and other more fluid systems have far less impact on the habit field than it has on them. But eventually, the ocean does carve its imprint on the shore; habit fields do evolve, impacted by the very energies they mold. In this chapter, you will be learning ways of accelerating their evolution to benefit your health and well-being.

Certain energy techniques can set your habit field like a thermostat. Suppose, for example, you have high blood pressure. What if you could alter the habit field that keeps your blood pressure elevated? If the habit field shifts, shifts in behavior, attitude, and actual physiological conditions follow.

"FIELDS" THAT PROMOTE YOUR HEALTH OR SUSTAIN YOUR AILMENTS

Since the 1920s, a handful of biologists have been using the term *morphogenetic fields* to refer to fields that give form to biological systems, and the concept has recently been expanded, refined, and popularized as "morphic fields" by the biologist Rupert Sheldrake.[1]

The premise is simple: Form follows field. Biological structures pattern themselves after invisible organizing fields. An acorn carries the blueprint of the oak not only in its genes but *also* in its morphic field. But, you might think, don't its genes carry all the coding necessary to produce the towering tree? Why add another concept? Because genes engineer parts, not wholes. No one has found a gene that puts all the parts together into the whole. No DNA code has been discovered that instructs an acorn to transform itself into an oak, a tadpole into a frog, or an egg into an ostrich. No one has explained how genes get a termite's head at the right end or cause a termite colony to cooperate in building a thirty-foot nest. As Sheldrake notes, "Properties are projected onto [genes] that go far beyond their known chemical roles."[2] These biologists have concluded that the embryo carries an informational field that guides it into its adult form and governs its instinctive behavior. They believe that these form-generating fields are as integral to biological development as genes themselves.

Harold Burr, a neuroanatomist at Yale in the 1940s, provided some of the earliest empirical evidence of the existence of an energy field surrounding the body. Using a conventional voltmeter, he measured the electrical field around a sprout. It closely resembled the field that would be there when the plant was fully grown. Similarly, the energy field surrounding a young salamander was shaped roughly like an adult salamander. As Burr mapped progressively earlier stages of salamander development, he was amazed to find that the electrical axis that would later be aligned with the brain and the spinal cord in the adult is present in the unfertilized egg.[3]

In another experiment using salamanders, primitive, unspecialized tissue cells from an amputated foreleg, transplanted near the tail, regrew as a tail; transplanted near the hind leg, they grew into another hind leg. Reviewing these findings, the orthopedic surgeon and Nobel nominee Robert Becker concluded, "It was obvious that chemical messengers were totally unable to convey this much organizational complexity. . . . There had to be something else—something similar to the morphogenetic field—that could contain within itself the entire organizational plan."[4]

Not surprisingly, this concept is controversial, but it accounts for some uncharted mechanisms of inheritance, and its implications are profound. If you could transform a deep organizing field that is detrimental to your health, you would be giving your health a significant boost.

Distinguishing the Aura from the Habit Field. For many years I sensed, within the layers of the aura, a second energy surrounding the body, but it was a long time before I felt sure. This second energy seemed to have little movement other than a steady rhythm that didn't fluctuate, and it had virtually no color. Next to the aura's rich, pulsating colors, this second field was often difficult to distinguish.

I found that when I worked with the triple warmer meridian, however, this field was noticeably affected. Because triple warmer controls the body's habits, I began to call this less obvious field "the field of habit." Changes in it seemed to correlate with changes in physiological patterns. Then, in 1991 at a conference in Prague, I was in the audience when Sheldrake presented his theory of the morphic field. Bingo!

Fields of habit seem to be an energy system of a different order from the meridians, chakras, basic grid, aura, Celtic weave, triple warmer, or strange flows. Those all fluctuate with your mood, health, and circumstances. Your habit field, however, like your primary rhythm, is more steadfast and more pervasive. My best guess, in fact, is that the primary rhythm is not a separate system from the habit field but rather is the reverberation of the habit field through each of the other energy systems.

Sometimes I see the habit field around the person. More often I feel it; sometimes I taste it. Because I register these fields with my senses, I am certain that the impassioned debate about morphic fields that has been waged in the scientific literature[5] will ultimately resolve in favor of the morphic field biologists. But the dispute will go on for some time, because morphic fields are generally too subtle to be detected by standard measuring instruments, and the whole concept seems a bit metaphysical compared with conventional science. It really isn't any more metaphysical than gravity, however, another field we cannot see directly and know only by its effects. But we've had three hundred years to get used to that idea.

Habit fields seem to interact directly with information in the environment. Like other subtle energies, they appear to be nonlocal—that is, their effects are not necessarily limited to their location.[6] Many studies have shown that human intention can influence distant biological events. Perhaps fields that organize physiological processes are shifted by thought. For instance, after being instructed in how to use visualization to inhibit the breakdown of red blood cells in a test tube located in a different room, experimental subjects achieved statistically significant results in their efforts to slow the rate of cell deterioration.[7] In another experiment, two healers used prayer to send love to subjects two hundred miles away. The subjects—as contrasted with controls who did not receive prayers—showed significant EMG changes that correlate with a reduction of tension and an increase in relaxation.[8] Of 131 studies that investigated the effects of prayer on healing published up to 1993, 77 reported statistical significance.[9]

How to Relieve Pain in Your Hands, Wrists, or Elbows

The spindle cell mechanism (page 283) reestablishes neurological equilibrium in a disturbed area. Carpal tunnel syndrome, stiff joints, sore arms, and other related problems often respond to tiny pinches because the spindle cell at the belly of each muscle will reset. Make tiny light pinches up and down the inside of your arm for about 20 seconds. The lighter the pinch, the better. This will improve the condition. In fact it is good to energy test because otherwise ***you won't believe it.*** *Suppose you have been typing at the computer for an hour. To reset the spindle cells (time—about a minute):*

1. *Begin with a general indicator test (pages 65–67).*

2. *If you test strong, open and close your hands several times and retest. Repetitive movement such as typing or opening and closing your hands can stop energy from flowing off your fingers.*

3. *If you tested weak, reset the spindle cell mechanism by pinching up and down your arm.*

4. *Retest. If you now test strong, the pinching has freed energies and is helping.*

SHIFTING A FIELD OF HABIT

Your habit field is designed to conserve what evolution has created. Its purpose is to keep you from changing! Habit fields always resist innovation. But because our bodies evolved by adapting to a world that no longer exists, and because circumstances continue to change at a bewildering rate, it has become critical that we learn to alter the fields of habit that keep us locked in outmoded adaptations. A habit field is actually a constellation of many fields that influence specific habits. You have already been working with your habit field. Chapters 8 and 9 provided techniques for changing deep patterns in your immune system and other processes related to illness. It is possible to shift your habit field to promote any change you wish.

Your habit field can also keep you stuck. It is easy for some people to tell others to just get beyond their plan or sorrows or wounds, and it is a great idea. But I have had many people come to me who have worked very hard and sincerely, though unsuccessfully, to resolve physical or emotional injuries from long ago. The issue for many of these people is not that they wallow in self-pity, want sympathy, long to hold on to the past, need to bond with others over their misfortunes, or lack courage. The invisible but decisive factor in many of these cases is a field of habit that is outside their awareness and thus their control. Human will and intention are powerful, but if all your energies are going in the opposite direction of your will, best focus your will on changing those energies.

Certain glands and energy systems, such as the hypothalamus, the spleen meridian, the triple warmer, and the basic grid, seem to have a particularly strong influence on habit fields. While habit fields resist change, they can and do change, and energy work is often the key. This chapter will walk you through a series of steps you can take to consciously and deliberately create shifts in your habit field.

CLEARING YOUR HABIT FIELD

Build it, and they will come, was the enigmatic advice in the movie *Field of Dreams.* Like the field of dreams, a new "field of habit" can begin with a vision and a belief that it can exist. Build a new field of habit, one you have carefully envisioned, and *they*—powerful energies to support that vision—*will come.* To secure lasting changes in your habit field, especially when you cannot get a clear vision or concentrate your mind, simple techniques for unscrambling your energies and releasing stress can be followed with sophisticated methods for reprogramming your field of habit. Begin by experimenting with each of the following ways of clearing your energies:

1. Do the Daily Energy Routine (time—5 to 8 minutes). If you are simply using the six-part Daily Energy Routine you learned in Chapter 3, you are already doing much to keep debris from accumulating in your habit field. If you are working with a partner, add one more step to the spinal flush (pages 79–80). After clearing your spine, as you are lying on your stomach, have your partner place one hand on your sacrum, a couple of inches below your waist, and the other hand on your upper back, while rocking you for two to three minutes. Just as rocking a baby soothes and calms, so does rocking an adult. The rocking also dispels stress that is building up, and, most important, it inhibits new defense responses from entering the field of habit. When you stop rocking, energy keeps moving through your body, and most people feel a delightful buzz and release of tension. Stresses are dissipated before they act to further lock in old habits.

2. Vigorously Rub Your Neck and Head (time—less than a minute): Use the fingers of both hands to briskly massage your neck and scalp. This gets your blood flowing and your energy moving, and it prepares your system to gain the greatest benefit for holding the neurovascular points.

3. Hold Your Neurovascular Points (time—3 to 8 minutes): Open the space by stretching your forehead to the sides. Lay one palm on your forehead and the other on the back of your head. Hold them there for at least three minutes, breathing deeply and comfortably. Again, holding the front neurovascular points returns blood to your forebrain and balances the circulation of blood throughout your body. Holding your other hand behind your head sedates the fear points and calms your hypothalamus. It also creates an energy polarity between your hands while linking the energies throughout your brain. If you are in stress, this allows better communication between the front part of the brain as it is trying to figure things out and the back part as it is mobilizing for the fight-or-flight response. Your body relaxes and your habit field softens its grip.

ENVISIONING A NEW HABIT FIELD

The following guidelines for using your mind to set your field of habit concern themselves with intention, focus, use of repetition, choice of imagery, and willingness to experiment.

Begin with a Clear Intention. Setting a more auspicious habit field begins with your intention to do so. While the powers of the mind have been overrated by some health practitioners (e.g., the well-meaning but shortsighted and harmful outlook that *blames* a person's illness on unexpressed anger, a bad attitude, or the need to please), the role of the mind in health is often underrated by conventional medicine. The evidence that mental stress negatively impacts health is abundant, and the evidence that peace of mind, meditation, and directed imagery can enhance health is growing.[10] Your body *minds* your mind, responding to what you feel, think, and want. But if your mind does not *mind* your body—if your will, intention, and actions are not aligned with your body's requirements—you pay a toll in health and happiness.

Ride the Power of Your Mind. Sheldrake believes that morphic fields influence not only biological characteristics but also mental activity, behavioral patterns, and social organization. But even as your habit field shapes your thoughts, your thoughts can reshape your habit field, and that can be an empowering insight. Mental devices such as self-hypnosis, guided imagery, and focused intention may be particularly powerful toward this end. It has been repeatedly demonstrated that the mind can affect inanimate objects,[11] and it is not surprising that prayer, spontaneous imagery, and focused imagination also impact the body and its health.[12] In my work, I have used guided imagery to alter people's fields of habit. Consider a phobia—say, a fear of elevators. The emotional response is coded in the person's habit field. I might have the person imagine being in an elevator, which evokes the fear. At the same time I hold the person's neurovascular points, which defuses the flight response, relaxes the body, and desensitizes the person to the feared event. This works. Habit fields change when such methods are applied. Beliefs, intention, and imagery can be directed to alter the field that supports any given habit.

Employ Repetition. Repetition increases the strength of an organizing field.[13] By frequently evoking a desired field in your imagination, you are building its potency. A woman who wanted to change her compulsive eating was hypoglycemic, and her panic was fueled by experiences of having gone into convulsions during low blood sugar attacks. The only way she knew to eliminate her panic was to indulge her craving for sweets. In her imagery, she was able to "see" the feeling of panic that was laced throughout her energy body. Though she didn't understand why, the image of a house came to mind, and it became clear that the house represented panic. As in the fairy tale of the Three Little Pigs, sometimes the houses of panic were made of stone, sometimes they were made of logs, but the vast majority of them were made of straw. She found after working with this imagery that when she was overcome with panic, she could "huff and puff" at the panic and the craving, and it usually proved to be a straw house that would just blow away. When that happened, she was left with an image of herself as thinner, in control, and safe. Repeatedly holding that image built it into her habit field. At other times, when the house was made of stone or wood, she could not simply blow it away. She came to trust that at these times the desire was a real need, not a panicked craving, and she learned to give her body what it required.

Find the Imagery That Is Yours. You can influence your habit field with directed imagery, but no one can tell you what your field of habit will look or feel like in your imagination. You may *see* it, *feel* it, see *and* feel it, experience it through other sensory channels such as sound, taste, or smell, or know it without knowing how you know. It may appear to you as an area in space with little detail; at other times, particularly as you work with it, you may experience its elements as complex symbols, like the house or a face or an animal or a wall. The way you sense your habit field at any particular moment is the right way for you.

A ten-year-old girl began having terrible pain in her left side. It was intermittent, but the frequency had increased to more than once each day, and she was missing a great deal of school. She had twice been rushed to the hospital screaming in agony. At first, hearing the symptoms, the doctors guessed that it was appendicitis, a bowel obstruction, or an infection, but extensive medical examinations and laboratory tests revealed no organic basis to the problem. Yet the pains kept returning. When she was brought to me, I found a blockage in the ileocecal valve that sits between the small and large intestines. Energy work relieved her pain and the blockage, but the pain always returned in a day or two. I gave her mother my home number and asked her to call me the next time the daughter had an acute attack. When she did, I immediately went over to their house.

The girl was writhing in pain. I noticed that she had posters of the 1980s rock star Boy George on her walls. I asked her to imagine Boy George singing "Karma Chameleon" and chasing the pain away. One of the lyrics in the song is "You come and go, you come and go," and we changed the line to "You go away, you go away." She was able to see Boy George with a big smile, dancing and singing on the trail of the pain, until the pain simply dissolved. The pain was never again a problem, because any time she started to feel it, she invoked Boy George, and he was immediately able to make the pain *go away.* After a few of these encounters, the pain stopped returning.

You of course might surmise, "Well, it must have just been hysterical pain." Perhaps, but it was no less real to her pain receptors. By relieving the pain and rechanneling the energies, she was also preventing a buildup that might have progressed into a more serious condition. And even if the pain was a hysterical reaction, another way to look at persistent hysterical pain is that most chronic pain is a habit that doesn't have to be. The body doesn't need to keep sending the pain message after the cause has subsided. Whatever the original cause of chronic pain, the body learned the pain response, the nervous system grabbed on to it, and it became more and more deeply entrenched in the habit field. Treating even hysterical pain at the level of the energies and the underlying field that maintain it is usually more effective than simply telling the sufferer, "There is no medical reason for your excruciating abdominal pain, so stop experiencing it!"

Do an Experiment. There are infinite variations to an imagery approach. Sometimes it is useful to add words—an affirmation—that describes the new field as already existing. "I am bathed in energies that keep me calm and peaceful." Say the words fervently. Some people have responded well to the image of removing a computer disk or a CD that holds the old field of habit and replacing it with one that holds the new field. Another computer image that one of my clients suggested is to imagine deleting the old field from your hard drive and, with deliberation, typing in a new file.

Select a physical, emotional, or behavioral condition you would like to change. Vividly imagine that the change has already occurred. See it, feel it, hear it, smell it, taste it. Use all the sensory channels you can. Then breathe deeply and immerse yourself in the energy field that is created by this vision.

Imagery techniques can, in themselves, begin to shift your habit field. They are not only psychological. They impact your neurons, and they shift the energy patterns in your brain. Physical techniques that are attuned to your energy system can also make your efforts more effective. They "lock in" or at least turn up the volume on your intentions. You can, for instance, reinforce your new field by doing the Zip Up (page 82) simultaneously with imagining you are immersed in the energy of a new field. The Zip Up traces the central meridian, which affects the parts of your brain that are involved in autosuggestion and hypnosis, and it opens the energy so that the field you are imagining becomes more deeply embedded.

How to Relieve Low Back Pain

Each of the following addresses low back pain. If one does not provide relief, go to the next:

1. *Tape the north side (see pages 300–301) of a small magnet over the area.*

2. *Hold the kidney sedating points (see Figure 26L, page 121).*

3. *Low back pain may indicate that toxins are backing up in your large intestine so that waste products are not being removed. If you feel pain in your back slightly above your waist, you can release toxins and get them moving out of your system by stretching the area or massaging the large intestine lymphatic points on your back (Figure 9e).*

4. *If the pain persists, hold the large intestine sedating points (see Figure 26X, page 123).*

REPROGRAMMING YOUR FIELD OF HABIT

The need to reprogram our fields of habit has never been more urgent. No longer can we keep up in today's world by thinking as our parents thought, by believing what they believed, or by acting as they acted. No longer, in fact, can you flourish by using adaptations that were at your own cutting edge ten years ago.

Many people know the kinds of changes they want to make but find themselves unable to make them. Trapped in habits from the past, they may put tremendous effort into willing a thinner body, a relationship with a healthy partner, or vocational success. Guided by the popular notion that if your intention is strong enough you will succeed—"Energy flows where attention goes"—they only gather discouragement, cynicism, and self-loathing. But the opposite is also true: "Attention goes where energy flows." To change your mind, change your energy. You can't will happiness. You can't always will pain to stop. You cannot even will inspiration. But you can shift your energies in ways that support your happiness, diminish your pain, and increase your inspiration.

In my work, I rarely begin with a person's attitude or state of mind. If your energies are freed for healing and creativity, your mind will follow. A person doesn't need to believe in energy healing for the techniques to work. I, in fact, enjoy skeptics in my classes and in my practice. Perhaps I am their last grasp or their spouse coerced them to see me. It is always rewarding and somehow validating for them when they start to feel better as energy is moved through their body with techniques that seem absolutely bogus to them.

It is particularly difficult to simply will yourself to change patterns that are emotionally locked into your body because of trauma, and the following techniques address this issue. Three approaches for reprogramming your habit field include: (1) defusing traumatic residue that is already in your habit field, (2) programming positive feelings into your habit field, and (3) causing your habit field to embrace a new physiological or psychological pattern.

1. Defusing Traumatic Residue. Like land mines left in the ground after a war, the defensive responses you mobilize during trauma need to be defused and cleared for life to be fully restored. A primary kind of restriction I often see in the field of habit is based on early trauma. The trauma may have been physical or emotional. We all have been wounded in one way or another, and, in trying to protect us, the hypothalamus and triple warmer often lock into outdated defensive strategies that tend to rigidify the habit field. This situation continues to limit us, though we may have found our way out of the original traumatizing situation decades earlier. Self-talk is often not enough to change such a primal response, but a number of techniques can mobilize the hypothalamus and triple warmer's energies to open your habit field for reprogramming.

You may not be aware of a past trauma that is holding you back, or now may not be the time to deal with it, but if and when that time comes, return to the following two techniques.

A. HOLDING YOUR NEUROVASCULARS TO DEFUSE A TRAUMATIC MEMORY. You have already worked with your neurovascular points to reprogram your stress response patterns (Chapter 3) and learned variations on that method in Chapters 4, 7, and 9. The following discussion summarizes the approach. It is one of the most powerful and important techniques I know for changing dysfunctional responses that are maintained in your habit field. It is not possible for you to think clearly when the blood has left your forebrain and moved into your limbs for fight or flight. Each time, however, that you are able to keep the blood from leaving your brain as you deal with a stressful problem, you are showing your brain a new strategy, and your habit pattern begins to change. As your body learns to tolerate increasing amounts of stress before invoking the fight-or-flight response, your general level of health will increase as well.

A friend of mine had a childhood that was among the more painful of any I've encountered, and her rage against her parents was overwhelming. Each evening she would, in the bathtub, hold her neurovascular points to defuse a different memory. She did this every night because it worked; her inner torment was diminishing. Eventually she found that her energies were no longer being drawn into daily tirades of hatred toward her parents, and she was able to stay focused on what was happening in the present moment rather than continually being pulled into her past.

Select a memory from the past or a situation from the present that is painful to you or has an emotional grip on you. Hold your neurovascular points for at least three minutes (don't forget that this is also a powerful technique to use at any moment you are feeling overwhelmed). You can choose among the following ways to hold the points. All three are effective, but you may sense that one works better for you than the others (time—at least 3 minutes):

- Place the fingers of both hands on your forehead and stretch the skin. Bring your fingers back to the neurovascular points on your forehead (page 90) and rest.
- Using these same points, cross your hands one on top of the other, fingers on the *opposite* neurovascular points, stretching the skin apart. Crossing your hands helps your energies cross over from one side of the body to the other. With your hands crossed, breathe deeply and relax. Or,
- Lay one hand on your forehead and the other on the back of your head. The polarity between your hands creates an energetic link between the front and back parts of the brain, brings energy to the hypothalamus, and sedates the kidney meridian fear points.

You may also have a friend hold your neurovasculars in one of these ways so you can totally relax into the experience. Replay your bad memory on a movie screen in your mind, or tell it as a story, while continuing to hold the points. Don't try to change the story, don't try to be positive about it. Sink into it. Eventually, the blood *will* return to your brain, the fight-or-flight response *will* disengage, and the emotional impact of your story *will* diminish.

The technique not only removes the charge from the memory but also breaks up the constellation of memory, feeling, and defensive response in your habit field.

The next time you think about the scene, you will probably notice that it has less of a hold on you. Your body won't automatically respond as it has in the past. If the memory continues to provoke a traumatic response or if you have any negative residue, repeat the steps. By proceeding in this way, you will eventually get to a place where the memory is no longer debilitating.

B. EYE PATTERN RELEASE (time—about 2 minutes). While thinking of the stressful memory, cluster the thumb and first three fingers of either hand and hold them six to twelve inches in front of the bridge of your nose and make sideways figure 8 patterns. Move the fingers upward and to the right, as high and wide as you can while still keeping them in your field of vision. Follow them with your eyes, while keeping your face straight ahead. Follow your fingers with your eyes—up to the right, around, and under, and then to the left—up, around, and under (see Figure 47). Keep your mind on the stressful memory.

If you notice that the stress gets to you more when your eyes are fixed in a specific direction, keep looking at your fingers in that direction and, with your other hand, hold your frontal neurovascular points as you continue to think of the memory. The eye position can substantially augment the neurovascular technique. Eye desensitization procedures for releasing traumatic memories have recently become popular and quite sophisticated.[14] I have, however, heard of versions of this technique being passed down in various cultures for thousands of years.

Combining this procedure with energy testing yields an extremely potent neurological correction. Have your partner do the figure 8 with one hand as you keep the memory in your mind and follow the figure 8 with your eyes. The tester stops every few inches and does the general indicator test. At any eye position that you test weak, hold your neurovasculars for at least twenty seconds while continuing to look at the tester's fingers. Eventually the memory loses its emotional charge and you will test strong.

2. Programming in a Positive Feeling (time—less than a minute). The next time you feel fabulous or are having a wonderful memory, you can

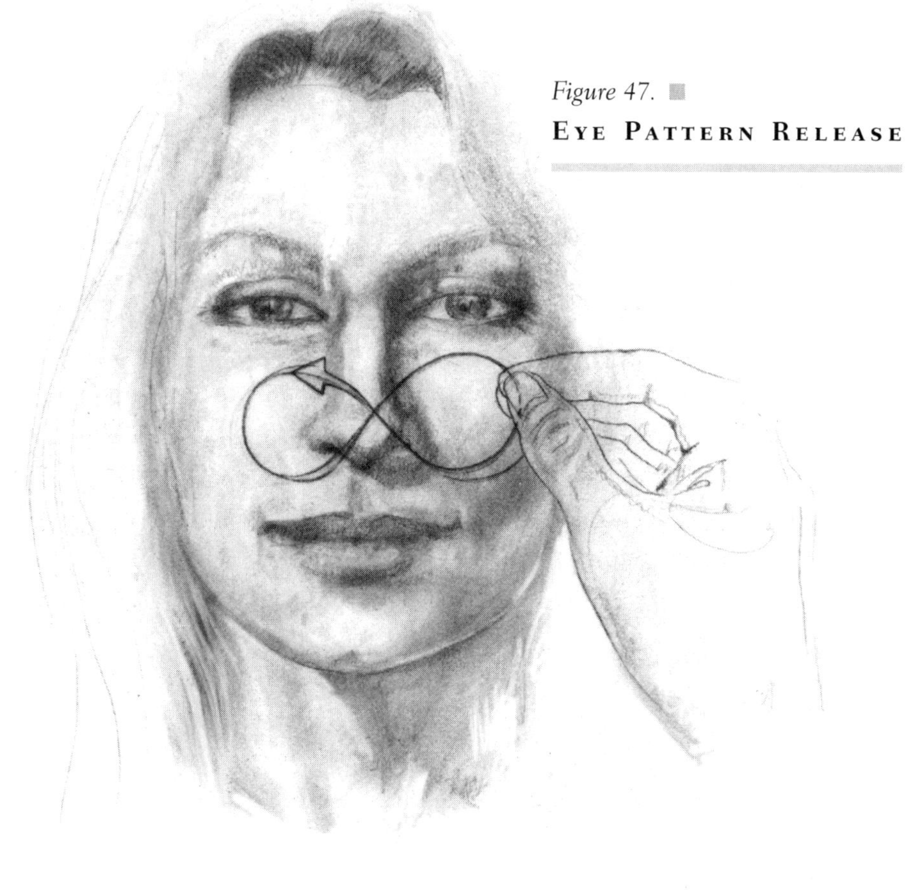

Figure 47.
Eye Pattern Release

imprint these feelings into your habit field. If you never feel fabulous and do not have access to wonderful memories, find a way to mentally thank someone or something. As the feeling of gratitude begins to expand within you, focus on that feeling. With your middle finger, "tap the feeling" into the spot between your eyebrows, the third eye point. Use a steady, light, firm tap. The third eye sits at the beginning of the bladder meridian. All the nerves coming down your spine are on this meridian. To tap there is to send a steady message, a pulsing beat, into your nervous system and into your habit field. In the legendary torture where water is dripped on the forehead, the prisoner's inevitable fears reverberate throughout the nervous system, culminating in insanity; it's not just the steady drip-drip-drip. But tapping can send happiness as well as terror through the nervous system. Tapping in a positive feeling is a simple, natural, and lovely way to instill more joy into your field of habit.

3. Changing a Physiological or Psychological Habit. The "temporal tap" is an ancient technique that recently has been rediscovered. It originally was used for pain control in the Orient, but it is also enormously effective for breaking old habits and simultaneously establishing new ones. Tapping around the temporal bone—beginning at the temples and travelling around the back side of the ear—makes the brain more receptive to learning while temporarily suspending other sensory input. It also sedates triple warmer because you are tapping in the opposite direction of its natural flow, and it is triple warmer that most decisively governs your body's habits. By calming the part of the nervous system that fights to maintain your habits, you can more easily slip in a new habit.

In the 1970s, the founder of applied kineseology, George Goodheart, discovered that by tapping along the cranial suture line that starts between the temporal and spenoidal bones, you can temporarily shift the mechanisms that filter sensory input.[15] If you introduce a self-suggestion or spoken affirmation while you are tapping, your mind will be particularly receptive to it. The temporal tap is also attuned to the differences between the left and right hemispheres of the cerebral cortex. Because the brain's left hemisphere is in most people more critical and skeptical, statements made with a negative wording are in closer accord with the way it functions and are more likely to be assimilated. Such statements are tapped into the left side of the head. Likewise, because the right hemisphere is highly receptive to favorable input, statements that are worded positively are tapped into the right side. This pattern is reversed in some left-handed people, and you can energy test to be sure. If you stay strong while tapping a negative statement into the left side, stay with the instructions; if you become weak, swap the words "right" and "left" in the instructions.

Begin by identifying a habit, an attitude, an automatic emotional response, or a health condition you would like to change. Describe the change you would like to bring about in a single sentence, stated as an affirmation in present time, that is, as if the desired condition already exists. It does not have to be a truth yet, but rather a statement that you would like to be a truth in the future. For instance, you might say, "Under pressure, I stay calm and centered." It usually helps to write your statement.

Then restate it, keeping the same meaning, but with a negative wording (i.e., "no," "not," "never," "don't," "won't," etc.). Thus the "calm and centered" statement may be worded negatively as "I no longer get stressed under pressure." Notice that the meaning is still positive, even with the negative wording. Another example: "I eat for health and fitness and I enjoy my food" can be negatively worded as "I don't eat from anxiety or compulsion." "My fingernails are growing long and healthy" can be negatively worded as "I no longer bite my nails" (time—about a minute):

1. Starting at the temple, tap the left side of your head from front to back with the three middle fingers of your left hand (see Figure 48). State the negatively worded version of your statement in rhythm as you tap. Tap hard enough to feel a firm contact and a bit of a bounce. Tap from the front to the back about five times, making your statement with each pass.
2. Repeat the technique on the right side, tapping with your right hand, but this time using your statement in its positively worded form.
3. Repeat the procedure several times per day. The more you tap in the affirmation, the quicker and stronger the effect on your nervous system and your field of habit.

The temporal tap combines a variety of powerful elements, including repetition, autosuggestion, and neurological reprogramming. It affects not only the brain but also each meridian, so the message of your intention is carried to every system in your body. It is a disarmingly simple way to change many patterns that cannot be overcome by willpower alone.

The parallels between our physical and our emotional difficulties are often striking.[16] Sometimes an organ will literally begin to behave in a way that reflects the way *you* behave. I'll use myself as an example. I sometimes have difficulty setting clear boundaries with people. Given the nature of my work, this can be an overwhelming problem. At one point some years ago, I became very vulnerable to infections. It was as if my thymus gland, which is responsible for protecting the body from infection, was modeling *its* behavior after

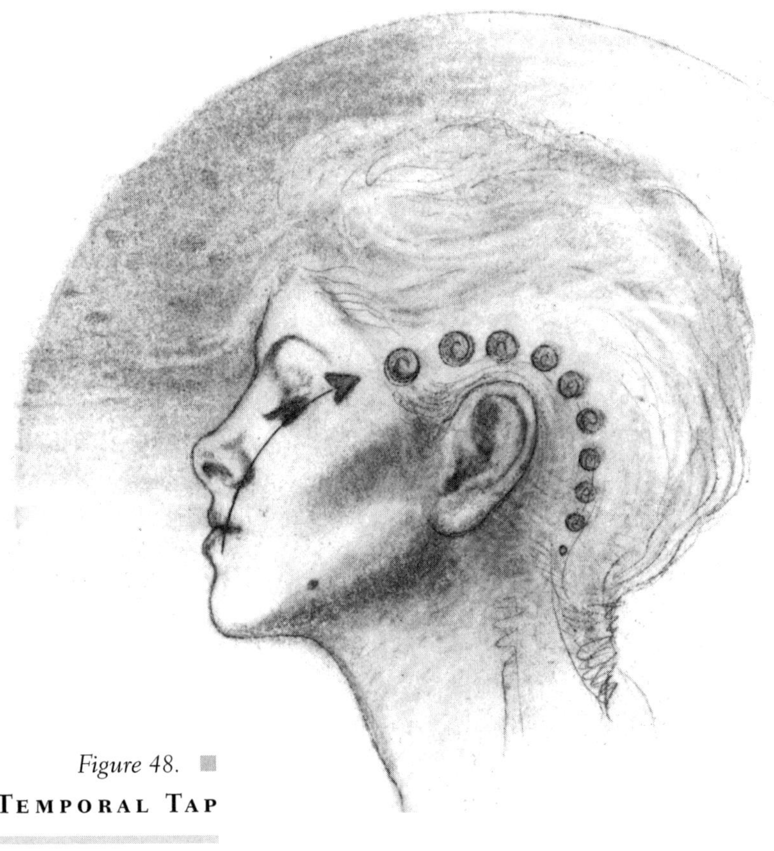

Figure 48.
Temporal Tap

mine. I began to tap into my left temple the statement, "My thymus gland no longer allows foreign invaders to enter my system," and into the right, "My thymus gland vigilantly maintains strong boundaries that keep out foreign invaders." Not only did my susceptibility to infections improve, my ability to set clear boundaries improved as well. Part of the beauty of the temporal tap is that in funneling your intention down to two short statements, you can sometimes get to the nub of a complex and puzzling relationship among body, mind, and soul.

Another use of the temporal tap points involves wordlessly *smoothing* the points instead of tapping them. This can be both calming and nurturing. Place your fingers at the temples on both sides of your head and *smooth* both sets of fingers along your temporal points as you breathe deeply. When you want to nourish yourself, perhaps to satisfy a craving, smooth the temporal points on both sides of the head simultaneously, while breathing deeply. If, for instance, you are feeling a panicked craving to eat or indulge in some other addiction, smooth your temporal points. Many people feel nurtured from this simple procedure, and the sense of panic leaves.

I am so certain that the temporal tap is one of the best tools available for taking greater control of your life that I want to relate a number of stories where *it didn't work,* as these stories are instructive for getting it to work. I have found that when the temporal tap fails, it is almost always because of how the statements are worded. The words need to be in a language that is comfortable for you. Sometimes it is as simple as translating the statements into the simplest words possible, rather than leaving them in moralistic, religious, New Age, or other highfalutin' terms that are a bit alien to the way you think. It is also important that the words be aligned with your values and congruent with your feelings, that you not be saying one thing while thinking another, and that you are not attempting to negate a primal need.

A woman who wanted to lose twelve pounds gained eighteen using this approach. On the left side, she tapped in the words, "I no longer hold on to extra weight." On the right side: "My set point is dropping to 134." These wordings seemed reasonable, but the tapping was having the reverse effect. I asked her to notice if her thoughts were following her words as she tapped, if her mind was wandering, or if any images were entering her awareness. It turned out that what came into her mind *every* time she tapped was, "Oh, hell, I've got a Slavic body, I'm always going to have a Slavic body, and I'm going to end up looking just like my [fat] Aunt Sophie." She was tapping this in five times every day. And it was working! The thought and image were a far more primal self-suggestion than her carefully worded statements. She did, by the way, lose a good deal of weight after tapping into her left side, "I did not inherit Aunt Sophie's body" and "I inherited a body that can be thin and lithesome" into her right side.

A man whose job was eliminated after twenty-four years with a company was sent back to college for a year, at the firm's expense, to upgrade his skills. Five weeks into the first quarter, he had pulled Cs and Ds on the midterms of several of his courses, threatening his future re-employment. He was trying very hard, but he found himself unable to concentrate during any sustained period of study. Having heard that I have helped people with learning disabilities, he scheduled a session with me. I could not find a learning disability, but I did sense that after working for a quarter of a century in a people-oriented position, his habit field had no juice for inwardly oriented, concentrated study. I explained this to him, and he was highly enthusiastic after I taught him how to use the temporal tap for changing the habit field that was affecting his ability to study. But he came back a week later deeply discouraged. He had followed my instructions to the letter and devoted all the time required, but he could detect no change in his study habits.

He told me the statements he had been tapping in, and I watched him do the tapping. The words sounded good to me, but his energies weren't receiving them very well. As I explained how important it was that the words must sound right to him, I learned that he had come to the United States at age six and that English was his second language. After doing the temporal tap for another week, this time using his native language to make the same statements, he reported respectable improvements in his ability to read and concentrate.

A friend wanted to quit smoking after fighting a chronic cough for more than a year. I taught her the temporal tap. As she tapped in her statements—"I no longer smoke" on the left side and "I enjoy my cigarette-free existence" on the right—she started to smoke like crazy, and she felt miserable. Something about the tapping was making the habit stronger, and it was also making her extremely anxious. But since the tapping was having an undeniable effect, even if it was the opposite of the desired one, she knew it was potent and decided to figure out how to use it to her benefit. She sensed that her anxiety was a key to the problem. She figured out that saying "I no longer smoke" was triggering her anxiety, in part because smoking was her major method of relaxation. Playing her intuition, she found new words. She began to tap into the left side "Anxiety no longer causes me to smoke" and into the right side "I only smoke now for my highest health and pleasure." The anxiety dramatically receded, and she went down from a pack per day to about three cigarettes per week. She kept it at that level for several months, using her occasional cigarette as a mantra, an act of meditation. Her cough diminished. Eventually she quit smoking completely.

As with energy testing—as with much in this book, for that matter—the procedures often require more finesse than simply following a pat formula. To summarize some of the considerations to keep in mind, your temporal tap statements will be more effective if: (1) the wording of the statement tapped into the left side of your head contains a negative word and the statement tapped into the right side contains only positive words, (2) the statements are in harmony with the way you naturally talk and think, (3) as you say the statements, you keep your attention on the words and their meaning, and (4) the statements do not instruct you to do something that contradicts a core value or a more basic need.

I have seen the temporal tap be effective for an enormously wide range of problems. It has helped people stop smoking, drinking, overeating, and scratching compulsively. It has assisted them in building confidence, optimism, and self-esteem. It has aided them in stimulating their immune systems when fighting a serious illness, improving their metabolisms when trying to lose weight, and enhancing their coordination when trying to learn a new skill. It has been a factor in reducing tumors, clearing eczema, and lowering blood pressure.

Select a target behavior, emotion, or physiological condition. Craft a negatively worded statement to tap around your left ear and a positively worded statement to tap around your right ear. You can test whether your statement is in harmony with your energies by tapping with one hand while making your statement and using the other hand as a friend energy tests you. Temporal tap your statements four or five times a day for at least a week. With deeply entrenched habits, I have seen it take up to thirty days before the results became evident. Even if you are skeptical, judge for yourself based on the results you attain.

Three closing thoughts for achieving optimum health, joy, and vitality: drop out, tune in, turn on.

- *Drop out* of your busy world and regularly focus your awareness on your body. Bring it compassion for its struggles rather than distancing yourself from them, like a citizen who looks away to not notice when bad things happen. Your body fights many battles for you every day. When it is overwhelmed and becomes ill, as it will from time to time, do not treat it as your enemy. Cherish it as your beloved until death do you part.
- *Tune in* to the subtle cues your body provides about its well being and its needs. With curiosity and concern, intuition and energy test-

ing, read the energies of your body, for they are the vocabulary of its native language.

- *Turn on* and nourish your body's energies by tracing meridians, circling chakras, weaving currents, holding points, and setting fields as ways of interacting, communicating, and unraveling the mysteries of illness and health.

Epilogue

Journeys to the Outer Realms

THERE ARE MORE THINGS IN HEAVEN AND EARTH,
HORATIO, THAN ARE DREAMT OF IN YOUR PHILOSOPHY.

—WILLIAM SHAKESPEARE
Hamlet

WE DON'T STOP AT OUR SKIN.

—DOLORES KRIEGER
Alternative Therapies in Health and Medicine

I have personally had countless experiences that have caused me to add to my vocabulary things not "dreamt of" in many people's philosophies, such as the existence of "spirit guardians," past lifetimes, and messages from people who have died. These experiences may seem strange or even inconceivable to you, and you don't have to come to the same conclusions I have drawn about them in order for the techniques presented in this book to work for you. I do hope, however, that these claims will pique your curiosity about the mysterious relationship of body, mind, and soul, and that you will take them in as data. For me, these experiences place energy medicine into a larger context. It is not just about technicians who know how and when to trace this meridian or spiral over that chakra to remove a symptom. It is also through the bridge of energy work that a practitioner walks into a client's life at the soul level.

When you consciously begin to work with energy, unexpected realms of existence may become apparent. Sometimes this may be no more than the simple blessing of moving invisible energies and hearing a moment later that the person feels better. Sometimes, because it is possible to tap into a person's inner healer, I can stand back and watch as the person's body goes into an amazing spontaneous healing. Occasionally a client has said something like, "My grandmother is here and wants you to listen to her. She's trying to tell you what to do." And then it is as if a presence comes into the room and instructs me to carry out a procedure that never would have occurred to me but that proves effective. I have also had the experience of feeling that the middle of my back was being unzipped as "someone" stepped in and conducted a profound healing session. While an inner gyroscope is assuring me that we are engaging in a positive process, I never know in advance the outcome in any of these situations, and I am often amazed by the unexpected sequence of events. Again and again, I meet such experiences with a sense of wonder. They have never become matter-of-fact to me.

Relationships with spirit guardians, guides, angels, or whatever you call them, have been reported throughout history. Socrates openly spoke of an entity whom he believed had guided him throughout his life. He referred to it as "the oracle," and he credited it for having protected him in numerous ways.[1] More than two thousand years later, the discriminating mind of Helen Keller disclosed: "My most loyal and helpful friends are those of the spirit. As I wander through the dark, encountering difficulties, I am aware of encouraging voices that murmur from the spirit realm."[2] In 1993, a *Time* magazine/CNN poll on angels revealed that 69 percent of the population, or an estimated 144 million adult Americans, said they believed in spiritual beings with special powers to act on Earth, and 32 percent of them, or an estimated 46 million, reported having personally felt such an "angelic presence" in their own lives.[3]

I do not try to open people to their spiritual guardians or personal angels. My instinct is that it is best for such developments to occur organically. By keeping your mind clear, your heart open, and your energies balanced, you are doing your part in inviting such experiences if they are meant to happen. But since people do stumble into these realms, it is an issue worth addressing. If such experiences are totally outside your belief system, or if you believe they occur only as a manifestation of evil, you are more likely to be terrified if they do happen, and you are less likely to keep the experience in perspective or to use it constructively. I know many people who regularly communicate with inner guidance that to them feels different from their ordinary mind. I can only observe that while this "communication" is sometimes a vehicle for the ego and the imagination to run away in a fling of "psychic" communication, at other times it provides sound instruction.

I was in my twenties before I knew that everyone didn't make decisions the way I do, which is to listen to the voices inside and around me. While these voices felt as though they were from a different realm, they did not seem strange. It was more like they were coming from precious friends. They felt so natural, in fact, that it never even occurred to me to discuss them with anyone in order to make sense of them. My mother was quite psychic, but I didn't know this was unusual when I was young. Sometimes she would respond to my questions by saying, "Donna, you have your angels, ask them." I thought my father was the unusual one because he never had experiences like these. I have, in fact, so often been blessed by experiences of benevolent forces that I have refused to believe that evil may dwell on the "other side."

ENCOUNTERING DARK FORCES

When entering deeply into another's being, you sometimes open to forces you never invited, wanted, or expected. Jim was deeply distressed when he scheduled an appointment with me. He inexplicably kept losing control of his car. His experience was that someone invisible was in the car struggling with him for control of the steering wheel and trying to make the car go off the road. He feared for his life, so much so that he had left the area, gone to Alaska, and had lived peacefully for a year on a fishing boat. When he returned to Oregon, however, the "spirit" was in his life again, and he wanted me to help him get rid of it.

We agreed to work together. I felt I could help him, although I didn't really believe he was dealing with an "evil spirit." Psychological explanations for his account seemed far more probable, and besides, I am a person who had and has a very hard time recognizing evil. I can be in denial even when I am faced with direct experiences of its presence.

At the time, my office was in my home. After Jim lay down on my table, the session quickly became very strange. When I went to touch him, it seemed that my hands couldn't get any closer than about four inches from his body. There was some kind of barrier around him. I would try to touch him, but it was as if an invisible force kept me from bringing my hands any closer. I heard words from within that urged caution, but I disregarded them. This was someone who had come to me for help. I found the warning to be a distraction, and I didn't let it in. Jim was lying there with his eyes shut, not knowing what I was going through.

I finally gathered my will and forced my hands through the barrier. When I did, I could feel his skin, and I could feel that his body was solid. But I couldn't make an energetic connection. The experience had an ominous feel. I was pushing against something I didn't understand, and I soon started to feel vaguely ill. Finally, I said, "Jim, I don't know what's wrong with me, but I've been working on you for forty-five minutes, and I still can't connect with your energy. I could stop now and set up another time for you to come back." He said, "Forty-five minutes! Didn't I just come in?" But then to his surprise he saw the clock, and he said it seemed as if he had been drugged. But he agreed to end the session, and he left.

The whole room felt thick and gloomy. I found my husband, David, and told him I was going for a bike ride. He offered to go along with me. It was early evening. I noticed that when I pedaled more slowly, I began to have the sensation of something hitting the back of my sacrum. It was a thud that came with a rhythm that hit hard and seemed to be getting worse. If I rode fast, it wasn't so bad, but when I slowed down, it became awful. David and I stopped at a cafe for a cup of tea, and I was in a very strange space. A friend came over to talk with us, and I couldn't even hear her. I went home to sleep and had a terrible, terrible night. Looking back, it felt like I was possessed, though I wasn't thinking in those terms at the time. It was as if something had me in its grip. The strange and very unpleasant pounding continued. After a sleepless night, I got up early, not knowing what to do. I had clients to see, but I felt I couldn't work on anybody. David suggested that I ask my guides for help. I canceled my clients, rode by bike to the park, lay down on the grass, and asked my guides what was happening to me.

What I heard was that by breaking through Jim's energy barrier, I had taken on the spirit that had been in him. It had left him completely and was now in me. Evil spirits were a foreign concept to me, but I couldn't deny that something real and painful seemed to be slamming against my sacrum. It was difficult to ride home. When I did get home, David was there, and I told him what was happening. He didn't miss a beat. He said, "Well, we'll have to exorcise it." I lay on the bed, and David began to speak. He has had some experience with both hypnosis and exorcism, and he knows something about the difference between the two. Eventually, I saw a grayish dark energy, with the faintest outline of a figure, lift up out of me. I fell asleep.

When I awoke, I was fine, but David was upset. His first question, after receiving assurance that I was okay, was "Where did it go?" I said, "Straight up, through the ceiling." Above our bedroom was an attic where David had his computer. He was deep in the throes of writing his first book. He discovered that the file he had been working on that morning before the exorcism suddenly could not be read by his machine. Later, technicians would be able to find only "garbage" in the file. The file contained two chapters from his book that atypically had not yet been backed up. They represented months of work. If the "spirit" was responsible for erasing them, it certainly knew where to strike.

When I heard from Jim, he was jubilant. His experiences of being pushed off the road had abruptly ceased with our session.

PROTECTING THE HEALER

I've always thought of evil as a consequence of human vanity, greed, and corruption, and I ignored my instincts when they conflicted with this notion. By far, the mistakes I've made that have hurt me the most have occurred when I was too invested in wanting to help. I have given treatments when I was exhausted, emotionally drained, or on the edge of illness myself. I have given treatments when my relationship with the person was off in some significant way, and those sessions have at times hurt me. I've also ignored the wee small voice from within telling me to back off, as with Jim, and I've rushed in where wisdom would fear to tread. It has been all the more difficult to learn this lesson because when I am so open and eager to help that I end up sacrificing my life force for the person, the session can be quite profound. The client may leave much improved, grateful, and wanting another treatment just like the one we just completed, but I leave in illness or pain.

When you fully open yourself to working with another's energies, you are involving yourself with that person intimately. If you are healing me, my energies are affecting you. I know no way around this. I personally don't want a healer who is overly careful not to take on my energies, to keep too much distance, or to maintain overly rigid boundaries. I am always personally involved during a session with a client. The distinction between forming an appropriate professional relationship and being able to open yourself to another's energies is critical. The professional relationship—the definition of the exchange, fees, and time structure—helps create a container within which it is safe to fully open oneself to the mystery of the moment. Almost always, even if disturbing energies move through my body during a session, by the end of the session I am restored, feeling energized and more robust than when I began. But I've also sometimes gotten myself and my health into trouble, so I want to share a few guidelines for protecting yourself if you "dispense" energy medicine, either professionally or to loved ones.

If I am not well, I am more vulnerable to picking up another's illness. You have to get to know yourself on this score. A different standard applies from other jobs. It is easier to get away with working in an office or a factory when you are feeling a bit ill than when you mingle your own and another's energies in a healing setting. Stay alert to your own internal gauge, and be prepared to learn by trial and error. Sometimes I can be dragging before a session and the session turns out to be restorative for us both. At other times it can be overwhelming for me. A rule of thumb is that if I cannot relatively easily get my own energies flowing by doing a few simple energy techniques, I should not be working on someone else.

Another way that healing work is different from other jobs is that you can fry your circuits if you hold on too tightly to the concept of the forty-hour week. The amount of time you can give to this work varies widely with your constitution, your heartiness on a given day, and what else is going on in your life. I need to have time between sessions to come back fully into myself, and I routinely do simple energy exercises on my body, such as the Three Thumps, the ear pull, and Separating Heaven and Earth. I also push up hard with my fingers on my cheek bones, which stimulates the first points on the stomach meridian, clearing stresses that have just been registered.

Not only can the very energy techniques that I offer usually reverse any difficulty I have gotten myself into during sessions, the process of doing healing work is almost always revitalizing for me. The room is almost always filled with a healing and nurturing atmosphere. The forces that appear from the other side are almost always benevolent.

SPIRIT GUARDIANS

Though always unexpected and a bit mystical, the appearance of spirit guardians occurs often enough in my work that I am forced to come to terms with their sometimes conspicuous presence. Reports of help and guidance from another realm are increasing, and there are sometimes urgent reasons that cause otherworldly help to make itself known. Sometimes it can. More often our conditioning and cultural filters prevent us from becoming conscious of it, but I have many times experienced a presence in a session that provided an important key to the healing process. Here is one of those instances.

I was teaching a class at Palomar College, near San Diego, in the summer of 1981. A woman drew my attention because she never participated in the class, and she generally left just before the sessions ended. She seemed to be in considerable physical pain. I later learned that to attempt this kind of healing without summoning Jesus conflicted with her religious faith, and she felt confused.

Eventually, however, the woman asked for an appointment. Betty had suffered with a great deal of pain on her left side since having had polio as a child. Now that she was in her early sixties, her bone structure and musculature weren't carrying her well, and the pain was intensifying, particularly in her left shoulder and hip. She had been to numerous doctors, but none of them had been able to relieve the pain. She had been told, "This is just something you're going to have to learn to live with."

Betty was pleasantly surprised by substantial relief from her pain after our first and then our second sessions. During the third session, I kept having a sense of seeing a red energy in the corner of the room. This didn't compute. I was trying to dismiss it, but it was very demanding. The session was suddenly interrupted. A male, ghostly, almost human figure was standing in that corner. He reminded me of Yul Brynner in *The King and I,* even to the bald head. In a regal way, he commanded: "I am Balasheem. Stand at her shoulders!"

I stood at her shoulders and put my fingers on them. Suddenly it felt like this being stepped into my body. The middle fingers on each hand felt like they had become steel rods that were plugged into her shoulders. I couldn't move them away. Such a strong force of energy pulsed through my body and out of my hands that she began to vibrate, very hard, until it looked as if she were going into convulsions. I was stunned, and I couldn't pull my fingers away. She was moving up and down, up and down, something between vibrating and convulsing, until she fell off the table. Only then did we disconnect.

I rushed to her, thinking, "Oh my God, is she okay?" Betty was sobbing uncontrollably on the floor, and I could do nothing that made her stop. Finally, she said, "Look!" She showed me her increased range of motion and declared that she was having no pain. None whatsoever. It was all gone. She hadn't had full range of motion since prior to her polio, and she hadn't had relief from pain for many years. She began to excitedly wonder aloud if I were "Mary, Mother of Jesus." No other explanation fit for her. Meanwhile, I was saying, "Betty, it wasn't me! It was like a spirit came into me. He called himself Balasheem." "It was Mary and Jesus," she screamed excitedly. "No," I replied, "He said his name was Balasheem, and he looked like Yul Brynner."

She was perplexed. This didn't fit anything she believed, but she could not deny her improved condition. She began telling her church community that Mother Mary was using my body for healings. I was flooded with calls from people wanting me to heal them. At first I was trying to explain, "It wasn't like that; I'm not Mary." Finally, I began to just agree that some divine healing force had apparently come through. I was leaving soon and could not accept any new clients, but I did agree to see Betty one more time.

At that session, she was filled with questions. "Why me?" I didn't have any answers, but I sensed deeply that Balasheem was personally aligned with her, that he was not some generic force of the heavens who was available for everybody. I said that to her, and she responded, "Well, if he comes again, I want to know why he doesn't just talk to me directly!" Suddenly, I could sense him in the room. I said, "Betty, I believe he's here." She said, irritated, "Well, ask him!" I heard the reply, "I have been trying to reach Betty all her life. And now it is imperative that she believe in me." And that was it. He was gone. I didn't know what that meant, that it was *imperative* that she believe in him. She still grumbled, but she was fascinated and felt that regardless of what I wanted to call him—it might have something to do with my own strange spiritual beliefs—she knew that Mother Mary had graced her life.

I left for home. I had no contact with Betty for months. Then, one day as I was walking in the park, I was confronted with the apparition of Balasheem. He just appeared and told me to call Betty. He gave me specific dietary instructions I was to convey to her, having to do with nine vitamins and minerals, the amounts she needed, and what was out of balance. I went straight to the phone and said, "Betty, I saw Balasheem in the park just now, and he told me to call you." I described his dietary suggestions for her. There was a long silence. She finally asked me to repeat what I had said. She was just opening an envelope with the results of lab tests after a total checkup. What I said to her matched the lab report precisely, down to indicating that she was dangerously low in potassium. This was the event, more than even the healing, that made her accept Balasheem. She called him her personal angel and embraced him into her belief system.

My next trip to teach in San Diego was that summer. I was driving south on Highway 5, my daughters Tanya and Dondi were sleeping in the back seat, and it was raining heavily. We had just passed through Sacramento when suddenly I saw the figure of Balasheem outside the windshield. He told me to call Betty. "Tell her, that which has just happened is not her fault."

I anxiously drove to a phone and called. A man answered, and I asked to speak with Betty. He said, "I'm sorry, you can't speak with her right now." I said, "Please, I must speak with her." He said, "It's impossible! Please call back later." "No, I must speak with her now. Please tell Betty this is Donna, and I have spoken to Balasheem."

When he told her it was me, she came to the phone. I began, "Betty, don't say anything," and I told her how Balasheem had come in front of the windshield and told me to tell her "That which has just happened is not your fault." She let out a scream and began to sob. Someone was trying to take the phone away from her, and Betty was saying, "No, no!" Then a peace seemed to come over her. She asked if he had said anything more. I said, "Not yet, but I'm on my way to San Diego. Betty, what is it?"

"My son just killed himself."

I went straight to Betty's house when I arrived in San Diego. She was sleeping. Her husband told me she was having heart irregularities before my call, but she was calmer now. He spoke of their shock and guilt. Betty's love for this son was both boundless and agonizingly complicated, but the husband's grief was made worse because he and the son had been estranged. Betty had been placed in the middle, forcing herself to assume a "tough love" attitude that went against her maternal instincts.

I heard Betty stir and went in to see her. The room was filled with so much love and tenderness that, before I saw him, I knew Balasheem was present. Emanating from him was a phenomenal compassion and a reverence for the gravity of the situation. I felt privy to something extraordinarily intimate. I said, "Betty, he's here."

Remarkable information began to come forth. The son's body was in such bad shape from years of drug abuse that he would have, in any event, died within a couple of years of liver disease. He had the kind of body chemistry that simply could not afford to trifle with drugs. But by stepping out of his life at this time, he felt he was preventing harm from coming to those he loved as he recognized that he could not stop himself from doing horrendous things to get drugs. He chose suicide as the less terrible alternative. This blew me away. I do not condone suicide. It goes so strongly against my instincts that I had a hard time reconciling myself to Balasheem's words.

For the next three days, I saw Betty each day, and Balasheem appeared each time. He explained that for the entire year before the suicide, it appeared unstoppable. *That* was the reason Balasheem had first "tapped Donna on the shoulder" to make his presence known to Betty. Healing Betty's body was the vehicle to make contact and gain her trust as a larger story was unfolding. It became clear to Balasheem that Betty's heart would not survive her son's suicide. So Balasheem began to create conditions that would minimize the wider damage of the inevitable suicide.

During the third session with Betty, she gasped, "I see him! I SEE YOU!!!" She began to speak with and hear Balasheem, and I was no longer the intermediary. But it was her son's voice that she really wanted to hear. Her heart was so burdened from missing him that she could not stop the pain. Every day for six months she went to his grave, leaving him fresh flowers. One day, after she had spoken to him, cried for him, and put the flowers down, she turned to leave and heard his voice as distinctly as if he were standing behind her. He said, "Oh, Mom, take the flowers home and enjoy them." The terrible agony of her grief was lifted.

EVIDENCE OF OTHER REALMS

Personality theorists, from Freud onward, have often associated spirituality with psychopathology. Psychological research, however, has consistently identified correlations between "spirituality" and mental health. People reporting mystical experiences, for example, have scored lower on psychopathology scales and higher on measures of psychological well-being than others.[4] These spiritual experiences may take many forms.

While some of the most outrageous deceptions perpetrated on people involve "channeled spirits," when you are intimately involved in experiences such as Balasheem's impact on Betty's life, it may cause you to rearrange your belief system. Behavioral scientists are also finding increasing evidence that upsets a purely materialistic worldview.

Some children, for instance, from the time they can talk, make it known that they are experiencing themselves as someone who does not match the child's given name and local circumstances. In case after case, such children have spontaneously provided highly specific details about another person whose identity they think they hold. Ian Stevenson, a psychiatrist at the University of Virginia, has collected over twenty-six hundred reported cases of these "past-life memories." He and his colleagues have travelled around the world, interviewed hundreds of children who have reported such memories, and then investigated the details the children reported. In many of the cases, Stevenson and his team were able to confirm that a person with the name the child offered did exist and had died prior to the child's birth. Other facts also frequently matched: the individual lived in the town the child identified, had relatives and friends the child named, or worked in an occupation in which the child had an uncommon interest.[5] Measures to safeguard against fraud or hoax were employed.

While these findings can be interpreted in a number of ways, a more persuasive link between the child and the person the child "remembers" was discovered through a chance observation. In 35 percent of 895 of the cases Stevenson investigated, the children had birthmarks or birth defects. In many instances these matched wounds that the person being remembered had actually received! For instance, a three-year-old who recalled being shot in the abdomen claimed to be a person who, on investigation, was found actually to have died in that manner. Moreover, the boy had a small round birthmark in his abdomen and a larger, irregular birthmark on his back, matching the way a bullet would enter and exit. Cases involving such birthmarks are significant because they rule out other paranormal explanations, such as that the child is telepathically picking up the information or that the spirit of the deceased individual has stepped in and "possessed" the child.[6]

These findings have been used to account for inexplicable phobias, some troubled child-parent relationships, gender dysphoria of early childhood, unusual childhood interests, and inexplicable skills that are sometimes evident in early childhood. A question many people ask regarding past lives is: "If we have lived before, why don't we remember?" My own belief is that children have generally forgotten their previous lifetime by the time they can talk. If they did not, such memories would, among other things, confound their ability to form an identity. Stevenson observed that most of the children he investigated stopped talking about the lives they "remembered" some time between the ages of five and eight. It is also very interesting that in more than half the children whose recollection of a past lifetime has been verified, the previous lifetime ended in a violent death.[7] Perhaps when there wasn't an opportunity to reach a sense of completion in the previous lifetime, the soul is born into the present incarnation with something incomplete.

I do not try to evoke memories of past lifetimes during healing sessions, but when they emerge, as they do when you traffic in subtle energies, you have to deal with them in one way or another. There is good reason to be skeptical about "past life" readings (may investigators have found no corroboration when they have gone from one past life reader to another), but when a vivid memory arises during a session, sometimes nothing makes more sense than to consider that the client is working out an issue that had not been completed in a previous lifetime. A small clinical literature is developing about such cases,[8] and past life memories have several times provided a key in my own work.

Jolee was one of the first employees of a land investment firm that flourished in the San Diego real estate boon of the early 1980s. Hired as a typist, by the time she had been with the company for four years she was managing the division that helped companies relocate to California. At this point, the firm moved its own offices to showcase a beautiful rural area that still had good access to commercial services. Suddenly Jolee's latent and thoroughly neurotic fear of spiders surfaced. Garden spiders were often seen in the new location, and Jolee would run out of the room screaming in terror. She was actually afraid of tarantulas, though she'd never seen one, and she became delusional in her fear that the small harmless creatures she would see walking along the carpet or on the wall were tarantulas and were life-threatening. She was inconsolable after these incidents, her behavior made everyone uncomfortable, and her supervisor didn't know how to handle the situation. She had proven herself to be a competent, motivated, and loyal employee, but her periodic outbursts were unacceptable. After disrupting a meeting with a major client, she was fired.

Coming to me, I began to work with her basic grid. As the grid energies started to connect, she had a past life "memory," entering a scene from about sixty years ago in upper Mexico, near the Texas border, out in the middle of nowhere. She recalled seeing tarantulas cover the roads, so much that the road could not be seen. One day her youngest child was found dead on the road with tarantulas covering his body. She almost died from horror and grief, and she lived the rest of her life in utter terror every night of tarantulas covering her or one of her remaining children. I held her neurovascular points as she related this story amidst her sobs. As she calmed, I had her continue to talk about it and, while I continued to hold her points, also to tell me about her present heart-palpitating anxiety around spiders. After three treatments like this on consecutive days, her fear of spiders was simply gone, and the firm was happy to rehire her.

I have many times experienced a presence in a session that, once I've described how it feels or sounds or looks, the client has recognized as a relative or friend who has died. Useful information often follows. I believe, in fact, that many more people intuit help from "the other side" than has commonly been acknowledged, although the wide publicity and impressive information provided by channels such as James Van Praagh and George Anderson may be opening people's minds to these possibilities.[9]

Jean was a beautiful and spirited thirty-eight-year-old woman with low self-esteem. Her relationships with men did not last. She would put on an act with them, trying to be who she thought they wanted, certain that if they knew her, they would leave. In fact, they left because they didn't know her.

Jean had done quite a bit of therapy, and she was not interested in dealing with her past. She wanted me to "just do the energy thing" so her pattern with men would stop. On her fourth visit, she had the flu and only wanted to get well quickly, so what happened was a complete surprise. It was as if the room lifted the sick energy out of her and turned it into a happy quality. She and I both felt it, and we began to giggle. We couldn't stop. I cannot tell you how strange this was. She said, "What are you doing to me? I feel so happy!" I said, "I don't know."

At that point, I felt a presence in the room and heard the name, "Weldon." I said, "All right, who's Weldon?" She replied with surprise, "My dad." I said, "Well, you have a very strong psychic connection because I am picking him up in your energy." She said, "No way!" but gave me no further explanation, and she became a little agitated. Suddenly I heard myself saying, "Forgive me, Jeannie, for leaving you. You were the only good thing I left behind. Forgive me. Forgive me." I felt the words were being spoken through me.

Jean broke into tears. Finally her story began to spill forth. Her father adored her, and she knew he would never leave her. But he was an alcoholic, and one day he never came back. As children do, Jean figured it was her fault. She hadn't done things right. She decided that she needed to be prettier, nicer, and more perfect so that her daddy would "see her from behind a bush" one day and come home. He never came, and she never got over performing for the men who did come along.

I said, "Jean, he's no longer alive [she didn't know this, although it was later verified]. But he is connected to you, profoundly so, and he wants to guide you through the labyrinth of your relationships." Happiness again filled the room. This session was a turning point in her self-understanding and in her pattern of performing for and obsessing about men. The happiness that filled the room somehow stayed with her after the session, as did her sense of her father's presence. Within two years, I was invited to her wedding.

REFLECTIONS

Poignant encounters with invisible spirits, previous lifetimes, and voices of the deceased who provide verifiable information tend to open one's metaphysical perspective. I am certain that the soul passes through other realms that we can only glimpse. These glimpses, however, seem to offer a natural if unconventional source of information. Such information has again and again provided me with a key for unlocking a healing process, occasionally in individuals who have been abandoned as beyond hope by Western medicine, its true marvels notwithstanding.

So how do I make sense of bizarre encounters such as those with the likes of Balasheem? There is no question in my mind that these exceptional realms are just as "real," just as valid, as anything in the ordinary physical world, and that the principles governing them are just as lawful. Tuning into these more elusive zones has provided me with a deeper understanding about healing by placing illness into the larger context of the soul's journey, and it has consistently made my own life a more magical voyage.

I, in fact, love being tuned into that realm. From it I feel in touch with deeper truths and truer perceptions. I see beauty and perfection. I'm always ready for the unexpected, and it may appear in the most interesting and gratifying ways. At such moments I enter a state of deep appreciation of life and its many wonders. I become more accessible to a Balasheem and more open to subtle knowledge about the nature of things. To simply imagine this mystical zone is a knock on its door. To actively pursue it is a turn of the key. If you frequent this world—pursuing and entering it through meditation, ritual, sacred plants, healing work, or spontaneous grace—you know that my attempt here to capture it with words is inadequate. If you do not, knock at its door from time to time as you go through your week, and be receptive to the advent of mystery.

While that is another topic from this book, it is not unrelated. In attuning yourself to the realm of subtle energies, you have not only been paving an avenue toward better health and greater vitality, you have stepped into a world where your soul's purpose penetrates through the veil of your busy life. May you abide in its radiance. Blessed be.

Appendix

How to Find a Competent Energy Medicine Practitioner

WHEN WE ARE WELL, WE ALL HAVE GOOD ADVICE FOR THOSE WHO ARE ILL.

—TERENCE
The Woman of Andros

The following list of resources is designed to lead you to a competent practitioner of energy medicine. Finding the right practitioner in the healing arts can be even trickier than finding the right contractor to remodel your home. At least as important as the person's professional specialty is the person. Besides skill level and intuitive abilities, a certain alchemy needs to exist between your energies and the practitioner's.

The common denominator among energy kineseology, acupuncture, Touch for Health, Ayurveda, applied kineseology, qi gong, homeopathy, Reiki, Therapeutic Touch, and other forms of energy medicine is that their interventions are explicitly concerned with the energy body as well as with the physical body. In homeopathy, for example, where minute quantities of the healing agent are ingested, it is the energetic rather than the dietary effects of the remedy that are being sought. While the strategies for influencing the client's energy differ widely among these approaches, the map of the energy body presented in this book is not confined to any particular technique, and the energy tests can be used to assess the results of any given treatment.

The program you are now completing has limited itself to offering tools for self-treatment, but skilled and knowledgeable practitioners of any of the approaches listed here can help you go to the next level when a health condition requires more intensive treatment than you yourself can provide. The following professional associations can tell you of practitioners in your area who are qualified according to the organization's professional standards. But don't be afraid to shop carefully. A list of websites that are pertinent to energy medicine also follows.

ACUPRESSURE

Acupressure Institute
1533 Shattuck Avenue
Berkeley, CA 94709
510-845-1059
www.healthy.net/acupressure

ACUPUNCTURE

American Academy of Medical Acupuncture
5820 Wilshire Boulevard
Los Angeles, CA 90036
213-937-5514
www.medicalacupuncture.org

American Association of Oriental Medicine
433 Front Street
Catasqua, Pa 18032
610-266-1433
www.aaom.org

APPLIED KINESEOLOGY

International College of Applied Kineseology
6405 Metcalf Avenue, Suite 503
Shawnee Mission, KS 66202
913-384-5336
www.icakusa.com

AYURVEDA

The Chopra Center
7630 Fay Avenue
La Jolla, CA 92037
619-794-2425
www.chopra.com

Maharishi University
1000 4th Street, DB-1155
Fairfield, IA 52557
515-472-7000
www.vedic-health.com

EDGAR CAYCE APPROACH

Association for Research and Enlightenment (A.R.E.)
P.O.B. 595
Virginia Beach, VA 23451
800-333-4499
www.are-cayce.com

A.R.E. Clinic
4018 N. 40th Street
Phoenix, AZ 85018
602-955-0551
www.starlighter.com/are

ENERGY KINESEOLOGY

Innersource
P.O.B. 213
Ashland, OR 97520
541-482-1800
www.innersource.net

GUIDED IMAGERY

Academy for Guided Imagery
P.O.B. 2070
Mill Valley, CA 94942
800-726-2070
www.healthy.net/agi

HEALTH KINESEOLOGY

Health Kineseology, Inc.
R.R. 3
Hastings, Ontario K0L 1Y0
Canada
705-696-3176
hk@subtlenergy.com

HOLISTIC MEDICINE

American Holistic Health Association
P.O.B. 17400
Anaheim, CA 92817
714-779-6152
www.ahha.org

American Holistic Medical Association
6728 Old McLean Village Drive
McLean, VA 22101
703-556-9728
www.ahmaholistic.com

American Holistic Nurses' Association
P.O.B. 2130
Flagstaff, AZ 86003
520-526-2196
www.ahna.org

HOMEOPATHY

National Center for Homeopathy
801 N. Fairfax Street, Suite 306
Alexandria, VA 22314
703-548-7790
www.homeopathic.org

LEARNING ENRICHMENT THROUGH KINESEOLOGY

Educational Kineseology Foundation
P.O.B. 3396
Ventura, CA 93006
800-356-2109
www.braingym.com

Three in One Concepts
2001 West Magnolia Blvd, Suite B
Burbank, CA 91506
818-841-4786
www.onebrain.com

MAGNET THERAPY

Bio-Electro-Magnetics Institute
2490 West Moana Lane
Reno, NV 89509
702-827-9099
www.bemi.org

MASSAGE THERAPY

Massage-One
4379 North Capistrano
Dallas, TX 75287

888-251-6793
www.massage-one.com

NATUROPATHY

American Association of Naturopathic Physicians
2366 Eastlake Avenue, Suite 322
Seattle, WA 98102

800-206-7610
www.naturopathic.org

QIGONG

Qigong Association of America
2021 NW Grant Ave.
Corvallis, OR 97330

541-752-6599
www.qi.org

REFLEXOLOGY

International Institute of Reflexology
P.O. 12462
St. Petersburg, FL 33733

813-343-4811
ftreflex@concentric.net

REIKI

International Center for Reiki Training
29209 Northwestern Highway, Suite 592
Southfield, MI 48034

800-332-8112
www.reiki.org

SHIATSU AND OHASHIATSU

Ohashi Institute
12 West 27th Street
New York, NY 10001

212-684-4190
www.ohashi.com

Shiatsu School of Canada
547 College Street
Toronto, Ontario, Canada M6G-1A9
1-800-263-1703
www.shiatsucanada.com

THERAPEUTIC TOUCH

Nurse Healers Professional Associates
1211 Locust Street
Philadelphia, PA 19107
215-545-8079
www.therapeutic-touch.org

THOUGHT FIELD THERAPY

Callahan Techniques
45-350 Vista Santa Rosa
Indian Wells, CA 92210
760-345-9216
www.tftrx.com

TOUCH FOR HEALTH

Touch for Health Kineseology Association
3223 Washington Blvd., Suite 201
Marina del Rey, CA 90292
800-466-8342
www.tfh.org

WEBSITES

Alternative Medicine Connection
www.arxc.com/hotlinks.htm
Alternative Medicine Network
www.sonic.net/~nexus/links.html
Alternative Medicine Yellow Pages
www.alternativemedicine.com
American Academy of Pain Management
www.aapainmanage.org
Barbara Brennan School of Healing
www.barbarabrennan.com
Bio-Electro-Magnetics Institute
www.bemi.org

Brain Wave Stimulation
www.flexyx.com
Commonweal Cancer Help Program
www.commonwealhealth.org
Environmental Medicine
www.healthy.net/aaem
HealthWorld Online
www.healthy.net
Institute of HeartMath
www.webcom.com/hrtmath
Institute of Noetic Sciences
www.noetic.org
International Society for the Study of Subtle Energies and Energy Medicine
www.nekesc.org/~issseem
Kineseology (Applied, Clinical, Specialized)
www.kineseology.net
MedScape
www.medscape.com
National Center for Health Statistics
www.cdc.gov/nchswww
National Council against Health Fraud
www.quackwatch.com
National Library of Medicine (includes MEDLINE)
www.nlm.nih.gov
Office of Alternative Medicine, National Institutes of Health
http://altmed.od.nih.gov/oam
Therapeutic Touch Training Videos
www.haelauworks.com
Touch for Health Education
www.touch4health.com
Traditional Chinese Medicine
www.acupuncture.com
Wellness Web
www.wellweb.com
Women's Health
www.4womenshealth.com

Notes

INTRODUCTION

1. H. D. Thoreau, entry for January 5, 1856, *A Writer's Journal,* Laurence Stapleton, ed. (New York: Dover, 1960).
2. Many of the pioneering research studies that have accompanied the development of energy medicine are nicely summarized in three accessible books: *Vibrational Medicine* by Richard Gerber, rev. ed. (Santa Fe, N.Mex.: Bear, 1996); *Subtle Energy* by William Collinge (New York: Warner Books, 1998); and *Cross Currents: The Promise of Electromedicine, The Perils of Electropollution* by Robert O. Becker (Los Angeles: Tarcher, 1990).
3. Among the professional journals that regularly include articles on energy medicine are *Alternative and Complementary Therapies, Alternative Therapies in Clinical Practice, Alternative Therapies in Health and Medicine, American Journal of Natural Medicine,* and *Journal of Alternative and Complementary Medicine.* Alternative medicine also has spawned a watchdog journal, *Scientific Review of Alternative Medicine,* which is emphatically skeptical of the field's claims.

CHAPTER 1

1. P. Wang, X. Hu, and B. Wu, "[Displaying of the Infrared Radiant Track Long Meridians on the Back of the Human Body]," *Chen Tzu Yen Chiu Acupuncture Research* 18(2), 1993, 90–93.
2. Applying systems theory to the concept of energy, Gary E. Schwartz and Linda G. Russek have presented a cogent theoretical framework for understanding the lawful ways by which energy and information interact in biological systems and healing. Their paper "Dynamical Energy Systems and Modern Physics" appeared in *Alternative Therapies,* May 1997, 3 (3), 46–56.
3. Lewis Thomas, *The Lives of a Cell: Notes of a Biology Watcher,* rev. ed. (New York: Penguin, 1995).
4. Claire Sylvia, William Novak, and Bernie S. Siegal, *A Change of Heart* (New York: Little, Brown, 1997).
5. Names and other identifying information in the cases presented in this book have been changed unless explicit permission to reveal them has been received. Where conversational exchanges are placed in quotation marks, the words shown represent my best recollection.
6. William Collinge, *Subtle Energy* (New York: Warner Books, 1998), pp. 2–3.
7. Jerome H. Barkow, Leda Cosmides, and John Tooby, eds., *The Adapted Mind: Evolutionary Psychology and the Evolution of Culture* (New York: Oxford University Press, 1992).
8. Cited in Anthony Stevens, *The Two-Million-Year-Old Self* (College Station, Tex.: Texas A&M University Press, 1993), p. 3.
9. Robin Fox, *The Search for Society* (New Brunswick, N.J.: Rutgers University Press, 1989), p. 29.
10. Rupert Sheldrake, *Seven Experiments That Could Change the World* (New York: Berkley, 1996).
11. Collinge, *Subtle Energy,* p. 1.
12. Hiroshi Motoyama, *Measurements of Ki Energy, Diagnosis and Treatments* (Tokyo: Human Science Press, 1998), available through the American center for Motoyama's research: California Institute for Human Science, 701 Garden View Court, Encinitas, Calif. 92024, 760-634-1771.
13. On the ion flow: Motoyama, *Measurements of Ki Energy.* On light emissions: I. Dumitrescu and Julian Kenyon, *Electrographic Imaging in Medicine and Biology* (Suffolk, England: Neville Spearman, 1983).
14. Dumitrescu and Kenyon, *Electrographic Imaging.*
15. William A. Tiller, "A Gas Discharge Device for Investigating Focused Human Intention," *Journal of Scientific Exploration* 1990 (4), 225–271. See also William A. Tiller, *Science and Human Transformation: Subtle Energies, Intentionality and Consciousness* (Walnut Creek, Calif.: Pavior, 1997).

16. Roger D. Nelson, G. Johnston Bradish, York H. Dobyns, Brenda J. Dunne, and Robert G. Jahn, "FieldREG Anomalies in Group Situations," *Journal of Scientific Exploration* 1996 (10), 111–41.
17. Dean Radin, *The Conscious Universe: The Scientific Truth behind Psychic Phenomena* (San Francisco: HarperCollins, 1997).
18. C. Norman Shealy, "Clairvoyant Diagnosis," in T. M. Srinivasan, ed., *Energy Medicine around the World* (Phoenix, Ariz.: Gabriel Press, 1988), pp. 291–303.
19. Russell Targ and Jane Katra, *Miracles of Mind: Exploring Nonlocal Consciousness and Spiritual Healing* (New York: New World Library, 1998); Larry Dossey, *Healing Words: The Power of Prayer and the Practice of Medicine* (San Francisco: HarperCollins, 1995).

CHAPTER 2

1. Rupert Sheldrake, *Seven Experiments That Could Change the World* (New York: Berkley, 1996).
2. Ronald Melzack, "Phantom Limbs," in *Mysteries of the Mind* (a special issue of *Scientific American,* September 1997, pp. 84–91.
3. Report of an interview with Ronald Melzack in *Discover,* February 1998, p. 20.
4. Sheldrake, *Seven Experiments,* p. 127.
5. Larry Dossey, *Healing Words: The Power of Prayer and the Practice of Medicine* (San Francisco: HarperCollins, 1995); Dean Radin, *The Conscious Universe: The Scientific Truth Behind Psychic Phenomena* (San Francisco: HarperCollins, 1997); Russell Targ and Jane Katra, *Miracles of Mind: Exploring Nonlocal Consciousness and Spiritual Healing* (New York: New World Library, 1998).
6. Valerie Hunt, *Infinite Mind: The Science of Human Vibrations* (Malibu, Calif.: Malibu, 1995).
7. Dean Radin and Jannine M. Rebman, "Lunar Correlates of Normal, Abnormal and Anomalous Human Behavior," *Subtle Energies* 1996 (5), 209–38.
8. Michael A. Persinger and Stanley Krippner, "Dream ESP Experiments and Geomagnetic Activity," *Journal of the American Society for Psychical Research* 1989 (83), 101–6.
9. Rollin McCraty, "The Electricity of Touch," paper presented at The International Society for the Study of Subtle Energies and Energy Medicine, Sixth Annual Conference, Boulder, Colo., June 24, 1996, available through the HeartMath Institute, www.webcom.com/hrtmath.
10. Jacques Lusseyran, *And There Was Light,* translated by Elizabeth R. Cameron (New York: Parabola Books, 1987).
11. Larry Dossey, "The Healing Power of Pets: A Look at Animal-Assisted Therapy," *Alternative Therapies in Health and Medicine* 1997, 3(4), 8–16.
12. Summarized in William Collinge, *Subtle Energy* (New York: Warner Books, 1998), p. 99.

13. Alan G. Beardall, *Clinical Kineseology,* vols. 1–5 (Lake Oswego, Oreg.: Alan G. Beardall, 1980–1985).
14. Dean I. Radin, "A Possible Proximity Effect on Human Grip Strength," *Perceptual and Motor Skills* 1984 (58), 887–88.
15. Personal communication, January 16, 1998.
16. Grace E. Jacobs, "Applied Kineseology: An Experimental Evaluation by Double Blind Methodology," *Journal of Manipulative and Physiological Therapeutics* 1981 (4), 141–45; J. J. Kenney, R. Clemens, and K. D. Forsythe, "Applied Kineseology Unreliable for Assessing Nutrient Status," *Journal of the American Dietetic Association* 1988 (88), 698–704.
17. Arden Lawson and Lawrence Caleron, "Interexaminer Agreement for Applied Kineseology Manual Muscle Testing," *Perceptual and Motor Skills* 1997, (84), 539–46.
18. Chang-Yu Hsieh and Reed B. Phillips, "Reliability of Manual Muscle Testing with a Computerized Dynamometer," *Journal of Manipulative and Physiological Therapeutics* 1990 (13), 72–82.
19. G. Leisman, R. Zenhausern, A. Ferentz, T. Tefera, and A. Zemcov, "Electromyographic Effects of Fatigue and Task Repetition on the Validity of Estimates of Strong and Weak Muscles in Applied Kinesiological Muscle-Testing Procedures," *Perceptual Motor Skills* 1995 (80), 963–77.
20. G. Leisman, P. Shambaugh, and A. H. Ferentz, "Somatosensory Evoked Potential Changes During Muscle Testing," *International Journal of Neuroscience* 1989 (45), 143–51.
21. Andrew Abarbanel, "Gates, States, Rhythms, and Resonances: The Scientific Basis of Neurofeedback Training," *Journal of Neurotherapy,* 1995, 1 (2), 15–38.

CHAPTER 3

1. These findings come from a series of studies conducted at the Institute for HeartMath in Boulder Creek, California, and many of their reports can be found at www.heartmath.org.
2. Rollin McCraty, Institute of HeartMath, Boulder Creek, Calif.; personal communication, March 31, 1998.
3. The series of studies that produced these findings is summarized in William Collinge, *Subtle Energy* (New York: Warner Books, 1998), p. 256.
4. Paul E. Dennison and Gail E. Dennison, *Brain Gym* (Ventura, Calif.: Edu-Kinesthetics, 1986).
5. Carl H. Delacato, *The Diagnosis and Treatment of Speech and Reading Problems* (Springfield, Ill.: Charles C. Thomas, 1963).
6. John Zimmerman, "Laying-on-of-Hands Healing and Therapeutic Touch: A

Testable Theory," *Newsletter of the Bio-Electro-Magnetics Institute* 1990, 2(1), 8–17.

7. Both published and unpublished reports of these studies are available as a packet, "Educational Kineseology Foundation Research Reports," from the Educational Kineseology Foundation, P.O. 3396, Ventura, Calif. 93006; 800-356-2109. Other literature and resources are also available through this foundation.
8. Jan Irving, "The Effect of PACE on Self-Reported Anxiety and Performance in First-Year Nursing Students," Ph.D. diss., Oregon State University, Corvallis, 1995.

CHAPTER 4

1. Reported in S. Rose-Neil, "The Work of Professor Kim Bong Han," *The Acupuncturist* 1967(1), 15.
2. P. Wang, X. Hu, and B. Wu, "[Displaying of the Infrared Radiant Track Long Meridians on the Back of the Human Body]," *Chen Tzu Yen Chiu Acupuncture Research* 1993, 18(2), 90–93.
3. Robert O. Becker, *Cross Currents: The Promise of Electromedicine, The Perils of Electropollution* (Los Angeles: Tarcher, 1990), p. 47.
4. Z. Yan, Y. Chi, P. Cheng, J. Wang, Q. Shu, and G. Huang, "Studies on the Luminescence of Channels in Rats," *Journal of Traditional Chinese Medicine* 1992 (4), 283–87.
5. Reported in Richard Gerber, *Vibrational Medicine,* rev. ed. (Santa Fe, N.Mex.: Bear, 1996), p. 187.
6. John Thie, *Touch for Health,* rev. ed. (Marina del Rey, Calif.: DeVorss & Co., 1996).

CHAPTER 5

1. Study by Valerie Hunt summarized in Richard Gerber, *Vibrational Medicine,* rev. ed. (Santa Fe, N.Mex.: Bear, 1996), p. 133.
2. Hiroshi Motoyama and R. Brown, *Science and the Evolution of Consciousness: Chakras, Ki, and Psi* (Brookline, Mass.: Autumn Press, 1978).
3. Nancy Ann Tappe's audiotape, "Color Potpourri," is a good introduction to life colors. The tape, along with Tappe's books on the subject, is available from Starling Publishers, Carlsbad, Calif. 92018; 760-729-5950.
4. Glen Rein, Mike Atkinson, and Rollin McCraty, "The Physiological and Psychological Effects of Compassion and Anger," *Journal of Advancement in Medicine,* 8(2), 1995, 87–105.
5. Stanislav Grof and Christina Grof, *Spiritual Emergency: When Personal Transformation Becomes a Crisis* (Los Angeles: Tarcher, 1989).

CHAPTER 6

1. Valerie Hunt, *Infinite Mind: The Science of Human Vibrations* (Malibu, Calif.: Malibu, 1995).
2. Dorothy Huntley Gundling, "Electrophysical and Psychotronic Correlates of Music," Ph.D. diss., Saybrook Institute, San Francisco, 1977.
3. Personal communication, February 5, 1998.
4. Reported in William Collinge, *Subtle Energy* (New York: Warner Books, 1998), p. 28.
5. William G. Braud, "Human Interconnectedness: Research Indications," *ReVision* 14 (1992), 140–48.
6. This research is available through the Educational Kineseology Foundation, www.braingym.com.
7. Martin E. P. Seligman, *Learned Optimism* (New York: Simon & Schuster, 1990).

CHAPTER 7

1. Diane M. Connelly, *Traditional Acupuncture: The Law of the Five Elements,* 3rd ed. (Columbia, Maryland: Center for Traditional Acupuncture, 1987), p. 11.
2. This section is particularly indebted to Nancy Post's insights and training materials, "Systems Energetics" (616 West Upsal Street, Philadelphia, PA 19119).
3. Caroline M. Myss, *Why People Don't Heal and How They Can* (New York: Harmony Books, 1997).
4. A superb audiotape course that teaches about meeting pain with loving-kindness is Stephen and Ondrea Levine's "To Love and Be Loved: The Difficult Yoga of Relationships" (Boulder, Colo.: Sounds True, 1997).
5. Gordon Stokes, *Without Stress, Children Can Learn* (Burbank, Calif.: Three in One Concepts, 1996).

CHAPTER 8

1. Angelika Buske-Kirschbaum, Clemens Kirschbaum, Helmuth Stierle, Hendrick Lehnert, and Dirt Hellhammer, "Conditioned Increase of Natural Killer Cell Activity in Humans," *Psychosomatic Medicine* 54 (1992), 123–32.
2. Glen Rein, Mike Atkinson, and Rollin McCraty, "The Physiological and Psychological Effects of Compassion and Anger," *Journal of Advancement in Medicine* 8(2), 1995, 87–105.
3. Herbert Benson, *Beyond the Relaxation Response* (New York: Times Books, 1984).

4. Anne Manyande, Simon Berg, Doreen Gettins, and S. Clare Stanford, "Preoperative Rehearsal of Active Coping Imagery Influences Subjective and Hormonal Responses to Abdominal Surgery," *Psychosomatic Medicine* 1995, 57(2), 177–82.
5. Colin Wilson, *Beyond the Occult: A Twenty-Year Investigation into the Paranormal* (New York: Carroll and Graf, 1988), pp. 66–68.
6. Candace Pert, *Molecules of Emotion: Why You Feel the Way You Feel* (New York: Scribner, 1997).
7. Norman Cousins, *Anatomy of an Illness as Perceived by the Patient* (New York: Bantam, 1991).
8. Martin E. P. Seligman, *Learned Optimism* (New York: Simon & Schuster, 1990), pp. 172–78.
9. David McClelland and C. Kirshnit, "The Effect of Motivational Arousal through Films on Salivary Immunoglobulin," *Psychology and Health* 2 (1988), 31–52.
10. R. A. Martin and J. P. Dobbin, "Sense of Humor, Hassles, and Immunoglobulin A: Evidence for a Stress-Moderating Effect," *International Journal of Psychiatry in Medicine* 18 (1988), 93–105.

CHAPTER 9

1. D. Spiegel, J. R. Bloom, H. C. Kraemer, and E. Gottheil, "Effects of Psychosocial Treatment on Survival of Patients with Metastatic Breast Cancer," *The Lancet* 1989 (2), 888–91.
2. Daniel P. Wirth, "The Effect of Non-Contact Therapeutic Touch on the Healing Rate of Full-Thickness Dermal Wounds," *Subtle Energies* 1990, 1(1), 1–20.
3. William Collinge, *Subtle Energy* (New York: Warner Books, 1998), pp. 30–31.
4. Brian H. Butler, *Your Breasts: What Every Woman Needs to Know Now* (Surbiton: T.A.S.K. Books, P.O.B. 359A, Surbiton, Surrey, England KT5 8YP, 1995).

CHAPTER 10

1. *Acupuncture: NIH Consensus Statement* (Washington, D.C.: National Institutes of Health, 1997).
2. Reported in William Collinge, *The American Holistic Health Association Complete Guide to Alternative Medicine* (New York: Warner Books, 1996), p. 31.
3. This discussion draws from a report, *Pain Management,* Washington, D.C.: Office of Scientific and Health Research, National Institute of Neurological Disorders and Stroke, National Institutes of Health, 1997.
4. David Bresler, cited in Gordon Stokes and Daniel Whiteside, *Structural Neurology* (Burbank, Calif.: Three in One Concepts, 1989), p. 29.

5. Susan M. Wright, "Validity of the Human Field Assessment Form," *Western Journal of Nursing Research* 1991, 13(5), 635–47.

CHAPTER 11

1. Robert O. Becker, *Cross Currents: The Promise of Electromedicine, The Perils of Electropollution* (Los Angeles, Tarcher, 1990), pp. 173–88.
2. K. Nakagawa, "Magnetic Field Deficiency Syndrome and Magnetic Treatment," *Japanese Medical Journal,* no. 2745, December 4, 1976.
3. Reported in William Collinge, *Subtle Energy* (New York: Warner Books, 1998), p. 74.
4. Becker, *Cross Currents.*
5. Marcia Marinaga, "Giving Personal Magnetism a Whole New Meaning," *Science* 256 (1992), p. 967.
6. Committee on the Possible Effects of Electromagnetic Fields on Biologic Systems, National Research Council, *Possible Health Effects of Exposure to Residential Electric and Magnetic Fields* (Washington, D.C.: National Academy Press, 1996).
7. "Cancer Near Power Lines," *Microwave News,* November/December 1996, 16(6), 1.
8. "Summary and Conclusions of EPA's EMF Cancer Report," *Microwave News,* May/June 1996, 16(3), 3.
9. John Zimmerman and Dag Hinrichs, "Magnetotherapy: An Introduction," *Newsletter of the Bio-Electro-Magnetics Institute,* 1995, 4(1), 3–6.
10. John Zimmerman, Bio-Electro-Magnetics Institute, 2490 West Moana Lane, Reno, Nev. 89509; 702-827-9099.
11. John Zimmerman, "Laying-on-of-Hands Healing and Therapeutic Touch: A Testable Theory," *Newsletter of the Bio-Electro-Magnetics Institute* 1990, 2(1), 8–17.
12. Becker, *Cross Currents,* p. 42.
13. Carlos Vallbona, Carlton F. Hazelwood, and Gabor Jurida, "Response of Pain to Static Magnetic Fields in Postpolio Patients: A Double-Blind Pilot Study," *Archives of Physical Medicine and Rehabilitation* 1997 (78), 1200–1203.
14. Graham Spence, President, International Council of Magnetic Therapists, P.O.B. 72, Inglewood, 6052, West Australia.
15. Sarada Subrahmanyam, "Pulsed Magnetic Field Therapy," in T. M. Srinivasan, ed., *Energy Medicine Around the World* (Phoenix, Ariz.: Gabriel Press, 1988), pp. 191–203.
16. Many of the dilemmas involved in crises involving a difficult or unanticipated "kundalini opening" are addressed in Stanislav and Christina Grof's anthology, *Spiritual Emergency: When Personal Transformation Becomes a Crisis* (Los Angeles: Tarcher, 1989).

17. The most complete report of the vivaxis with which I am familiar is Judy Jacka's *Healing through Earth Energies* (Port Melbourne, Australia: Lothian Books, 1996). There is also a Vivaxis Energies Research International Society in Vancouver, Canada.
18. Bernard Grad, "Some Biological Effects of Laying on of Hands and Their Implications," in Herbert T. Otto, ed., *Dimensions of Holistic Healing: New Frontiers in the Treatment of the Whole Person* (Chicago: Nelson-Hall, 1979), pp. 199–212.

CHAPTER 12

1. Rupert Sheldrake, *The Presence of the Past: Morphic Resonance and the Habits of Nature* (New York: Random House, 1988).
2. Ibid., p. 158.
3. Harold S. Burr, *The Fields of Life* (New York: Ballantine, 1972).
4. Robert O. Becker, *Cross Currents: The Promise of Electromedicine, The Perils of Electropollution* (Los Angeles: Tarcher, 1990), p. 57.
5. S. Ertel, "Testing Sheldrake's Claim of Morphogenetic Fields," in E. W. Cook and D. L. Delanoy, eds., *Research in Parapsychology 1991: Abstracts and Papers from the Thirty-fourth Annual Convention of the Parapsychological Association* (Metuchen, N.J.: Scarecrow Press, 1994), 169–92.
6. David Feinstein provides a more detailed discussion of the relationship between morphic fields and quantum mechanics in "At Play in the Fields of the Mind: Personal Myths as Fields of Information," *Journal of Humanistic Psychology* 1998, 38(3), 71–109.
7. William G. Braud, "Distant Mental Influence of Rate of Hemolysis of Human Red Blood Cells," *Journal of the American Society for Psychical Research* 1990 (84), 1–24.
8. D. P. Wirth and J. R. Cram, "The Psychophysiology of Non-Traditional Prayer," *Proceedings of the International Society for the Study of Subtle Energies and Energy Medicine,* Fifth Annual Conference, Boulder, Colo., p. 20.
9. D. J. Benor, *Healing Research: Holistic Energy Medicine and Spiritual Healing* (Munich: Helix, 1993).
10. Herbert Benson and Eileen M. Stuart, *The Wellness Book: The Comprehensive Guide to Maintaining Health and Treating Stress-Related Illness* (New York: Simon & Schuster, 1993).
11. Dean Radin, *The Conscious Universe: The Scientific Truth Behind Psychic Phenomena* (San Francisco: HarperCollins, 1997).
12. Larry Dossey, *Healing Words: The Power of Prayer and the Practice of Medicine* (San Francisco: HarperCollins, 1995).
13. Sheldrake, 1988.

14. Francine Shapiro and Margot Silk Forrest, *E.M.D.R.: The Breakthrough Therapy for Overcoming Anxiety, Stress, and Trauma* (New York: HarperCollins, 1997).
15. The temporal tap is described in greater detail in David S. Walther, *Applied Kineseology, Vol. 1, Basic Procedures and Muscle Testing* (Pueblo, Colo.: SYSTEMS DC, 1981), pp. 109–14.
16. Fred P. Gallo, *Energy Psychology: Explorations at the Interface of Energy, Cognition, Behavior, and Health* (Boca Raton, Fla.: CRC Press, 1998).

EPILOGUE

1. Cited in Joel Martin and Patricia Romanowski, *We Don't Die: George Anderson's Conversations with the Other Side* (New York: Putnam, 1988), pp. 279–80.
2. Helen Keller, *Light in My Darkness* (West Chester, Penn.: Chrysalis Books, 1994), p. 160.
3. Nancy Gibbs, "Angels Among Us," *Time,* December 27, 1993, pp. 56–65.
4. R. W. Hood, "Differential Triggering of Mystical Experience as a Function of Self-Actualization," *Review of Religious Research* 1974, (18), 264–70; N. P. Spanos and P. Moretti, "Correlates of Mystical and Diabolical Experiences in a Sample of Female University Students," *Journal of the Scientific Study of Religion* 1988 (27), 105–16.
5. Ian Stevenson, *Children Who Remember Previous Lives* (Charlottesville: University Press of Virginia), 1987.
6. Ian Stevenson's *Reincarnation and Biology,* 2 vols. (Westport, Conn.: Praeger, 1997) is a 2,268-page study that presents detailed evidence of birthmarks that correspond with past life memories. It is summarized in a 203-page book by the same author: *Where Reincarnation and Biology Intersect: A Synopsis* (Westport, Conn.: Praeger, 1997).
7. See Stevenson.
8. Carl Schlotterbeck, *Living Your Past Lives: The Psychology of Past Life Regression* (New York: Ballantine, 1991); Roger J. Woolger, *Other Lives, Other Selves: A Jungian Psychotherapist Discovers Past Lives* (New York: Bantam, 1985).
9. James Van Praagh, *Talking to Heaven: A Medium's Message of Life after Death* (New York: Dutton, 1997); Joel Martin and Patricia Romanowski, *We Don't Die: George Anderson's Conversations with the Other Side* (New York: Putnam, 1988).

Index

Page numbers in italics refer to illustrations.

A

Abs of Steel, 175
Acupuncture, 278–79
  sedating points. *See* Sedating points
  strengthening points. *See* Strengthening points
Adrenals, 231
Alarm points, *112,* 113–15
Allergies, 231–33
Allergy Headache, Rid Yourself of, 230
Anabolism, 156–57
Anatomy of an Illness (Cousins), 248
Angels, 340
Applied kineseology, 332
Aqua, 156
Arthritic Pain, Relieving, 304
Aura, 172
  detecting, 173–74
  distinguishing from habit field, 319–20
  fluffing, 178–79
  loosely knit or tightly bound?, 174–75
  massage of, 179
  scanning, 177–78
  seeing, 176–77
  weaving, 179–80, *181,* 182–84, *185,* 186
Autoimmune disorders, 228–30, 239
Autonomic nervous system, 75
Autumn, 206–7

B

Bad Dream, Restoring Peace After, 275–76
Balance, 22, 24
Baraka, 16
Basic grid, 18, 186–93
  restoring
    by balancing your chakras, 193
    by cultivating inner peace, 193
    by using Celtic Weave, 193–95
Bathing to Restore Energies, 35
Baylor College of Medicine, 298–99
Beardall, Alan, 44
Becker, Robert, 319
Bed placement, 306–7
Beginner's mind, 57
Belt flow, 153, 268
Beta waves, 48
Bladder meridian, 32, 100, 103, *104,* 125, 127
Blood circulation, 273
Brain
  crossover of energy in, 69–73, *70*
  hemispheres of, 332
  reptilian, 88–90
Breasts, safeguarding, 263–64, *264*
Breathing out pain, 282

Bresler, David, 279
Broken Arm or Leg, Mending, 301–2
Burr, Harold, 318

C

Calcium, 158–59
Calming:
an Angry Spirit, 208
Your Kids (or Yourself), 164–65
Cancer, 224, 268
Catabolism, 156–57
Celtic Weave, 179–80, *181,* 182–84, *185,* 186, 265
Central meridian, 82–83, 100, *101,* 244
Chakras, 15, 34
awakening of (case study), 134, 136–37
clearing, balancing, strengthening, 167–69
connecting two, 170
defined, 133–34
energy test, 167, *168*
fifth (throat), 156–59
first (root), 43, 144–48
fourth (heart), 36, 154–56
holding and cradling, 170
mechanics of, 142–43
overview of, 137–38
second (womb), 148–52
seven layers of, 140–42
seven major, *135*
seventh (crown), 77, 163–64
sixth (third eye), 159–62
third (solar plexus), 152–54
A Change of Heart (Sylvia), 16
Chi, 16
Chinese medicine. *See also* Strange flows; Triple warmer meridian
five elements, 196–97
Chronic
fatigue syndrome, 229
pain, 279–80
Circulation-sex meridian, *105,* 106
Clockwise motion, 169
Coherent electrocardiogram, 61
Collinge, William, 20, 262
Color, Send Through Your Body, 252
Concentrating Amid Fear or Stress, 214
Connelly, Dianne, 197
Consciousness, 225
Cook, Wayne, 73, 75
Core rhythm, 209
Counterclockwise motion, 143, 169, 307
Cousins, Norman, 248
Cross Crawl, 69–73, *70,* 234
Crown Pull, 77–78, *79*
Crown (seventh) chakra, 77, 163–64

D

Daily Energy Routine, 86–87, 262, 265, 312, 322
Dark forces, 341–43
Deal, Sheldon, 175
Deceased, voices of, 351
Delta waves, 48
Dental chart, 284, 287
Despair, Overcoming Through Spiritual Connection, 21–22
Diaphragm, freeing, 266, 268–69
Diffuse pain, Hopi healing technique for, 285, 288
Digestive Problems, Overcoming, 281
Directed imagery, 225, 324–26
Direction faced in work and sleep, 309–11
DNA, 24
Dust, 231
Dying, 207
Dyslexia, 75

E

Eating Smarter, 53–54
Educational kineseology, 86
Ego, 160
Einstein, Albert, 15–16
Electromagnetism, 38, 295
creating environment supporting health, 296–98
magnets
fighting magnetism with, 298–302
in pain management, 302–4
using to counter electromagnetic fields, 304–14
Electromyograph, 173
Electropollution, 296
Emergency response loop, 88–89
reprogramming, 90–91
Emotional
overload, 240–41
pain, 292–94
Endocrine system, 166
Endorphins, 280
Energy. *See also* Energy testing; Subtle energies
body, 30–34
detecting, 36
effect of outside energy on energy within, 37–39
kineseology, 9, 45
language of body, 25–27
scrambled, 75
wearing field of, 39–41
work and the soul, 19–22
Energy field
out of balance, 60–61
and physical environment, 54–56
Energy medicine, 259–61
list of resources, 355–60
websites, 360–61
Energy stations. *See* Chakras
Energy testing, 7, 29, 44–46
biological basis for, 47–48
and chakras, 167, *168*
Cross Crawl, 73
Crown Pull, 78
and energy absorption of body, 307–8
fine-tuning, 56–59
and the five rhythms, 212–13
and food, 52–53, 232

with a friend, 49–50
and ileocecal valve, 266
of infant, pet, or comatose person, 55–56
K-27 thump, 65–67, *66*
learning to, 48–49
and meridians, 111, *112*, 113–15
reliability of, 46–47
spleen thump, 69
thymus thump, 68
Wayne Cook posture, 77
without a partner, 50, 52
Zip Up, 83, 86
Environmental illnesses, 231–33
Environmental Protection Agency (EPA), 297
Evil spirits, 341–43
Expectations, disengaging from, 57
Eye pattern release, 330, *331*

F

False memory syndrome, 192
Fear, 293
Blowing Out the Candle, 219
Escaping Grip of, 39
Fibromyalgia, 285
Fifth (throat) chakra, 156–59
Fight-or-flight response, 226, 229
First (root) chakra, 43, 144–48
Five rhythms, 200
autumn: ending, 206–7
balance among, 214–22, *217*
and energy testing, 212–13
Indian summer: solstice/equinox, 204–5
and meridians, 223
seasons wheel, *210*
spring: new growth, 202–3
summer: fulfillment, 203–4
winter: embryonic possibility, 201–2
Food energy testing, 52–53
Forebrain, 75, 88, 228
Forgiveness, Send Through Your Body, 252
Fourth (heart) chakra, 36, 154–56
Freeing the Diaphragm, 165
Frequencies, 47–48

G

Gait reflexes, 35
Gall bladder meridian, 106, *106*, 108
General indicator test, 65–67, *66*
Geomagnetic field, 295
Getting Kids Out of Bed in the Morning, 87
Gold, 155
Goodheart, George, 6–7, 44, 332
Governing meridian, 100–101, 244
Grad, Bernard, 314
Gratitude, Send Through Your Body, 252
Green, 143, 155
Greeting Kids After School or Spouse After Work, 127–28
Grief, Human Touching Divine, 222
Grounding, 163
Gundling, Dorothy, 173

H

Habit field
clearing, 322–23
defined, 317
distinguishing aura from, 319–20
and early evidence of morphogenetic fields, 318–19
reprogramming, 327–38, *331*, *334*
shifting, 321–22
and spindle cell mechanism, 320–21
using mind to set, 323–26
Headache, 77
Isometric Press, 138–39, *139*
points, 57
Healer, 8
protecting, 343–44
within, awakening, 24
wounded, 8, 150
Healing and the Mind (TV program), 279
Heart, 155–56
disease, 199
(fourth) chakra, 36, 154–56
meridian, 100, *102*
Heaven Rushing In, 21–22, *23*
Holy Spirit, 16
Homolateral crossover, 233–35, 265
Hook Up, 119, 176–77, 265
Your Yin and Your Yang, 251
Hopi healing technique, 285, 288
Humor, 253
Hunt, Valerie, 173
Hyoid bone, 158
Hypoglycemia, 44
Hypothalamus, 208, 226
Hysteria, 239
Getting Back in Control, 236

I

Iatrogenic illness, 45
Ileocecal valve
energy testing, 266
opening, 265–66, 267
Illness
balancing chakras and keeping strange flows flowing, 276
Energy Bathing Unborn Child, 270–71
and energy medicine, 259–61
freeing the diaphragm, 266, 268–69
Meridian Flow Wheel, 271–76, *274*
neurolymphatics, 269–70
Nipping in the Bud, 19
opening ileocecal valve, 265–66, 267

Illness (*cont.*)
safeguarding your breasts, 263–64, *264*
setting intention for healing, 265
strategy for getting well, 261–62
unscrambling energies, 265
Immune deficiency disorders, 228, 242
Immune system, 225–28
Indian summer, 204–5
Indigo, 151–52
Infants, 42, 71, 310
Intention
clearing, 323
focusing, 57–58
healing, 265

J

Jealousy, 218
Jet lag, 129, *130*, 131–32
Joy, Feeling More of the Time, 278
Joy, Send Through Your Body, 251–52
Judgment, 211–12
Jung, Carl, 24

K

Keller, Helen, 340
Ki, 16
Kidney meridian, 32, 103, *104*, 110
K-27 points:
energy testing, 65–67, *66*
tapping, 63–65
Kübler-Ross, Elisabeth, 207
Kundalini experience, 305

L

Large intestine meridian, 33, 108, *109*
Latissimus dorsi muscle, 52
Laughter, 248
Life
after-life experience, 199
color, 151
Liver meridian, *107*, 108
The Lives of a Cell (Thomas), 16
Long-distance medical diagnosis, 28
Low Back Pain, Relieving, 326–27
Lunar cycle, 37
Lung meridian, *107*, 108, 127
Lusseyran, Jacques, 37–38
Lymph, 80
Lymphatic
reflex points, 263
system, 79
Lymphocytes, 80

M

Magnetic field deficiency syndrome, 296
Magnetite, 296
Magnets, 129
basics in using, 300
contraindications for using, 302
fighting magnetism with, 298–302
in pain management, 302–4
testing polarity of, 300–301
using to counter electromagnetic fields, 304–14
Maroon, 155
Meditation, 193
Megbe, 16
Melzack, Ronald, 30
Meridian Flow Wheel, 124–25, *126*, 127–28, 271–76, *274*
Meridians, 15, 34, 49, 95–98. *See also* Meridian Flow Wheel; Muscle Meridian Chart
aligning with the earth's meridians, 124–25, *126*, 127–29, *130*, 131–32
balancing, 98
bladder, 32, 100, 103, *104*, 125, 127
central, 82–83, 100, *101*, 244
circulation-sex, *105*, 106
correcting imbalances in, 116–24
energy testing, 111, *112*, 113–15
flushing, 117–18
gall bladder, 106, *106*, 108
governing, 100, *101*, 244
heart, 100, *102*
inchworm trace, 273
kidney, 32, 103, *104*, 110
large intestine, 33, 108, *109*
liver, *107*, 108
lung, *107*, 108, 127
natural direction of, 61–62
sedating, 223
small intestine, 100, *103*
spleen, 44, 49, 100, *102*, 108, 110, 244–48
stomach, 108, *109*
strengthening, 223
tracing, 98–100
triple warmer, *105*, 106, 155, 226–44, 319
Metabolism, 156, 253
Metal, 199
Mind, power of, 323–24
Morphic fields, 318–20, 323
Moyers, Bill, 279
Mueller College of Holistic Studies (San Diego), 7
Multiple sclerosis, 229–30
Muscle Cramp, Relieving, 306
Muscle Meridian Chart, 284, *286*
Muscle testing. *See* Energy testing

N

National Institutes of Health (NIH), 11–12, 278
National Research Council (NAS), 297
Natural foods, 48

Neck and Shoulder Tension, Releasing, 198
Nervousness, Overcoming, 116
Neurofeedback, 48
Neurolymphatic reflex points, 68, 79, *84–85,* 214–15
  massaging, 118, 269–70, 272–73
Neurovascular points
  holding, 35, 89–90, *90,* 216, *217,* 294, 323, 328–30
  numbering, 273, *274,* 275
New Age guilt, 211
Nixon, Frances, 310
North pole (of magnet), 309, 314

O

Office of Alternative Medicine (NIH), 12
"Oh My God" points, 90
Orenda, 16
Outer realms, 339–41
  dark forces, 341–43
  evidence of, 348–52
  protecting the healer, 343–44
  spirit guardians, 345–48
Oxygen, 268

P

Pain, 277
  breathing out the pain, 282
  emotional and physical, 292–94
  in Hands, Wrists, or Elbows, Relieving, 320–21
  Hopi technique for diffuse pain, 285, 288
  nature of, 278–80
  pain chasing, 281–82, 289, 292
  sedating, 283–84, *286*
  siphoning off, 182, 283
  stretching area surrounding, 283
  taking teeth out of a toothache, 284, *287*
  tapping area of, 282–83
  zone tapping, 288–89, *290–91*
Pancreas, 44, 49, 253
Panic, Taking Down the Flame, 165, 220–21
Parathyroid glands, 158–59
Past-life memories, 349–50
Perceptual and Motor Skills, 46
Pericardium, 155
Phantom limb pain, 31–34, 280
Pick Me Up, 73
Pinching painful area, 283
Pink, soft, 155
PMS (premenstrual syndrome), 254–55
Polarity, 272, 295, 300–303
Positive Feeling, Programming in, 330–32
Posttraumatic stress disorder, 88
Power chakra, 152
Prana, 16
Princeton University School of Engineering, 27
Professional relationship, 344

R

Radin, Dean, 46
Red, 143
Repetition, 324
Reptilian brain, 88–90
Rhythmic 8s, 184, *185*
Rhythms. *See* Five rhythms
Room energy, 38
Root (first) chakra, 43, 144–48

S

Savants, 40
Scrambled energy, 75
Seasons. *See* Five rhythms
Second (womb) chakra, 148–52
Sedating points, 33, 118–19, *120–23,* 283–84, *286,* 293
Self-judgment, 211–12, 239
Seligman, Martin, 252
Separating Heaven and Earth, 248–49, *250,* 265
Seventh (crown) chakra, 77, 163–64
Sexual abuse, 190–93
Sexuality, 147
Shame, 218
Shealy, C. Norman, 28
Sheldrake, Rupert, 318–19, 323
Siegel, Bernie, 259
Siphoning pain away, 283
Sixth (third eye) chakra, 159–62
Skydivers, 175
Small intestine meridian, 100, *103*
Socrates, 340
Solar plexus (third) chakra, 152–54
Solstice/equinox, 204–5
South pole (of magnet), 309
Spinal
  Flush, 79–82, *81,* 118, 214–15
  Suspension, 119, 124, *125*
Spindle cell mechanism, 283, 320–21
Spirit guardians, 340, 345–48
Spirituality and mental health, 348–49
Spleen, 253
  meridian, 44, 49, 100, *102,* 108, 110, 244–48
  -pancreas energy test, 49–50, *51*
  points, 68–69
Spring, 202–3
Stevenson, Ian, 349–50
Stokes, Gordon, 57
Stomachache, 77
Stomach meridian, 108, *109*
Strange flows, 226
  activating when drying off, 249, 251
  jump-starting, 251–53
  keeping flowing, 248
  as psychic circuits, 243–45
  Separating Heaven and Earth, 248–49, *250*
  spleen meridian, 245–48

Strengthening points, 118–19, *120–23,* 293
Stress, 75
  emotion, 215–16, 218
  releasing built-up, 91
  reprogramming your response to, 87–91
Stretching painful area, 283
Stuttering, 75
Subtle energies
  in healing, 17–19
  measuring, 27–28
  seeing, 17
  sensing, 42–44
  soul as source of, 20
Summer, 203–4
Surrogate testing, 55–56
Sylvia, Claire, 16
Sympathy, Cradling the Baby, 221
Synesthesia, 42–43

T

Taking Down the Flame, 165
Tapping out pain, 282–83
Temporal
  smoothing, 334–35
  tap, 332–38, *334*
Tension
  Headache, Ridding Yourself of, 25
  Releasing, 62–63
Therapeutic
  massage, 7
  Touch, 262
Therapy localizing, 266
Thie, John, 6
Third
  chakra (solar plexus), 152–54
  eye (sixth) chakra, 159–62
Thomas, Lewis, 16
Thought, 27
Three Thumps, *51*
  K-27 points, 63–65
  spleen points, 68–69
  thymus gland, 67–68
Throat (fifth) chakra, 156–59
Thymus gland, 67–68
Thyroid gland, 156
Tiller, William, 27
Time, 340
Toothache pain, 284, *287*
Touch for Health, 6–7
Touching, 49
Toxins, 80, 269–70
Trauma, 188–95
  defusing residue of, 328–30
Triple warmer meridian, *105,* 106, 155, 244, 319
  and allergies, 231–33
  and autoimmune disorders, 229–30
  homolateral crossover, 233–35
  in overdrive, 228–29
  relation to immune system, 226–28
  reprogramming to stop attacking friendly forces, 236–37, 239
  reprogramming to stop attacking your body, 239–43
  sedating, 235–36
Triple warmer-spleen archetype, 253–55
Turquoise, 156

U

UCLA Energy Fields Laboratory, 173
Unborn Child, Energy Bathing, 270–71

V

Van Praagh, James, 351
Venom, Expelling, 219–20
Violent outbursts, 240–41
Vivaxis field, 310–11
  demagnetizing, 311–12, *313,* 314

W

Wakan, 16
Water, 49
  treating, 314
Wayne Cook posture, 73–77, *76*
Wellspring of Life points, 110, *111*
Wilson, Colin, 247
Winter, 201–2
Womb (second) chakra, 148–52

Y

Yang (masculine), 124, 147, 223, 245, 288
Yellow, 153
Yesod, 16
Yin (feminine), 124, 147, 223, 245, 288

Z

Zimmerman, John, 298
Zip Up, 82–83, 86, 326
Zone tapping, 288–89, *290–91*

ABOUT THE AUTHORS

A pioneer in the field of energy medicine, Donna Eden has been teaching people for more than two decades to understand the body as an energy system, to recognize their aches and pains as signals of energy imbalance, and to reclaim their natural healing capacities. She speaks to packed audiences throughout the United States, Europe, Australia, New Zealand, and South America, and is consulted frequently within traditional and alternative health care settings.

David Feinstein is a psychologist who has served on the faculties of Johns Hopkins University School of Medicine and Antioch College. His previous books include *Personal Mythology, Rituals for Living and Dying, Mortal Acts,* and *The Mythic Path.*

PROGRAMS BY DONNA EDEN AND DAVID FEINSTEIN

An optional six-hour video training program, *Energy Healing with Donna Eden,* illustrates methods presented in this book and is available directly through Innersource for $59.95 (list is $79.95). A separate 90-minute video program, *An Introduction to Energy Healing,* featuring Donna Eden teaching an introductory class, is available for $19.95.

The Mythic Path, David Feinstein and Stanley Krippner's twelve-week course for discovering and transforming your guiding mythology, is available as an integrated book and audiotape program for $24.95.

Please add $4 for shipping and handling per order ($6 for overseas airmail). Call 800-835-8332, or send name, address, and phone number with check, money order, or Visa/MasterCard number and signature to Innersource, PO Box 213, Ashland, OR 97520. A brochure describing other books and tapes as well as available training programs is included with each order or may be requested by mail, phone, or through www.innersource.net.